鲁班工坊数控技术专业系列丛书

金工实训

刘锐　苏晨　主编

天津大学出版社
TIANJIN UNIVERSITY PRESS

图书在版编目(CIP)数据

金工实训/ 刘锐，苏晨主编. — 天津：天津大学出版社，2022.6
（鲁班工坊数控技术专业系列丛书）
ISBN 978-7-5618-7193-5

Ⅰ.①金… Ⅱ.①刘… ②苏… Ⅲ.①金属加工—实习—职业教育—教材 Ⅳ.①TG-45

中国版本图书馆CIP数据核字(2022)第122872号

组稿编辑　胡小捷
责任编辑　胡小捷
装帧设计　逸　凡

JINGONG SHIXUN

出版发行　天津大学出版社
地　　址　天津市卫津路92号天津大学内(邮编:300072)
电　　话　发行部:022-27403647
网　　址　www.tjupress.com.cn
印　　刷　北京盛通商印快线网络科技有限公司
经　　销　全国各地新华书店
开　　本　185mm×260mm
印　　张　22.25
字　　数　645千
版　　次　2022年6月第1版
印　　次　2022年6月第1次
定　　价　50.00元

前言

《金工实训》一书的编写工作正是在贯彻落实天津职业教育与“一带一路”沿线国家的职业教育合作的基础上，为了配合马达加斯加“鲁班工坊”的理论和实训教学，开展交流与合作，提高中国职业教育的国际影响力，创新职业院校国际合作模式，输出我国职业教育优秀资源而开展的。

本书面向马达加斯加鲁班工坊项目，采用中文、英文两种语言编写。以“鲁班工坊”加工设备为载体，突出“以职业标准为依据，以企业需求为导向，以职业能力为核心”的教学理念，依据国家职业标准，结合企业实际和岗位需求，反映了新知识、新技术、新工艺、新方法在企业生产中的应用，并注重对职业能力的培养。本书共分 3 个项目，22 个任务，力图做到深入浅出，便于教学。

本书内容编排由浅入深，理实结合，并添加了精选案例，以任务驱动的形式将知识点与技能点有机融合；以科学性、实用性、通用性为原则，使教材符合现代职业教育机械类课程体系设置。金工实训是机械类各专业实践性很强的技术基础课，该课程为学生建立机械制造生产过程的概念，获得机械制造基本技能的训练奠定了基础。

本书作为实训课程教学用书，编写目的是使学生能初步接触机械制造的生产过程，了解机械制造常用材料和金属加工工艺基础知识，熟悉机械零件常用加工方法及所使用的主要设备和工具，初步掌握常用机床的基本操作技能并具有一定的操作技巧，为相关课程的理论学习及将来从事生产技术工作打下基础。本书可作为职业学校机械加工技术专业、机电技术应用专业及机电类相关专业的教学用书，也可作为有关行业的岗位培训教材及企业职工自学用书。

本书由天津市机电工艺技师学院刘锐、苏晨担任主编，参加编写的有天津市机电工艺技师学院杜暘、张治英、张金凤三位老师，在此表示感谢！限于编者经历及水平，书中难免有疏漏之处，恳请广大读者提出宝贵意见。

编者

目录

【项目描述】

钳工是利用虎钳和各种手工工具及设备，按技术要求进行零件加工、修整、机械装配、设备维修的工种。其特点是手工操作多，灵活性强，工作范围广，技术要求高，操作者本身的技能水平直接影响加工质量。钳工加工方法主要应用于机械加工方法不方便或难以解决的场合。钳工在机械制造中的历史悠久，对机械制造行业的发展做出了巨大的贡献，有着“万能钳工”的美誉。

随着生产技术的发展，机械加工逐步从使用各种手工工具制造发展到机械化制造和装配。但是，钳工依然是工业生产中一个独立的、不可缺少的工种，钳工工具多样，操作灵活，在自动化生产的今天，钳工工种依然重要。

【学习目标】

（1）了解钳工在工业生产中的主要任务和作用。

（2）了解钳工工作中各种工具和设备的操作及维护保养方法。

（3）熟悉实习场地和工厂的规章制度以及安全文明生产要求。

（4）掌握和理解钳工的基础知识和基本技能以及安全操作规程。

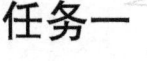

任务一　钳工一般知识

一、钳工的作用及主要工作内容

钳工是使用钳工工具，对工件进行加工、修整、机械装配、设备维修的工种。任何机械产品的制造过程都包括毛坯的制造、零件的加工制造、部件组装、整机装配和调试试运行等阶段。其中有大量的工作必须依靠钳工来完成。钳工加工方法是机械制造中不可缺少的一种方法，它的工作范围很广，主要包括以下几个方面。

1. 零件加工

钳工可以完成一些采用机械方法不适宜或不能解决的加工任务。如零件加工过程中的划线、精密加工（如刮削、研磨、锉削样板和制作模具等）以及检验和修配等。

2. 工具的制造和修理

钳工可以制造和修理各种工具、夹具、量具、模具及各种专用设备。

3. 机械的装配

钳工可以把零件按装配技术要求进行装配，并经过调整、检验和试车等，使之成为合格的功能部件或机械设备。

4. 设备的维修

当机械设备在使用过程中发生故障、出现损坏或长期使用后精度降低，影响使用时，可由钳工进行维护和修理。

随着机械工业的日益发展，许多繁重的工作已被机械加工所代替；但那些精度高、形状复杂零件的加工以及设备安装调试和维修是机械难以完成的。这些工作仍需利用钳工的精湛技艺去完成。因此，钳工是机械制造业中不可缺少的工种。

二、钳工必备的操作技能

1. 划线

划线作为零件加工的第一道工序，与零件的加工余量有着密切的关系。钳工在划线时，首先应熟悉图样，合理使用划线工具，按照划线步骤在待加工工件上画出零件的加工界限、各种孔的中心线，作为零件装夹、加工的依据。

2. 锯削

锯削用来分割材料或在工件上锯出符合技术要求的沟槽。锯削时必须根据工件的材料性质和工件形状，正确选用锯条和锯削方法，从而使锯削操作能顺利地进行，并达到规定的技术要求。

3. 锉削

锉削是利用各种形状的锉刀，对工件进行切削、整形，使工件达到较高的精度和较为准确的形状。锉削是钳工工作中的主要操作方法之一，它可以对工件的外平面、曲面、内外角、沟槽和各种形状的表面进行加工。

4. 錾削

錾削是钳工最基本的操作，是利用錾子和锤子这些简单工具对工件进行切削和切断的操作。錾削主要在零件加工要求不高或机械加工无法进行的场合采用。同时，錾削还要求操作者具有熟练的锤击技能。

5. 孔加工（包括钻孔、扩孔、锪孔和铰孔）

钻孔、扩孔、锪孔和铰孔是钳工对孔进行粗加工、半精加工和精加工的四种方法，应用时根据孔的加工要求和加工条件选用。其中，钳工钻孔、扩孔、锪孔是在钻床上进行的，铰孔可用手铰，也可以通过钻床进行机铰。掌握钻孔、扩孔、锪孔、铰孔的操作技能，必须熟悉所用刀具的切削性能，以及钻床和一些夹具的结构性能，合理选用切削用量。熟练掌握钻头的刃磨和手工操作的具体方法，是保证孔加工质量的关键。

6. 螺纹加工（包括攻螺纹和套螺纹）

攻螺纹是用丝锥在工件内圆柱面上加工出内螺纹的加工方法，套螺纹是用圆板牙在工件外圆柱面上加工外螺纹的加工方法。钳工所加工的螺纹，通常都是直径较小的三角螺纹或不适宜在机床上加工的螺纹。

7. 刮削和研磨

刮削是钳工对工件进行精加工的一种方法。通过刮削，不仅可以获得较高的几何精度、尺寸精度、接触精度和传动精度，而且还能利用刮刀在刮削过程中对工件表面产生的挤压，使表面组织紧密，从而提高材料的力学性能、耐磨性和耐腐蚀性。

研磨是最精密的加工方法。它通过磨料在研具和工件之间滑动、滚动产生微量切削，使工件达到很高的尺寸精度和很低的表面结构参数值。

8. 矫正和弯形

矫正和弯形是利用金属材料的塑性变形，采用合适的方法，对变形或存在某种缺陷的原材料和零件加以矫正，以消除变形等缺陷，或者利用专用工具将原材料弯成图样所需要的形状，并对弯形前的材料进行落料长度计算。

9. 装配和修理

装配是按照图样规定的要求，将零件通过适当的连接形式组合成部件或完整的机器。修理是对使用时间较长或由于操作不当而使精度和性能下降，甚至损坏的机器或零件进行调整，使之恢复到原来的精度和性能要求。

10. 测量

在生产过程中要保证零件的加工精度和要求，首先对产品进行必要的测量和检验。钳工在零件加工装配过程中，经常利用平板、游标卡尺、千分尺、百分表和水平仪对零件进行测量检查。这些都是钳工必须掌握的测量技能。

另外，钳工还必须了解和掌握金属材料热处理的一般知识，熟练掌握一些钳工工具的制造和热处理方法，如锤子、錾子、样冲、划针、划规和刮刀等。

三、钳工基本操作中常用工、量具

常用工具有划线用的划针、划线盘、划规、样冲、分度头和平板，錾削用的手锤和各种錾子，锉削用的各种锉刀，锯削用的锯弓和锯条，孔加工用的各类钻头、锪钻和铰刀，攻、套螺纹用的各种丝锥、板牙和绞杠，刮削用的平面刮刀和曲面刮刀以及各种扳手和旋具等。

常用量具有钢板尺、刀口形直尺、游标卡尺、千分尺、90°角尺、万能角度尺、塞尺、百分表等。

四、安全文明生产制度

1. 实习守则

（1）进入实习工厂实习，要穿好工作服、工作鞋、戴好工作帽、带好笔记和各种实习用品。

（2）上课前，学生以班级为单位在实习车间前列队，核查实到人数，提前 10 min 排队进入实习车间。不准穿拖鞋、戴手套、戴围巾进入车间，工作服保持干净整洁。

（3）进入实习车间，不准大声喧哗、打闹、随意跑动和滞留。不准吃零食、乱丢杂物。在教学实习场所行走时保持单列右行。

（4）进入实习工厂，听从实习指导教师及车间管理人员安排，接受实习指导教师的指导，认真听课，做好课堂笔记，勤学苦练，认真对待每一次技能实训。

（5）执行实习生产工艺标准，按实习生产计划进行实习生产。

（6）实习期间，要坚守自己的工位，不得窜岗。没有当班教师的批准，不准离开实习车间。

（7）严格遵守作息时间，不得迟到、早退、旷课，如确实有事，要先向当班教师请假。

（8）严格遵守各工种安全操作规程，按实习操作工艺标准进行实习操作。

（9）尊敬教师，有礼貌、虚心和教师交流，相互理解，增进友情。上课不得随意讲话，有问题先举手示意，教师同意后发言。

（10）实习工件、工具、仪器仪表、书籍等实习用品应摆放有序，保持实习台案、设备、周围地面的清洁卫生，不得有杂物、污渍和灰尘，按时清理自己的卫生责任区。

（11）未经教师批准，不得将与实习无关的物品带入实习车间，不得将公物带出实习车间。

（12）若需使用设备，必须在掌握设备性能、正确的操作规范和操作步骤后，经当班实习教师批准后，方可使用。设备使用后，应立即进行维护保养。

（13）实习过程中，遇到问题或仪器仪表、工具量具、机床设备等出现故障，要立即报告当班指导教师，说明情况，不得私自处理。

（14）遵守学校和实习工厂的各项规章制度。

（15）违反以上规定的，按有关管理规定处理，若造成损失，由责任人赔偿。

2. 钳工安全操作规程

（1）工作前，必须穿戴好工作服、工作帽和其他防护用品。

（2）工作前，必须检查使用的工具是否齐全、完整，锉刀、刮刀、手锤应有牢固的木把，冲子、錾子等工具的锤击处，不准有淬火裂纹、卷边和飞刺。

（3）使用手锤时，应选择好挥动方向，以防锤头脱落或铁屑飞出伤人，在錾切工件时，对面不准站人，固定操作处，应设防护网，握锤的手不准戴手套。

（4）使用手电钻、手砂轮及一切手提电动工具时，脚应踏在绝缘板上，并戴好绝缘手套和防护眼镜。

（5）使用手砂轮、软轴砂轮前，必须检查砂轮是否完好，必须仔细检查是否有漏电现象，不能在漏电情况下操作，并一定要等砂轮正常运转后，才可使用。

（6）在钻床上钻孔时，严禁戴手套操作，不准用手擦或嘴吹等方法清除切屑。

（7）在拆卸设备和调整设备运转部位前，必须切断电源，如果设备上的安全装置未修好，严禁试车；装修或调试设备后，必须认真检查，不准将工具或工件遗留在机床内，以防发生事故。

（8）未经电工准许，不准擅自拆装电器。

（9）合理使用工、卡、量、刃具，不准混放。

（10）使用砂轮、钻床、焊机和起重设备时，必须熟悉其操作规程，并严格遵守。

思考与练习

（1）钳工的主要工作内容包括哪些？

（2）钳工的主要操作技能有哪些？

（3）钳工操作过程中有哪些注意事项？

任务二　钳工的实训环境

一、钳工工作场地布局及要求

钳工工作场地是指钳工的固定工作地点。为工作方便、保证产品质量和安全生产，钳工工作场地布局一定要合理，符合安全文明生产的要求。

1. 合理布局主要设备

钳工工作台应安放在光线适宜、工作方便的地方。面对面使用钳工工作台时，应在两个工作台中间安装安全防护网。砂轮机、钻床应设置在场地的边缘，尤其是砂轮机一定要安装在安全、可靠的位置。

2. 正确摆放毛坯、工件

毛坯和工件要分别摆放整齐，并尽量放在工件搁架上，以免磕碰损坏。

3. 合理摆放工具和量具

常用工、量具应放在工作位置附近，不能任意堆放，以免损坏。在钳台上工作时，为了取用方便，左手取用的工、量具放在左边，右手取用的工、量具放在右边，各自排列整齐，不能把工具、量具、工件混放，且不能使其伸到钳台边以外。工、量具用后应及时清理、维护和保养，并且妥善放置。

4. 工作场地应保持清洁

训练后应按要求对设备进行清理、润滑，并把工作场地打扫干净。

二、钳工操作的常用设备及工具

钳工加工常用的设备大多比较简单，主要有钳台、台虎钳、砂轮机、台式钻床、立式钻床和摇臂钻床等。

1. 钳台

钳台主要用来安装台虎钳和存放常用手动工具、量具和夹具。钳台的样式有多人单排和多人双排两种。双排式钳台由于操作者面对面操作，中间必须设置防护板或防护网。钳台多由铸铁和坚实的木材制成，台面一般为长方形或六角形等形状，其长度和宽度尺寸由工作场地和工作需要确定，高度一般为 800~900 mm，如图 1-2-1 所示，装上台虎钳后，能够得到合适的钳口高度（一般以齐人手肘为宜）。

2. 台虎钳

它是用来夹持工件的通用夹具，其规格以钳口的宽度表示，常用的规格有 100 mm、125 mm、150 mm 等。常见的台虎钳结构类型为回转式，如图 1-2-2 所示。回转式台虎钳的钳体可以旋转，可使工件旋转到合适的工作位置。

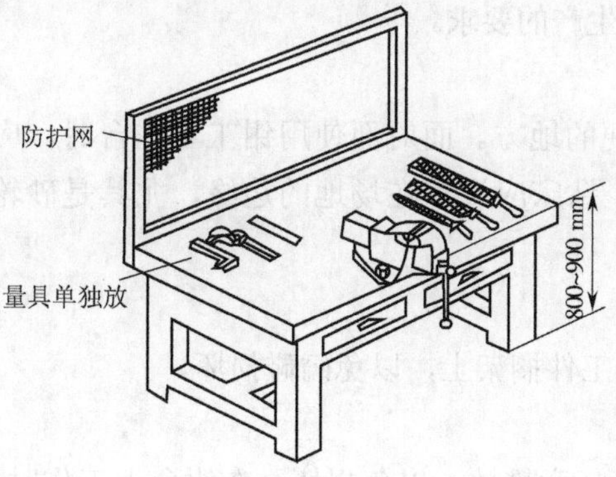

图 1-2-1　钳台

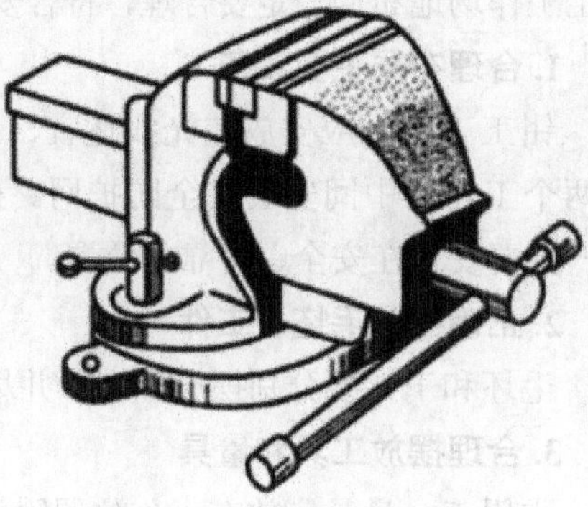
图 1-2-2　回转式台虎钳

3. 砂轮机

砂轮机用来刃磨钻头、錾子、刮刀等刀具或其他工具等，由电动机、砂轮、托架和机体组成，如图 1-2-3 所示。使用砂轮机必须注意安全，严格遵守砂轮机使用安全操作规程。

图 1-2-3　砂轮机

4. 钻床

钻床是钳工常用的孔加工设备，按结构的不同，可分为台式钻床、立式钻床和摇臂钻床三种。

台式钻床是一种用于加工孔的小型钻削机床，一般安装在钳台上。它以钻头等作为刀具。工作时，工件固定不动，刀具旋转作为主运动，同时拨动手柄使主轴上下移动，实现进给运动和退刀，如图 1-2-4 所示。台式钻床转速高，使用灵活，效率高，适用于较小工件的钻孔。其最低转速较高，故不适宜进行锪孔和铰孔加工。

图 1-2-5 所示为立式钻床的一种布局形式。加工时，主轴的旋转作为主运动，其轴向移动实现进给运动。利用操纵手柄可使主轴方便地实现手动快速升降、手动进给或机动进给。摇动工作台手柄，也可使工作台沿立柱导轨上下移动，以适应加工不同高度的工件。立式钻床适宜于单件或小批中型工件的钻孔、锪孔、铰孔和攻螺纹等加工。

图 1-2-4　台式钻床

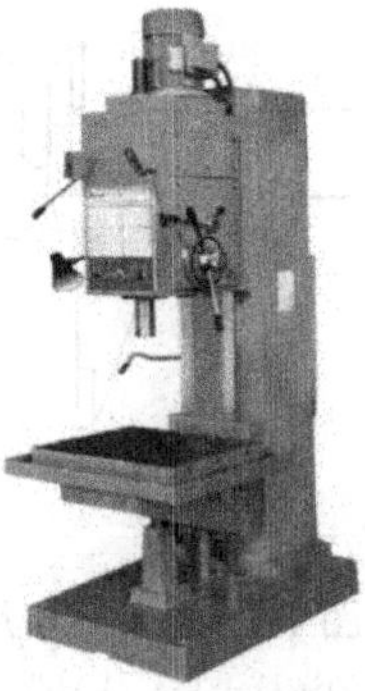

图 1-2-5　立式钻床

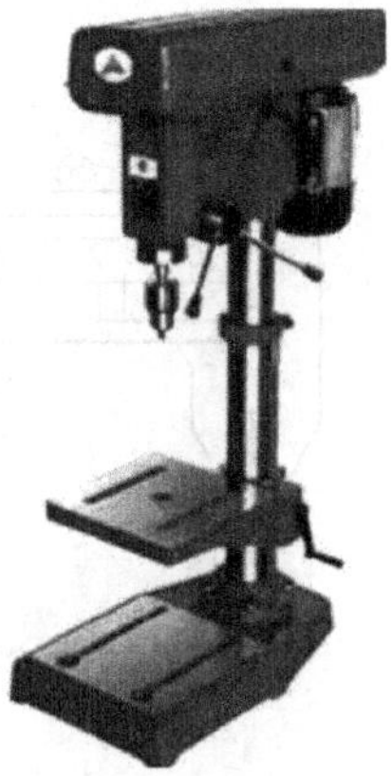

图 1-2-6　摇臂钻床

摇臂钻床如图 1-2-6 所示，其特点是：适用于单件、小批量和中等批量生产的中等件和较大件以及多孔件的各种孔加工。摇臂钻床能在很大的范围内工作，工作时工件可压紧在工作台上，也可以直接放在底座上，靠移动主轴来对准工件上孔的中心，使用时比立式钻床方便。摇臂钻床的主轴转速范围和进给量范围都很广，工作时可获得较高的生产效率和加工精度。

三、钳工常用量具

1. 钢板尺

钢板尺是常用量具中最简单的一种量具，如图 1-2-7 所示，主要用来测量工件的长度、宽度、高度和深度等，有时也可以作为画直线的导向工具。其常见规格有 150 mm、300 mm、500 mm、和 1 000 mm 四种。

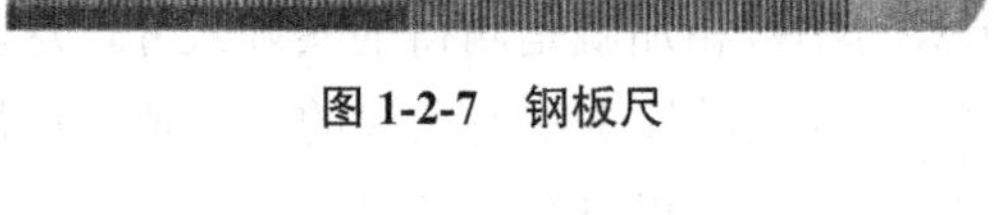

图 1-2-7　钢板尺

2. 游标卡尺

游标卡尺是一种中等精度的量具，可用来测量长度、厚度、外径、内径、孔深和中心距等。游标卡尺的精度有 0.1 mm（1/10）、0.05 mm（1/20）和 0.02 mm（1/50）三种，如图 1-2-8 所示。其中测量精度为 0.02 mm 的游标卡尺最为常用。

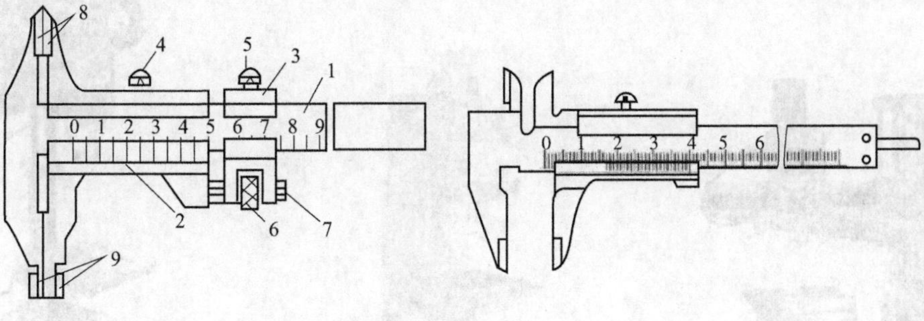

（a）　　　　　　　　（b）

图 1-2-8　游标卡尺

（a）可微动调节的游标卡尺　（b）带测深杆的游标卡尺

1—尺身（主尺）；2—游标（副尺）；3—辅助游标；4—锁紧螺钉；5—螺钉；

6—微调螺母；7—螺杆；8—外测量爪；9—内测量爪

测量外尺寸时，应将游标卡尺的两测量脚张开到略大于被测尺寸，先将固定脚的测量面贴靠工件，然后用大拇指轻轻推动游标，使活动量脚逐步紧靠工件后保持动作，并开始读数，如图 1-2-9 所示。

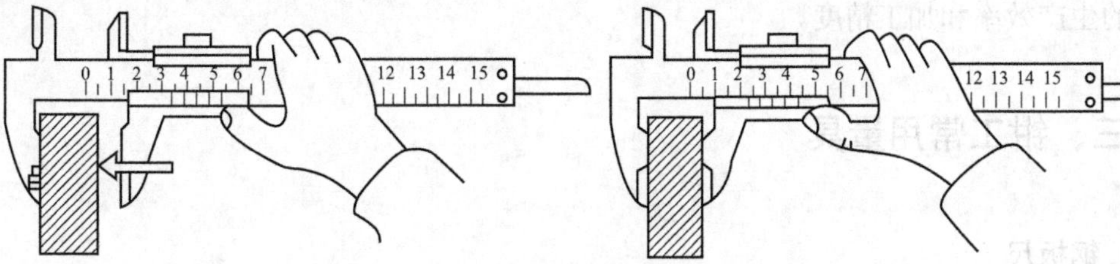

图 1-2-9　游标卡尺的使用方法

游标卡尺的正确读数方法如下。

（1）首先读出游标卡尺零刻线左边尺身上的整毫米数。

（2）再看游标卡尺从零线开始第几条刻线与尺身某一刻线对齐，其游标刻线数与精度的乘积就是不足 1 mm 的小数部分。

（3）最后将整毫米数与小数相加就是测得的实际尺寸，这里以 0.02 mm 游标卡尺的读数方法为例，读数时先读出游标零线前主尺的整数值，再加上游标与主尺重合线处的数值乘以精度值 0.02，二者之和即为所测尺寸。如图 1-2-10 所示，游标零线前主尺的整数值为 60，游标与主尺重合刻度线处的数值为 4，读数为 60 mm+24×0.02 mm=60.48 mm。读数时视线应垂直于游标刻度线，以免斜视引起读数误差。

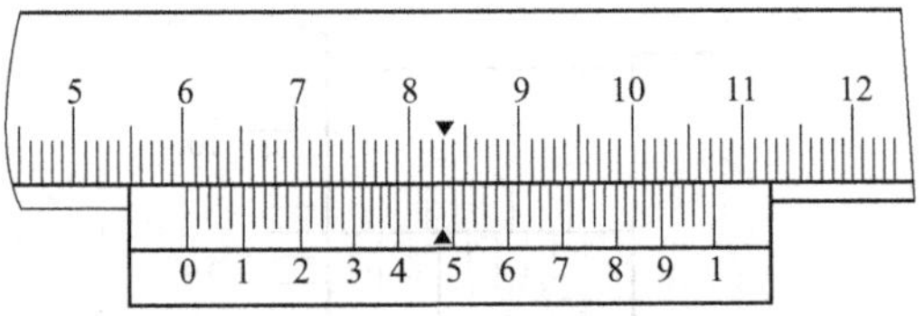

图 1-2-10 游标卡尺的读数方法

3. 千分尺

千分尺是测量中最常用的精密量具之一。千分尺的种类较多，按其用途不同可分为外径千分尺、内径千分尺、内测千分尺、深度千分尺、螺纹千分尺、公法线千分尺、线径千分尺、板厚千分尺等。按照测量对象的不同，常用千分尺有外径千分尺、内径千分尺和深度千分尺三种。

千分尺的测量范围在 500 mm 以内时，每 25 mm 为一挡，如 0~25 mm、25~50 mm、50~75 mm、75~100 mm 等；测量范围在 500~1 000 mm 时，每 100 mm 为一挡，如 500~600 mm、600~700 mm 等。千分尺按制造精度分为 0 级、1 级和 2 级，其测量精度一般为 0.01 mm。

外径千分尺的外观及结构如图 1-2-11 所示。它由尺架、测砧、测微螺杆、锁紧手柄、螺纹套、固定套管、微分筒、螺母、接头、测力装置、弹簧、棘轮爪、棘轮等部分组成。

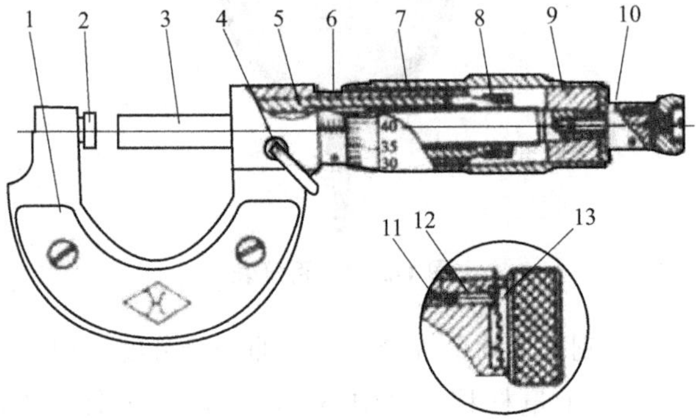

图 1-2-11 外径千分尺的外观及结构

1—尺架；2—测砧；3—测微螺杆；4—锁紧手柄；5—螺纹套；6—固定套管；7—微分筒；8—螺母；9—接头；10—测力装置；11—弹簧；12—棘轮爪；13—棘轮

外径千分尺的正确读数方法如下：

（1）先读出固定套管上露出刻线的整毫米及半毫米（0.5 mm）数；

（2）再确定微分筒刻线与固定套管基准线对齐的位置，读出不足半毫米的小数部分；

（3）最后将两次读数相加，即为工件的测量尺寸，如图 1-2-12 所示。

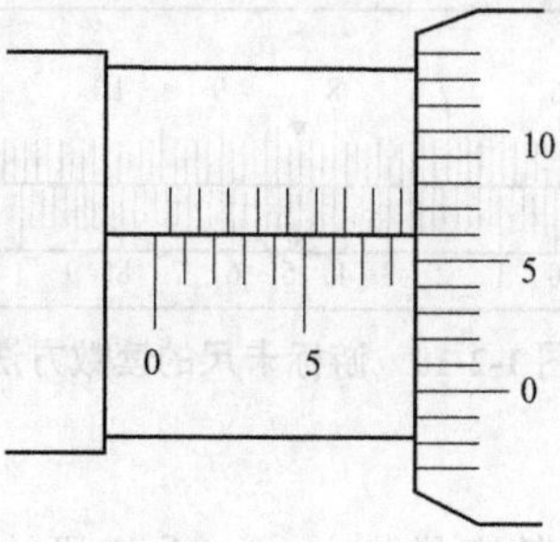

读图读数 8.56 mm

图 1-2-12　千分尺的读数方法

4. 百分表

百分表是一种指示式量仪，主要用来测量工件的尺寸、形状和位置误差，也可用于检验机床的几何精度或调整工件的装夹位置偏差。百分表的测量范围一般有 0~3 mm、0~5 mm 和 0~10 mm 三种。按制造精度不同，百分表可分为 0 级、1 级和 2 级，百分表的外形如图 1-2-13 所示。

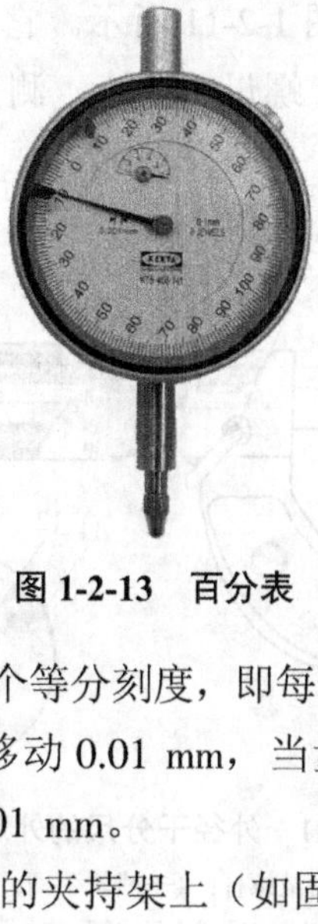

图 1-2-13　百分表

百分表的圆表盘上印制有 100 个等分刻度，即每一分度值相当于量杆移动 0.01 mm，所以大指针每转 1 格，表示量杆移动 0.01 mm，当量杆上移动 1 mm 时，大指针转 1 周，因此，百分表的测量精度为 0.01 mm。

测量时，必须把它固定在可靠的夹持架上（如固定在万能表架或磁性表座上），夹持架要安放平稳，测量杆必须垂直于被测工件表面，使测量杆的轴线与被测量尺寸的方向一致。当测头接触被测工件表面时，量杆被推向管内，量杆移动的距离等于小指针的读数（测出的整数部分）加上大指针的读数（测出的小数部分），即为测量结果。

5. 万能角度尺

万能角度尺是用来测量工件内、外角度的量具。其测量精度有 2' 和 5' 两种，测量范围为 0°~320°。其外观和结构，如图 1-2-14 所示。

测量时，直尺可通过扇形板上的小齿轮转动扇形齿轮，使基尺改变角度带动主尺沿游标转动。直角尺和直尺可以配合使用，也可以单独使用。

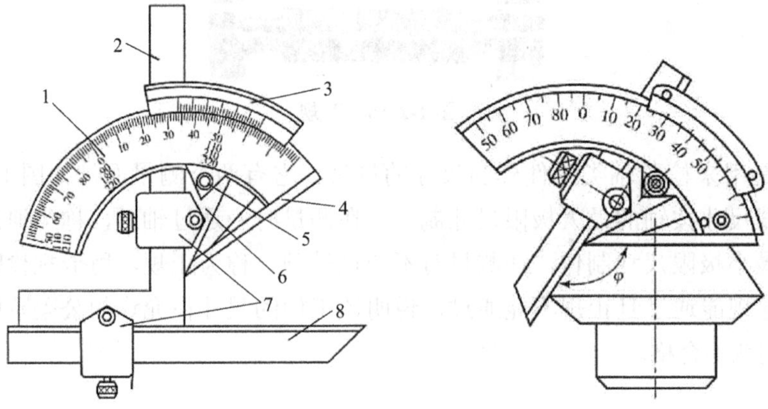

图 1-2-14　万能角度尺

1—主尺；2—直角尺；3—游标；4—基尺；5—制动器；6—扇形板；7—卡块；8—直尺

6. 塞尺

塞尺是用来检验两个结合面之间间隙大小的片状量规，是由一组厚度不等的金属薄片组成的量具，其长度有 50 mm、100 mm、200 mm 等多种，如图 1-2-15 所示。

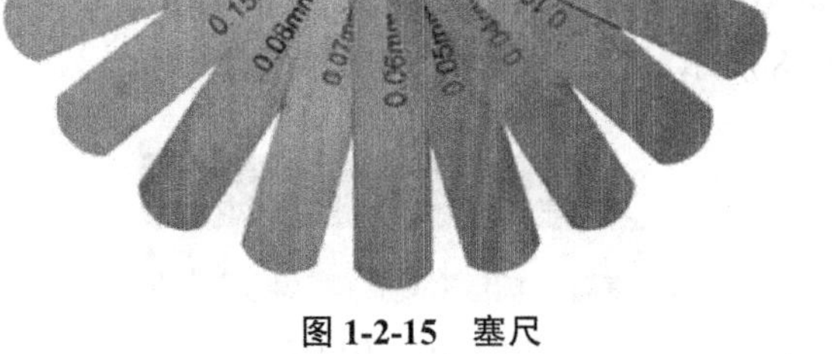

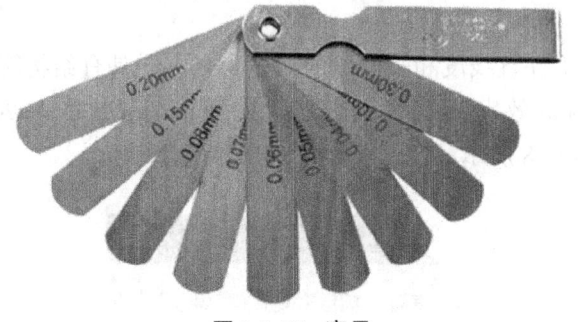

图 1-2-15　塞尺

使用塞尺时，应根据间隙的大小选择塞尺的片数，可用一片或数片重叠在一起插入间隙内。若塞尺塞入间隙后，移动时有轻微的阻滞感，则此时塞尺的厚度即为被测间隙的尺寸。

厚度小的塞尺片很薄，容易弯曲和折断，插入时不能用力太大。塞尺片用后应擦拭干净并及时合到夹板中。

7. 塞规和卡规

塞规是用来检验工件内径尺寸的量具。它有两个测量面，小端尺寸按工件内径的最小极限尺寸制作，在测量内孔时应能通过，称为通规；大端尺寸按工件内径的最大极限尺寸制作，在测量内孔时不通过工件，称为止规，如图 1-2-16 所示。

图 1-2-16　塞规

卡规是用来检验轴类工件外圆尺寸的量规。它有两个测量面，如图 1-2-17 所示。其中，大端尺寸按轴的最大极限尺寸制作，在测量时应通过轴颈，称为通规；小端尺寸按轴的最小极限尺寸制作，在测量时不通过轴颈，称为止规。用卡规检验轴类工件时，如果通规能通过且止规不能通过，说明该工件的尺寸在允许的公差范围内，是合格的，否则就不合格。

图 1-2-17　卡规

8. 刀口尺

刀口尺是检验工件直线度和平面度的测量工具，它具有结构简单、质量轻、不生锈、操作方便和测量效率高等优点。其常用规格有 125 mm、200 mm、300 mm 和 400 mm 等，如图 1-2-18 所示。

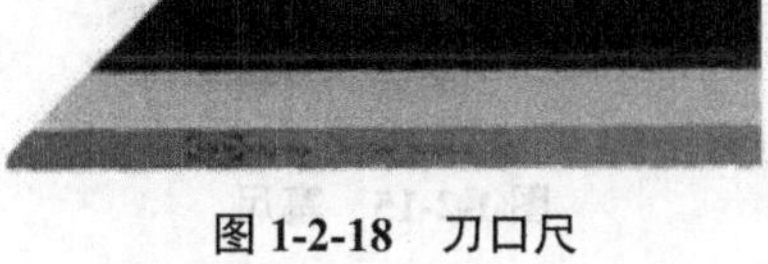

图 1-2-18　刀口尺

9. 直角尺

直角尺是一种专业量具，简称为角尺、靠尺。直角尺用于检测工件的垂直度及工件相对位置的垂直度，有时也用于划线，如图 1-2-19 所示。

图 1-2-19　直角尺

思考与练习

（1）钳工加工常用的设备有哪些？

（2）钳工常用量具有哪些？

（3）简述外径千分尺的正确读数方法。

（4）塞规和卡规的使用方法是什么？

任务三　划线

划线是机械加工中的重要工序之一，广泛运用于单件小批量生产。根据图纸和技术要求，在毛坯或半成品工件上利用划线工具划出加工界线，或者划出作为基准的点、线的操作过程称为划线。

划线分为平面划线和立体划线两种。平面划线就是根据图纸和技术要求，在工件的一个平面上划出加工界线的操作。如果需要在工件几个互成不同角度（一般是互相垂直）的表面上划线，才能明确表示加工界线的，称为立体划线。划线时要求线条清晰，定位要准确。

划线工序的作用主要有以下几点：

（1）确定工件的加工余量，并能及时地发现和处理不合格的毛坯或半成品工件；

（2）便于复杂工件在机床上的装夹，可按照划出的线条进行找正定位；

（3）当毛坯误差不大时，可以通过借料的方法进行补救，以提高毛坯的利用率；

（4）在板料上按划线下料，可以正确排料，合理利用材料。

所谓借料就是通过试划和调整，使各加工表面的余量互相借用，合理分配，从而保证各加工面都有足够的加工余量，从而使缺陷在加工后被排除。

一、划线工具

1. 划线平台

划线平台又称划线平板，它是用铸铁毛坯加工制成的，主要是用来作为划线时的基准平面，如图 1-3-1 所示。

图 1-3-1　划线平台

2. 高度划线尺

高度划线尺又称高度游标卡尺，它是一种比较精密的量具和划线工具，可以用来划线和测量高度，如图 1-3-2 所示。

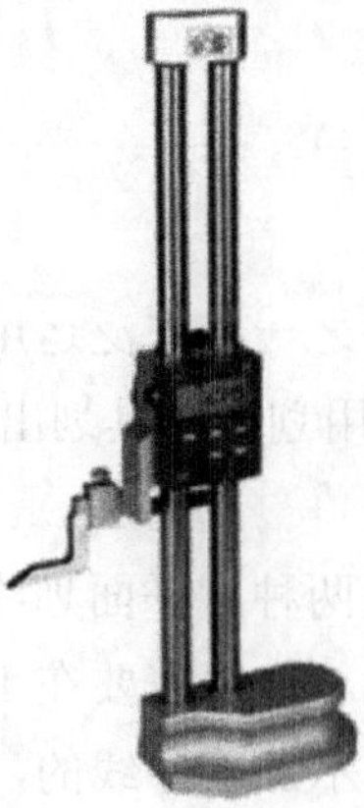

图 1-3-2　高度游标卡尺

3. 划规

划规是用来划圆和圆弧、等分线段、等分角度和量取尺寸的工具，它的用法与制图工具中的圆规相似，如图 1-3-3 所示。

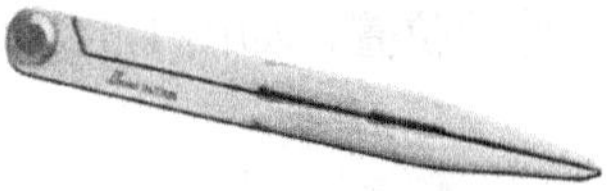

图 1-3-3　划规

4. 划针

划针是直接在工件上连线的工具，一般用钢丝或高速钢制成，如图 1-3-4 所示。

图 1-3-4　划针

划线时针尖要靠近导向工具的边缘，上部向外倾斜 15°~20°，向划线方向倾斜 45°~75° 并一次划出，如图 1-3-5 所示。用划针划线时不可以重复。为使划出的线条清晰准确，针尖要保持尖锐锋利，用钝后可用油石修磨。

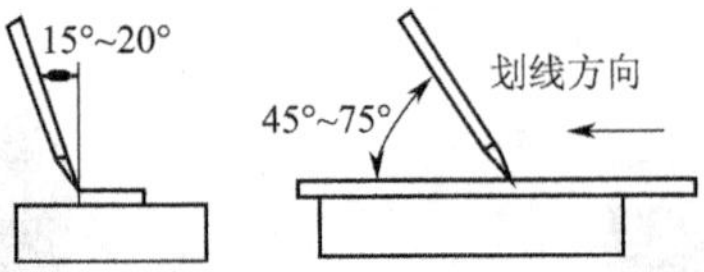

图 1-3-5　划针的使用

5. 样冲

样冲用来在工件所划的加工线条上打样冲眼，是加强标记的工具，如图 1-3-6 所示。

图 1-3-6　样冲

常在工件划线后，用手锤敲击样冲来打样冲眼，以防止工件上划好的线在加工中被磨掉，也用于划圆弧或钻孔时中心的定位。

6. 划针盘

划针盘是带有划针的可调划线工具，主要用于立体划线和校正工件的位置。划针盘由底座、立杆、划针和锁紧装置等组成。划针的两端常分为直头端和弯头端：直头端用来划线，弯头端用来找正工件的位置，如图 1-3-7 所示。

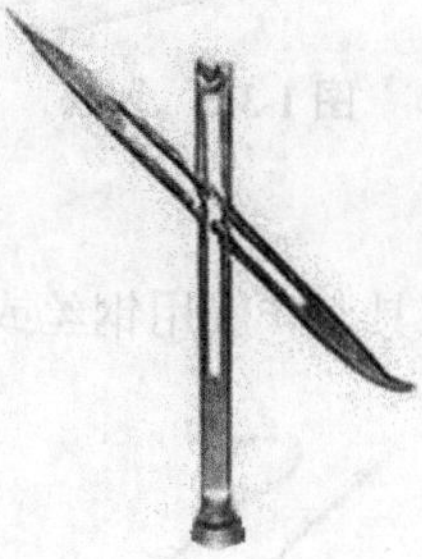

图 1-3-7　划针盘

7.V 形块和方箱

V 形块和方箱是用来支撑、夹持工件的，也作为薄形工件依靠的工具，如图 1-3-8 和图 1-3-9 所示。

图 1-3-8　V 形块

图 1-3-9　方箱

8. 千斤顶

千斤顶是划线或检测工件时的支撑工具，一般用来支撑形状不规则、带有伸出部分或较重的工件，以便工件进行校验、找正和划线，通常三个为一组使用。如图 1-3-10 所示。

图 1-3-10　千斤顶

二、平面划线前的准备工作和划线基准的选择

在划线之前，首先要看懂图样和工艺要求，明确划线的任务，检查工件是否合格，然后对划线部位进行清理、涂色，确定划线基准，选择好划线工具。

1. 划线前的准备

划线前的准备包括对工件或毛坯进行清理、涂色等。常用的涂料有石灰水（毛坯件）和蓝油（半成品件），目的是为了更好地看清楚所划线条。

2. 划线找正与借料

在对零件毛坯进行划线之前，一般都要先进行安放和找正工作。所谓找正，就是利用划线工具（如划线针盘、角尺、单脚规等）通过调节支承工具，使工件上有关的毛坯表面处于合适的位置。对于毛坯工件，划线前一般都要先做好找正工作。

当毛坯的尺寸、形状或位置误差和缺陷难以用找正划线的方法来补救时，就需要利用借料的方法来解决。借料就是通过试划和调整，使各待加工表面的余量互相借用，合理分配，从而保证各待加工表面都有足够的加工余量，使缺陷在加工后得以排除。

3. 划线基准的选择

基准是划线时用来确定工件上其他点、线、面之间位置关系所依据的点、线、面。而划线基准是在划线过程中采用的基准。一般划线基准选用设计基准为基准。

划线基准一般根据三种类型进行选择：

（1）以两个（或两条）互相垂直的平面（或直线）为基准，如图 1-3-11（a）所示；

（2）以两条（或两个）互相垂直的中心线（或中心面）为基准，如图 1-3-11（b）所示；

（3）以一个平面和一条（或一个）中心线（或平面）为基准，如图 1-3-11（c）所示。

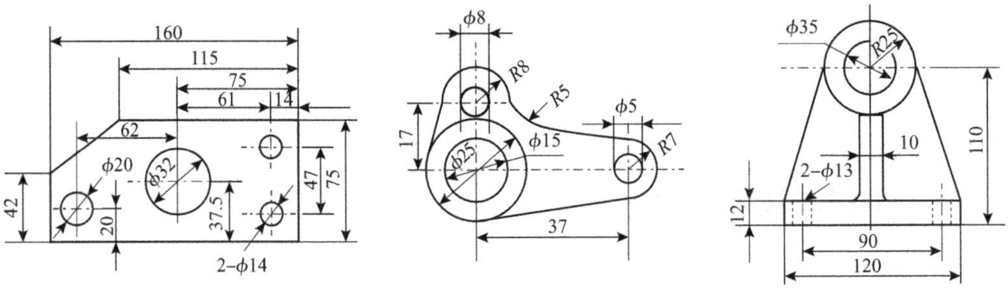

（a）　（b）　（c）

图 1-3-11　划线基准的选择

（a）以两个相互垂直的平面为基准　（b）以两条互相垂直的中心线为基准　（c）以一个平面和一条中心线为基准

三、划线的步骤

（1）看清、看懂图样，了解划线部位，明确划线要求和工艺。

（2）选择好划线基准。

（3）在工件上要求划线的部位涂色。

（4）根据图纸和工艺要求，先划基准线和位置线，再划加工线，即先划水平线，再划垂直线、斜线，最后划圆、圆弧和曲线，所划线条要求清晰均匀。

（5）立体工件按上述方法，进行翻转放置依次划线。

（6）划线结束后，按照图样和工艺要求，对工件依划线顺序从基准开始逐项检查，对错划或漏划应及时改正，保证划线的准确性。

（7）检查无误后在加工界线上打样冲眼；样冲眼必须打正。毛坯面要适当深些；已加工面或薄板件要浅些，密度小些；精加工表面和软材料上可不打样冲眼。

思考与练习

（1）简述划线的作用。

（2）划线常用的工具有哪些？

（3）划线分为哪两种形式？

（4）借料的作用是什么？

（5）简述划线的基本步骤。

任务四　方料锯削加工

一、图样与技术要求

如图 1-4-1 所示，工件毛坯尺寸为 150 mm×25 mm×25 mm，材料为 45 钢。评分表见表 1-4-1。

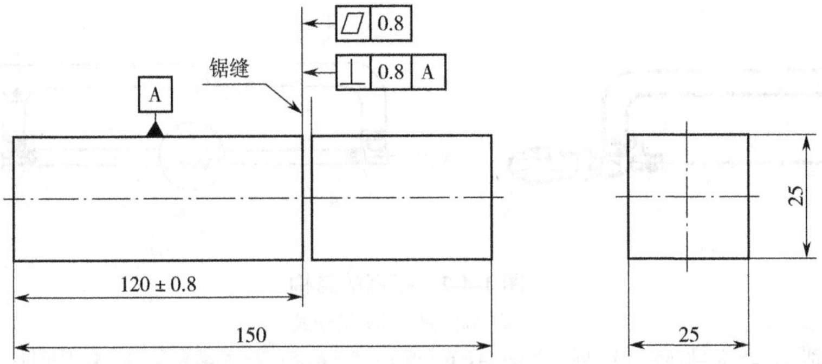

图 1-4-1 锯削方料

表 1-4-1 评分表

技术要求	配分	实测结果	得分
平面度公差为 0.8 mm	20		
垂直度公差为 0.8 mm	30		
尺寸为（120±0.8）mm	20		
姿势正确	15		
安全文明生产	15		

二、加工工艺分析

（1）检查来料尺寸。

（2）按图样要求划 120 mm 尺寸加工线。

（3）按锯削棒料方法锯下，达到尺寸精度、平面度和垂直度的要求，保证锯痕整齐，锯缝平直。

三、相关知识与技能

锯削是用手锯或机械锯（锯床）对材料或工件进行切断或锯出沟槽的加工方法。

1. 锯削工具——手锯

手锯由锯弓和锯条组成。

（1）锯弓是用来装夹并张紧锯条的工具，有固定式和活动式两种。其结构如图 1-4-2 所示。

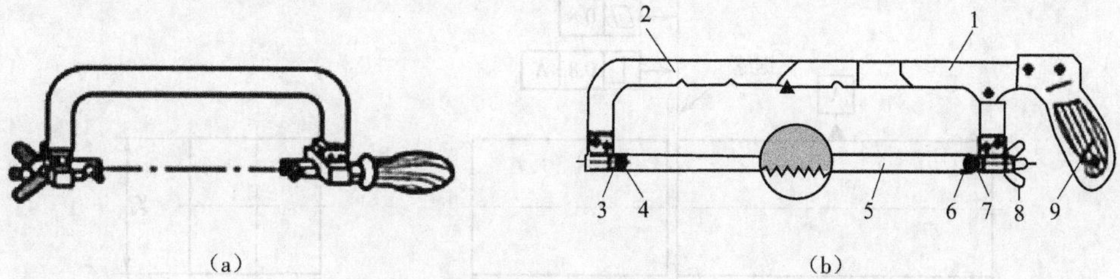

（a）　　（b）

图 1-4-2　手锯的结构

（a）固定式　（b）活动式

1—固定部分；2—活动部分；3—固定夹头；4、6—销子；5—锯条；7—活动夹头；8—蝶形螺母；9—手柄

（2）锯条是有齿刃的钢条片，是锯削的主要工具，如图 1-4-3 所示。锯条一般用渗碳钢冷轧而成，也有用碳素工具钢或合金工具钢经热处理淬硬制成的。锯条的长度是以两端安装孔的中心距来表示的，常用的锯条约长 300 mm、宽 12 mm、厚 0.8 mm。

锯条锯齿的粗细是以锯条每 25 mm 长度内的齿数来表示的，一般分粗、中、细三种，粗齿锯条适用于锯切软材料或较大的切面，细齿锯条适用于锯切硬材料或切面较小的工件，这样可以提高切削效率。

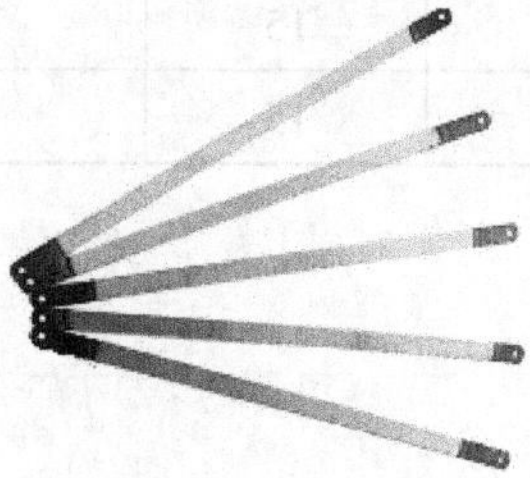

图 1-4-3　锯条

2. 锯削方法

1）锯条的安装

手锯只有向前推进时才有锯切作用，因此安装锯条时，应使锯条的锯齿方向向前（图 1-4-4 所示），锯齿朝下，锯条的松紧程度应调整适当，位置正确。其松紧程度以用手扳动锯条，感觉硬实并有一点弹性为佳。锯条安装调节后，还要检查锯条平面与锯弓中心平面是否平行，不得倾斜或扭曲，否则锯削时锯缝极易歪斜。

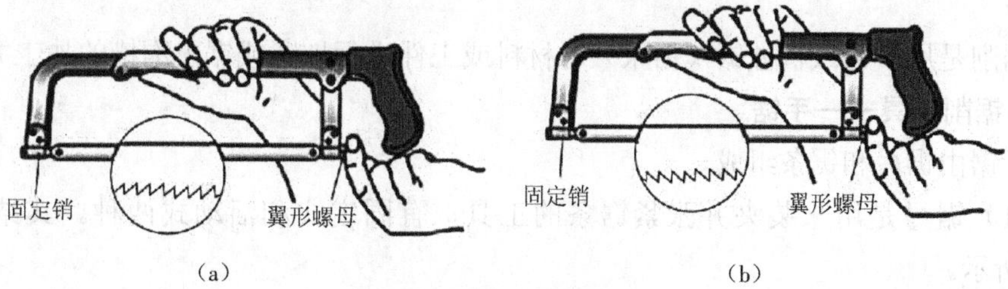

（a）　　（b）

图 1-4-4　锯条的安装

（a）正确的安装　（b）错误的安装

2）工件的装夹

工件要夹持在虎钳的左面，以方便操作，工件的锯缝离钳口约 15~20 mm，并与钳口侧面基本平行，工件装夹要牢固，以免锯切时工件松动而使锯条折断，如图 1-4-5 所示。

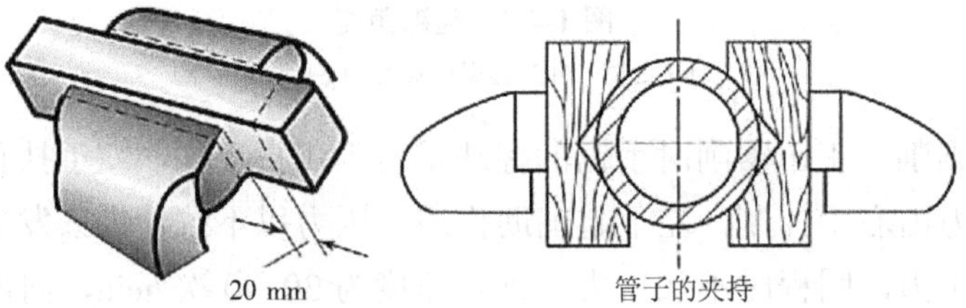

图 1-4-5　工件的装夹

3）锯削操作姿势

为了充分发挥较大的力量，操作者必须保持正确的站立位置，左脚跨前半步，两腿自然站立，人体重心稍微偏于后脚，视线要落在工件的切削部位，如图 1-4-6 所示。

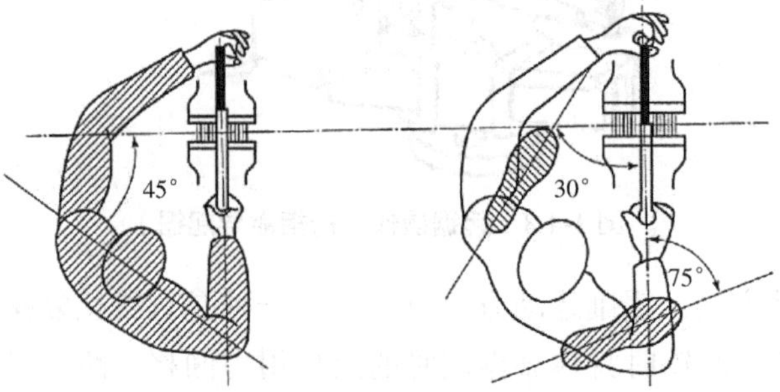

图 1-4-6　锯削站立姿势

4）锯削方法

（1）起锯。起锯方法有近边起锯和远边起锯两种，如图 1-4-7 所示。通常情况下采用远边起锯。因为这种方法锯齿不易被卡住，起锯时，左手拇指靠住锯条，如图 1-4-8 所示，使锯条能正确地锯在所需要的位置上，锯削行程要短，压力要小，速度要慢。无论用远边起锯还是近边起锯，起锯的角度都应在 15° 左右。如果起锯角太大，则切削阻力大，尤其是近边起锯时锯齿会被工件棱边卡住造成崩裂。起锯角太小，则不易切入材料，容易跑锯而划伤工件。

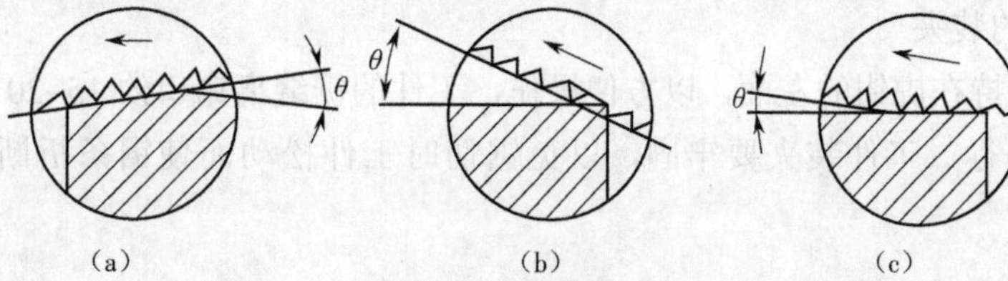

图 1-4-7　起锯角度

(a) 远边起锯　(b) 起锯角太大　(c) 近边起锯

（2）正常锯削。正常锯削时手锯的握法是右手满握手柄，左手扶住锯弓前端，锯削时推力和压力由右手控制，左手起辅助作用，压力以不致使锯条发生偏斜为准。锯弓向前推时加压力，回程时不加压力。锯削速度为 30~40 次 /min，向前推时要保持匀速，行程尽可能长，使锯条完全利用。

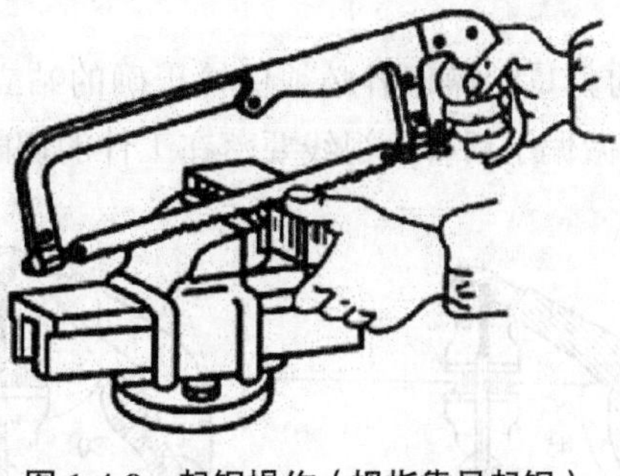

图 1-4-8　起锯操作（拇指靠导起锯）

（3）锯削运动方式。锯削运动方式有两种。一种是直线往复操作：推锯时身体与手锯同时向前运动；回锯时身体靠锯割时的反作用力回移，两手臂控制锯条做平直运动。直线往复操作适用于锯薄形材料和直槽。对锯缝断面要求平直的锯削，应采用此运动形式。另一种是摆动式操作：手锯推进时，身体略向前倾，双手随身体压向手锯的同时，右手下压，左手上翘；回程时右手上抬，左手自然跟回。这种运动形式，动作自然，不易疲劳，锯削时采用较多。

5）不同材料的锯削

（1）棒料的锯削。如果要求锯削面平整，则应从起锯开始连续锯削至结束。若对锯削面要求不高，则锯削时可以转过棒料已锯深的锯缝，选择锯削阻力小的地方继续锯削，以利于提高工作效率。

（2）管子的锯削。薄壁管子要用 V 形木垫夹持，以防夹扁和夹坏管表面。管子锯削时要在锯透管壁时向前转一个角度再锯，否则锯齿会很快损坏，如图 1-4-9 所示。

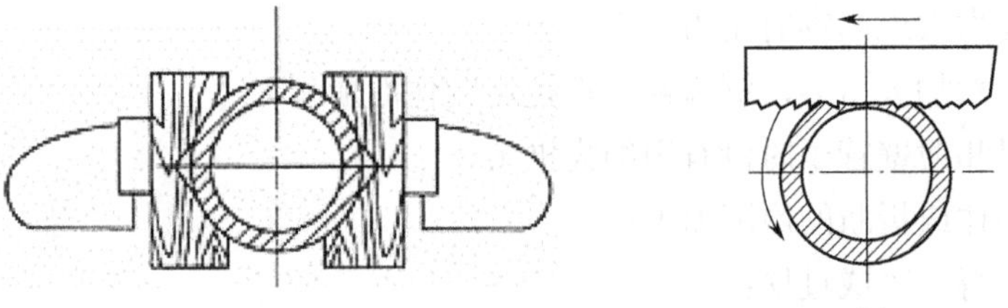

(a)　　(b)

图 1-4-9　管子的锯削

(a) 管子的夹持 (b) 转位后锯削

(3) 锯削薄料时，常将木板作为夹衬垫夹在台虎钳上，然后连木板一起锯切，或者采用横向斜推锯切的方法进行。锯切时，尽可能从宽的一面锯下去，这样同时锯削的齿数较多，锯齿不易被钩住和崩落，如图 1-4-10 所示。

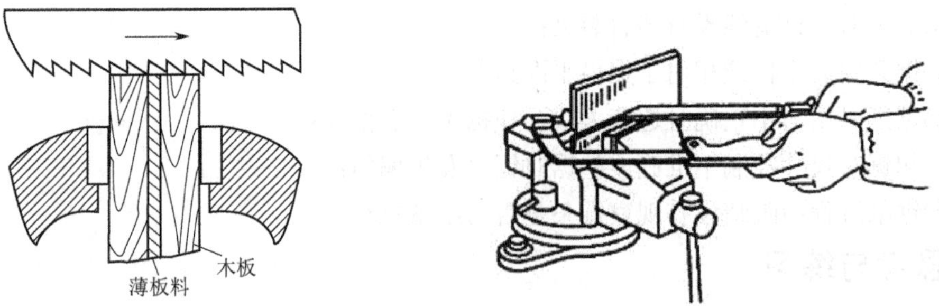

(a)　　(b)

图 1-4-10　薄板锯切

(a) 连木板一起锯切 (b) 横向斜推锯切

(4) 当锯缝深度超过锯弓高度的时候，应将锯条转过 90° 重新安装，使锯弓转到工件的一侧，当锯弓横下来其高度仍然不够时，可把锯条安装成锯齿向锯内的方向锯削，如图 1-4-11 所示。

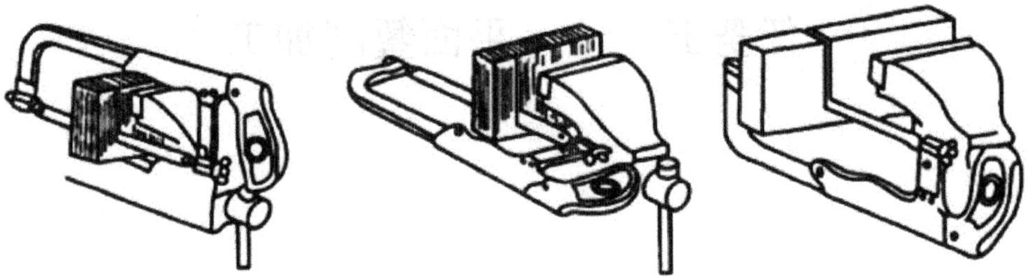

(a)　　(b)　　(c)

图 1-4-11　深缝锯切

(a) 正常锯切 (b) 锯条转过 90° 后锯切 (c) 锯条转过 180° 后锯切

6) 锯削时容易出现的问题及原因

(1) 锯齿崩齿的原因如下：

①起锯角度太大或起锯时用力过大；

②锯削时突然加大压力，锯齿被工件棱边钩住而崩裂；

③锯薄板料和薄壁管子时没有用细齿锯条。

（2）锯条卡住或折断的原因如下：

①锯条安装得过紧或过松；

②工件装夹不牢固或装夹位置不正确，造成工件松动或抖动；

③锯缝歪斜后强行纠正；

④运动速度过快、压力太大，锯条容易被卡住；

⑤新换锯条仍在老锯缝内锯削时易卡锯；

⑥工件锯断时没有减慢速度和减小压力，使手锯突然失去平衡而折断。

（3）锯缝不直或尺寸超差的原因如下：

①装夹时，锯缝线没有竖直装夹；

②锯条安装太松或相对于锯弓平面扭曲；

③锯削时用力不正确且速度太快，使锯条左右偏摆；

④起锯时尺寸控制不准确或起锯时锯路发生偏斜；

⑤锯削过程中眼睛没有观察锯条是否与线条重合。

思考与练习

（1）锯条松紧为何要调节适当？

（2）管子及薄板的锯削应注意哪些问题？

（3）简述锯条折断以及锯缝不直的原因。

任务五　平面錾削加工

一、图样与技术要求

如图 1-5-1 所示，工件毛坯尺寸为 100 mm×80 mm×45 mm，材料为 45 钢。评分表见表 1–5–1。

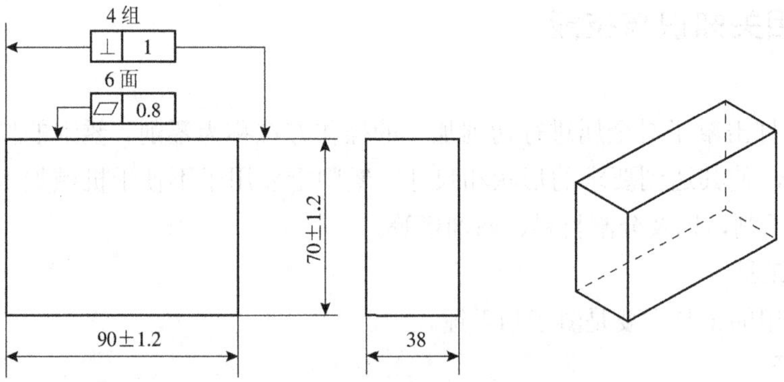

图 1-5-1 錾削平面

表 1-5-1 评分表

技术要求	配分	实测结果	得分
长度为（90 ± 1.2）mm	10		
宽度为 38 mm	10		
高度为（70 ± 1.2）mm	10		
平面度公差为 0.8 mm	18		
垂直度公差为 1 mm	24		
姿势正确	16		
安全文明生产	12		

二、加工工艺分析

（1）检查坯件尺寸。

（2）根据毛坯材料选择一个面，作为第一加工面（即基准）。粗錾后精錾，达到錾纹整齐，并用钢直尺检查錾削面，直至达到平面度 0.8 mm 要求后，即可作为六面体的加工基准面。

（3）按照图样要求，划线、錾削。按图纸要求进行尺寸錾削，直至达到技术要求。

（4）錾削时要注意工件夹紧，伸出钳口高度一般以 10~15 mm 为宜。同时下面加木垫块，台虎钳加软钳口保护工件。

（5）一次錾削量不宜过大，錾子后角要适宜。錾削大平面须开槽。

三、相关知识与技能

用手锤打击錾子对金属进行切削加工的操作方法称为錾削。錾削的作用就是錾掉或錾断金属，使其达到要求的形状和尺寸。錾削主要用于不便于机械加工的场合，如去除凸缘、毛刺，以及分割材料、凿油槽等。

1. 錾削工具

錾削所用的工具主要是錾子和手锤。

1）錾子

錾子由头部、錾身和切削部分组成，錾子是用碳素工具钢锻造而成的，经热处理使其硬度达 HRC 52~62，刃磨后便可使用。

（1）錾子的种类。根据用途不同，錾子一般分为扁錾、尖錾、油槽錾。

①扁錾。扁錾的切削刃较长，略带圆弧，切削面较扁平，如图 1-5-2（a）所示。其常用于錾削平面，切割，去凸缘、毛刺和倒角等。

②尖錾。尖錾的切削刃较短，两切削面从切削刃向整身逐渐狭小，切削刃与整身宽度方向呈“十字形”，如图 1-5-2（b）所示。其常用于沟槽，分割曲面、板料。

③油槽錾。油槽錾的切削刃较短，两切削面呈弧形，如图 1-5-2（c）所示。其主要用于錾油槽。

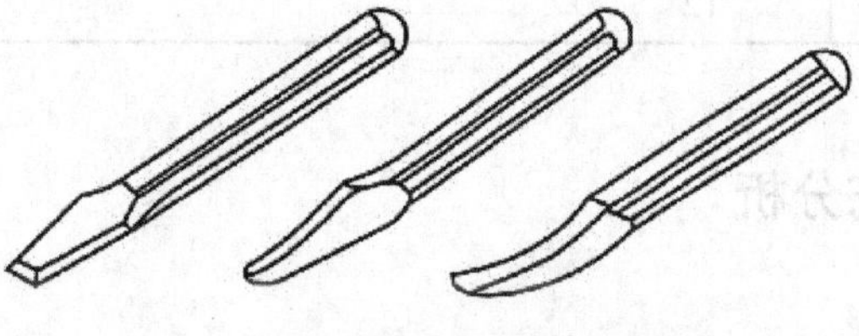

（a）　（b）　（c）

图 1-5-2　常用的錾子

（a）扁錾　（b）尖錾　（c）油槽錾

（2）錾子的几何角度。錾子一般用碳素工具钢 T7A 制成，其切削部分磨成楔形，錾子的几何角度如图 1-5-3 所示。

①楔角（β_o）。前刀面和后刀面之间的夹角称为楔角。楔角由刃磨形成，其大小取决于切削部分的强度及切削阻力的大小。楔角大时，刃部强度较高，但切削阻力也大。因此，在满足强度前提下应尽量选较小的楔角。

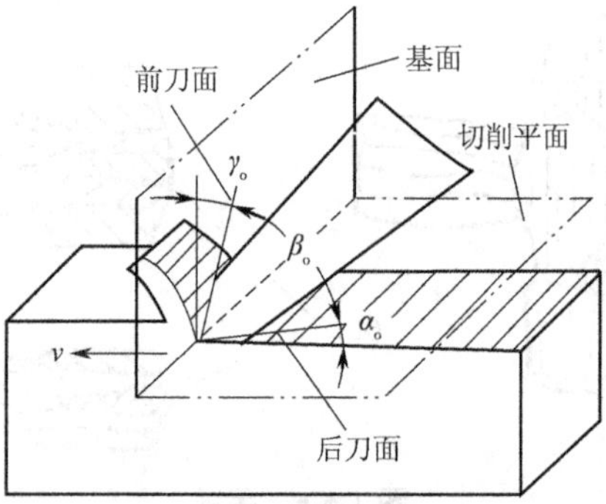

图 1-5-3　錾子的几何角度

②后角（α_o）。后刀面与切削平面的夹角称为后角。后角大小决定于錾子被掌握的方向，其作用是减小后刀面与切削平面的摩擦。后角大，切削深度大，切削困难；后角太小，易造成錾子从工件表面滑过。錾削时后角一般选 5°~8° 比较适宜。

③前角（γ_o）。前刀面和基面的夹角称为前角。前角对切削力、切屑变形都有影响，前角大，切削省力，切屑变形小。由于 $\gamma_o=90°-(\beta_o+\alpha_o)$，所以当楔角与后角确定之后，前角的大小也就确定下来了。

（3）錾子的修磨过程。

①磨平两斜面，并注意保持两斜面的对称性。

②磨平两侧面（位于斜面的两侧），并注意保持两侧面相互平行或对称。

③磨头部锋口，并注意保持两个平面的对称性。

修磨錾子的时候要经常蘸水来冷却，以防止切削部分退火硬度降低，錾子的高度需要略高于砂轮中心线，如图 1-5-4 所示。

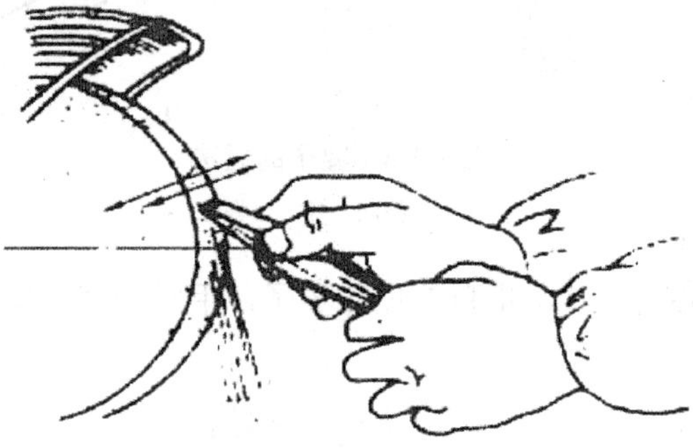

图 1-5-4　錾子的刃磨

2）手锤

手锤又称榔头，由锤头、木柄和楔子组成。手锤规格按锤头的质量分 0.25 kg、0.5 kg 和 1 kg 等多种。锤孔和木柄之间用楔子楔紧，以防锤头脱落，如图 1-5-5 所示。

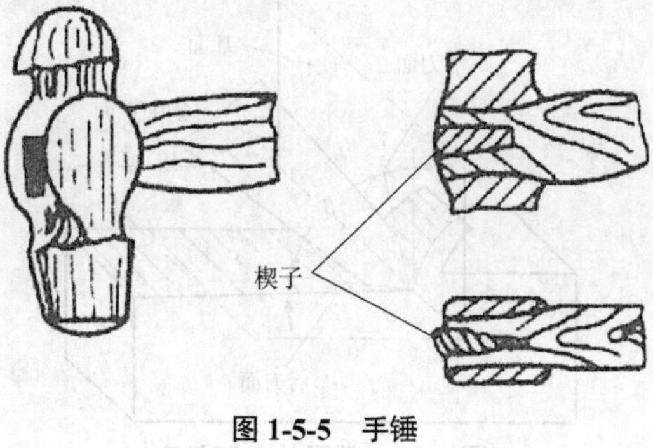

图 1-5-5 手锤

2. 錾削方法

1）錾子的握法

錾子主要用左手的中指、无名指握住，小指自然合拢，食指和大拇指自然接触，錾子头部伸出约 20 mm。轻松自如地握稳錾子，不能握得太紧，以免敲击时掌心承受的振动过大，或锤子打偏后伤手。錾削时握錾子的手要保持与小臂成水平位置，肘部不能下垂或抬高。

（1）正握法。手心向内，腕部伸直，用中指、无名指握住錾子，小指自然合拢，食指和大拇指自然伸直地松靠，錾子头部伸出约 20 mm，如图 1-5-6（a）所示。

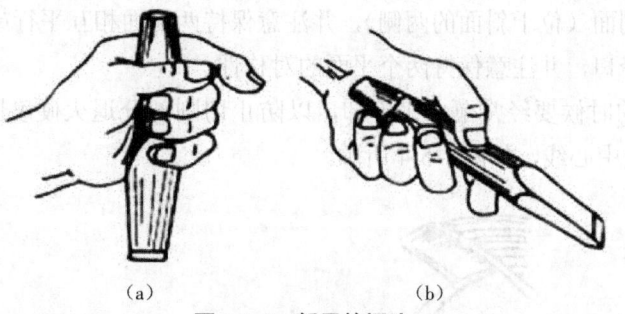

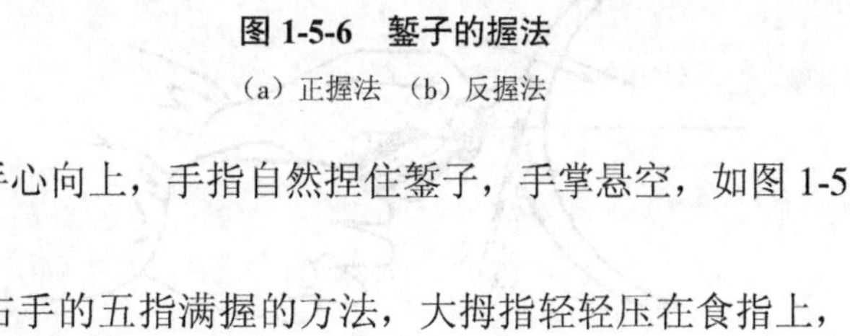

图 1-5-6 錾子的握法

（a）正握法 （b）反握法

（2）反握法。手心向上，手指自然捏住錾子，手掌悬空，如图 1-5-6（b）所示。

2）锤子的握法

锤子一般采用右手的五指满握的方法，大拇指轻轻压在食指上，虎口对准锤头方向，不能歪向一侧，木柄尾部露出 15~30 mm。在敲击过程中手握锤子的方法有两种：紧握法是五个手指从举起锤子至敲击都握紧并保持不变；松握法是在举起锤子时小指、无名指和中指依次放松，敲击时再依次收紧，如图 1-5-7 所示。

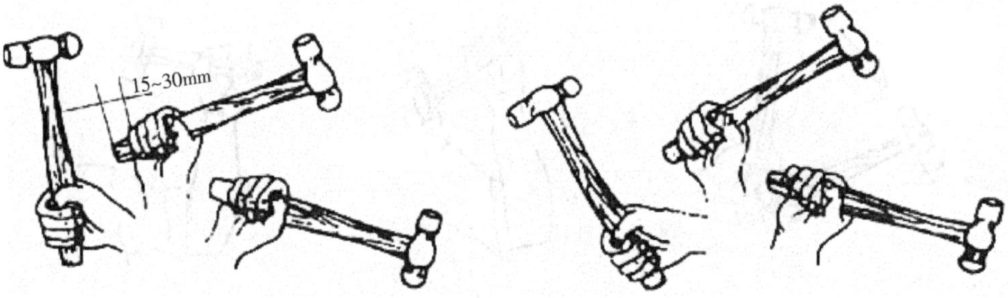

图 1-5-7　锤子的握法

（a）紧握法　（b）松握法

3）錾削姿势

錾削时身体与台虎钳中心线大致成 45° 角，且略向前倾，左脚跨前半步，膝盖处稍有弯曲，保持自然站立，右脚站稳伸直，不要过于用力。錾削的姿势如图 1-5-8 所示。

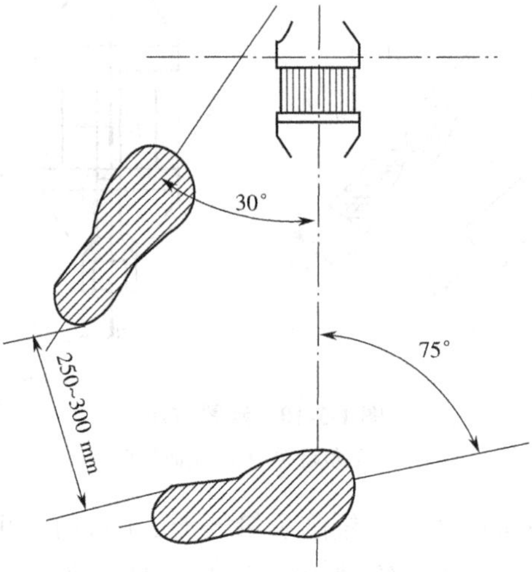

图 1-5-8　錾削姿势

4）挥锤方法

挥锤有三种方法，即腕挥、肘挥和臂挥，如图 1-5-9 所示。挥锤时要根据錾削余量的大小，选择不同的挥锤方法。当錾削余量较小及起錾和结尾时，錾削力较小，此时，应选用腕挥，即五指紧握锤，用手腕运动进行挥锤；当錾削余量较大，挥锤幅度也较大，锤击力较大时，应选用肘挥，即用手腕与肘部一起挥锤；臂挥是手腕、肘和全臂一起挥动，其锤击力最大，一般用于需要大力锤击的工作。

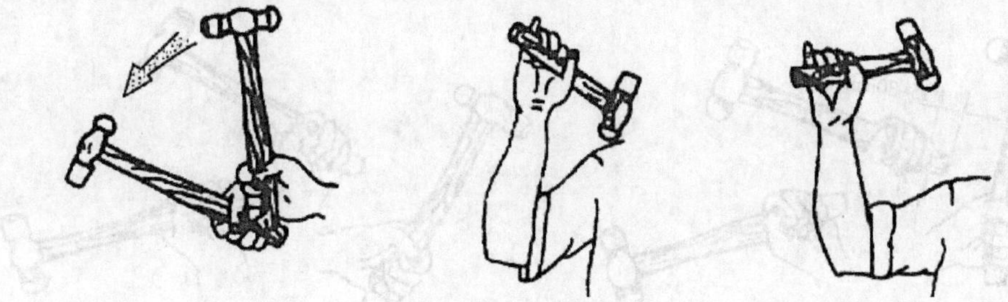

（a）　（b）　（c）

图 1-5-9　挥锤的方法

（a）腕挥 （b）肘挥 （c）臂挥

5）錾削方法

（1）起錾时，从工件的边缘尖角处着手，如图 1-5-10（a）所示；或者使錾子与工件起錾端面基本垂直，如图 1-5-10（b）所示；再用锤子轻敲錾子，即可准确和顺利地起錾。

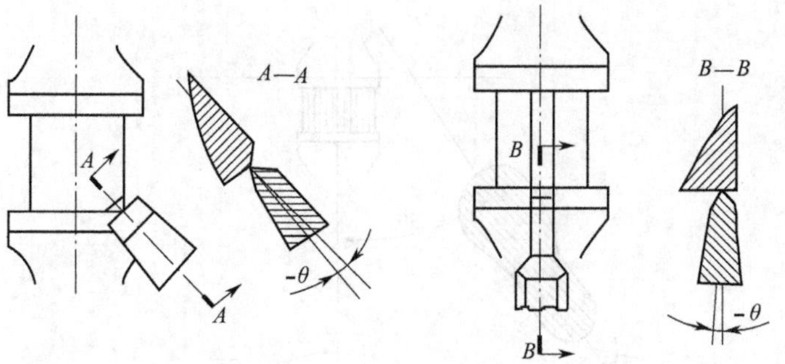

（a）　（b）

图 1-5-10　起錾方法

（a）斜角起錾 （b）正面起錾

（2）錾削深度选取 1 mm 为宜。錾削余量大于 2 mm 时，可分几次完成錾削。

（3）收錾时，每次錾削到距终端 10 mm 左右时，进一步錾削会使工件的边缘崩裂，应及时收錾，调转錾子从相反方向錾去剩余的部分，如图 1-5-11 所示。

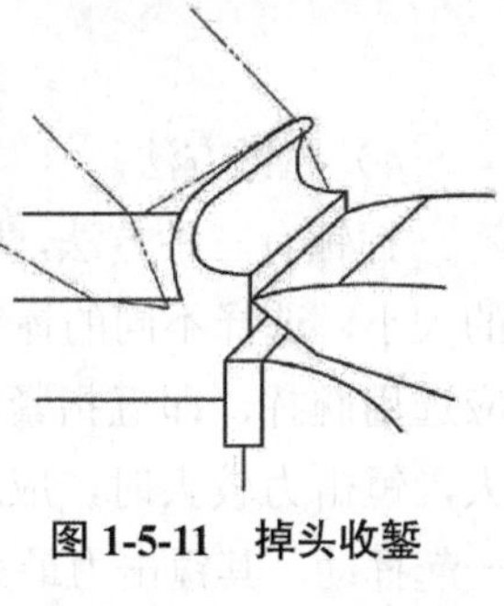

图 1-5-11　掉头收錾

6）錾削安全注意事项

（1）工件在台虎钳中央必须夹紧，伸出高度一般以离钳口 10~15 mm 为宜，同时下面要加木衬垫。

（2）发现手锤木柄有松动或损坏时，要立即更换或装牢。木柄上不应沾有油，以免使用时滑出。

（3）錾子头部有明显毛刺时，应及时磨去。

（4）手锤应放置在台虎钳右边，柄不可露在钳台外面，以免掉下伤脚，錾子应放在台虎钳左边。

思考与练习

（1）錾削时，挥锤的方式有哪几种？各有什么特点？

（2）錾削平面时，起錾和收錾各应注意什么问题？

（3）錾子的修磨过程包括哪些内容？

任务六　长方体锉削加工

一、图样与技术要求

如图 1-6-1 所示，工件毛坯尺寸为 120 mm×30 mm×30 mm，材料为 45 钢。评分见表 1-6-1。

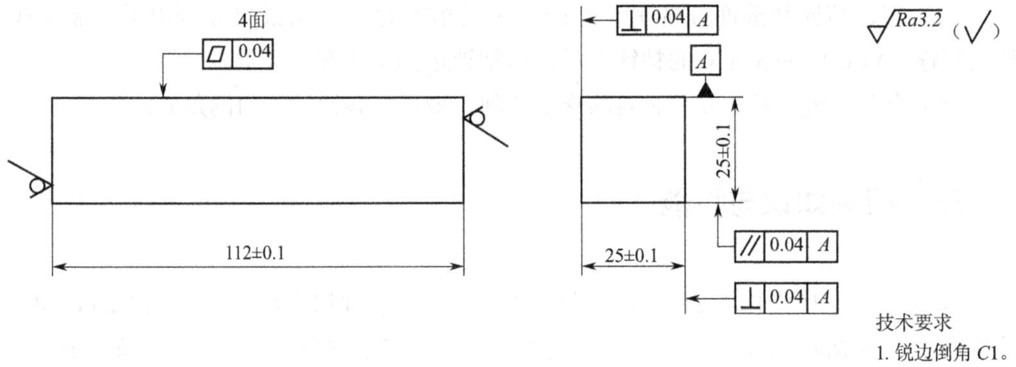

图 1-6-1　长方体零件图

表 1-6-1　评分表

技术要求	配分	实测结果	得分
长度为（112±0.1）mm	12		
宽度为（25±0.1）mm（2 处）	12		

续表

技术要求	配分	实测结果	得分
垂直度公差为 0.04 mm（2 处）	8		
平行度公差为 0.04 mm（1 处）	4		
平面度公差为 0.04 mm（4 处）	16		
表面结构参数 *Ra* 为 3.2 μm（4 处）	16		
锉纹整齐，倒角均匀（4 处）	16		
锉削姿势正确	10		
安全与文明生产	6		

二、加工工艺分析

（1）粗、精锉基准面 *A* 粗锉用 300 mm 粗板锉，精锉用 250 mm 细板锉。精度要求：平面度为 0.04 mm、表面结构参数 *Ra* 为 3.2 μm（表面结构参数用样块比较法目测检定）。

（2）粗、精锉基准面 *A* 的对面。用高度游标尺划出相距为 25 mm 的平面加工线，先粗锉，留 0.15 mm 左右的精锉余量，再精锉达到图样要求。

（3）粗、精锉基准面 *A* 的任一邻面。用 90° 角尺和划针划出平面加工线，然后锉削达到图样有关要求（垂直度用 90° 角尺检查）。

（4）粗、精锉基准面 *A* 的另一邻面。先以相距对面 25 mm 尺寸划出平面加工线，然后粗锉，留 0.15 mm 左右的精锉余量，再精锉达到图样有关要求。

（5）全部精度复检，并做必要的修整锉削。最后将各锐边均匀倒角 *C*1。

三、相关知识与技能

锉削是用锉刀对工件表面进行切削加工的方法。锉削精度可以达到 0.01 mm，表面结构参数可达 *Ra*0.8 μm。锉削的加工范围很广，它可以锉削平面、曲面、角度面、沟槽和各种形状复杂的表面等。

1. 锉削工具

锉刀是由碳素工具钢 T12、T13 或 T12A、T13A 制成的，经热处理淬火，切削部分的硬度达 HRC62~67，锉刀的结构如图 1-6-2 所示。

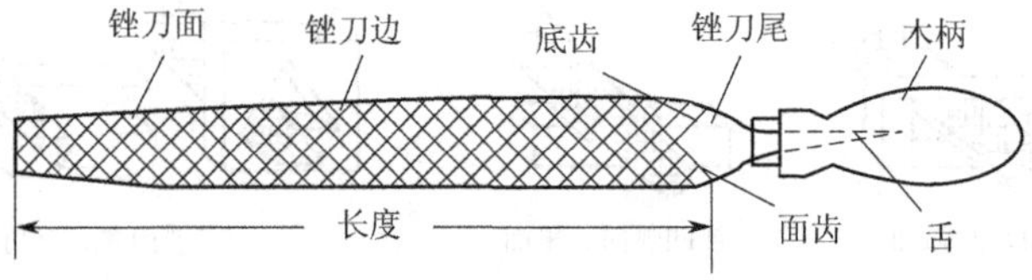

图 1-6-2　锉刀的结构

1）锉齿和锉纹

锉齿是锉刀面上用以切削的齿型，有铣制齿和剁制齿两种。齿纹是锉齿排列的形式，有单齿纹和双齿纹两种，如图 1-6-3 所示。

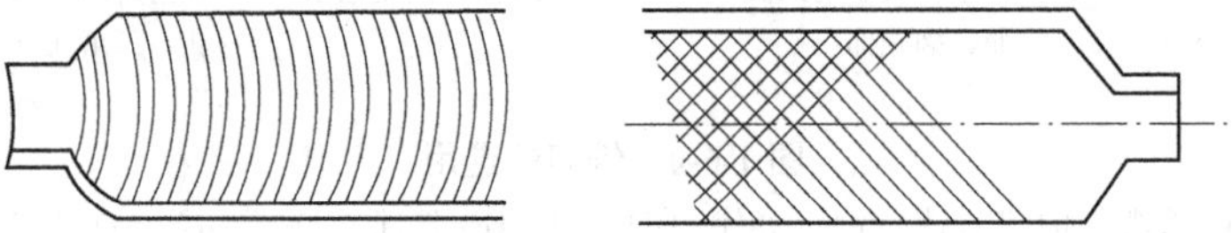

图 1-6-3　锉刀齿纹

（a）单齿纹（b）双齿纹

单齿纹锉的锉齿多为铣制齿，刀齿与轴线倾斜成一定角度，适用于锉削软质的有色金属。双齿纹锉的锉齿多为剁制齿：先剁上去的锉纹为底齿纹，其齿纹较浅；后剁上去的锉纹为面齿纹，其齿纹较深。面齿纹覆盖在底齿纹上交叉排列，起到分屑、断屑的作用，使锉削省力，适用于锉削硬金属材料。

2）锉刀的种类

（1）按锉刀齿的粗细进行分类，锉刀可分为粗锉刀、中锉刀、细锉刀和油光锉，一般粗锉刀齿数为 5.5~14 条/10 mm，适用于粗加工，还可以加工软金属材料；中锉刀齿数为 8~20 条/10 mm，适用于半精加工；细锉刀齿数为 11~28 条/10 mm，适用于精加工；油光锉齿数为 32~56 条/10 mm，适用于提高表面结构和修整尺寸。

（2）按锉刀剖面形状的不同，锉刀可分为板锉（平锉或扁锉）、方锉、三角锉、半圆锉和圆锉等。

3）锉刀的选用

锉刀的选用要根据加工零件的材料、加工表面质量和加工表面的形状来选择。具体选择情况如图 1-6-4 所示。

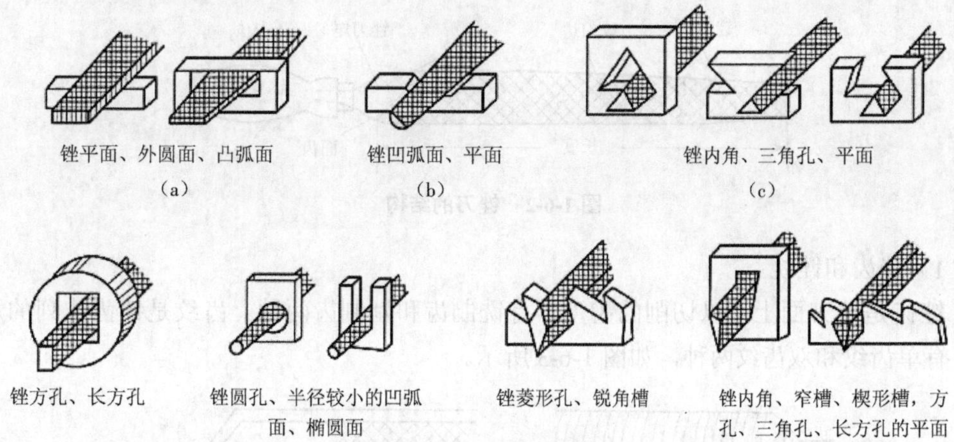

图 1-6-4　锉刀的选用

(a) 板锉　(b) 半圆锉　(c) 三角锉　(d) 方锉　(e) 圆锉　(f) 菱形锉　(g) 刀口锉

2. 锉削的操作

1）装夹工件

工件必须牢固地夹在台虎钳钳口的中部，需要锉削的表面应略高于钳口，但不能高得太多，夹持已加工表面时，应在钳口与工件之间垫上铜片或铝片。

2）锉刀的握法

根据锉刀的大小及使用情况不同，锉刀的握法也有所不同。较大锉刀的握法是右手紧握锉刀柄，柄端抵住在拇指根部的手掌上，大拇指放在锉刀柄上部，其余四指由下而上地握着锉刀柄；左手的基本握法是将拇指的根部肌肉压在锉刀头上，拇指自然伸直，其余四指弯向手心，用中指、无名指捏住锉刀前端。右手推动锉刀并决定推动方向，左手协同右手使锉刀保持平衡。大板锉的握法如图 1-6-5 所示。

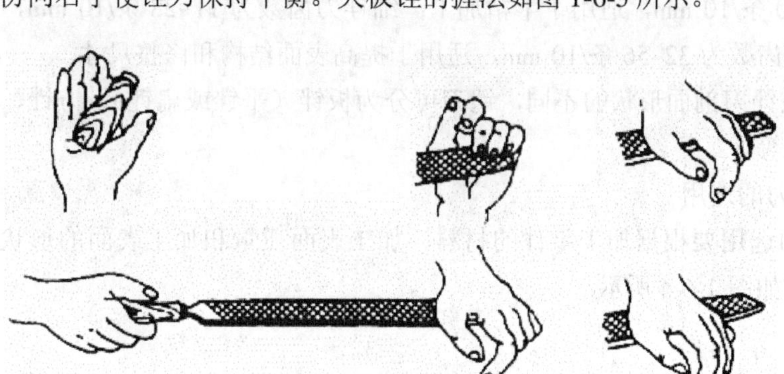

图 1-6-5　锉刀的握法

3）锉削的姿势和动作

锉削力的大小不同，锉削的姿势和动作也略有差异，因为正确的锉削姿势能够减轻疲劳，提高锉削的质量和效率。粗锉时，锉削力较大，所以姿势要有利于身体的稳定，动作要有利于推锉力的施加。精锉时，因为锉削力较小，所以锉削姿势要自然，动作幅度要小些，以保证锉刀运动的平稳性，使锉削表面的质量容易得到控制，如图1-6-6所示。

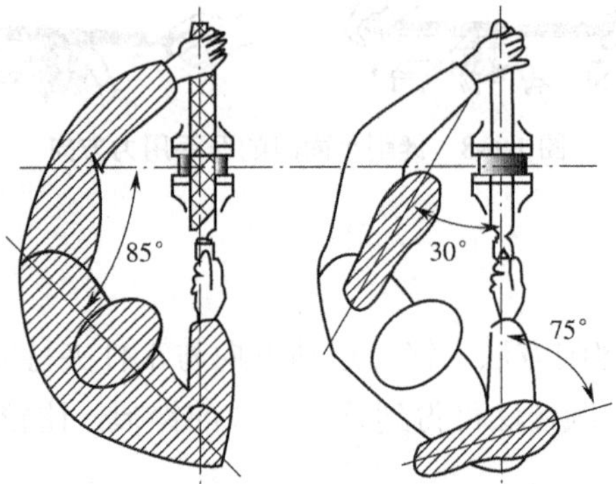

图 1-6-6　锉削时的站立部位和姿势

锉削时站立要自然，身体重心要落在左脚上，右膝伸直，左膝随锉削的往复运动而曲伸。在锉刀向前锉削的动作过程中，身体和手臂运动。开始时，身体向前倾斜10°左右，右肘尽量向后收缩；最初1/3行程时，身体前倾到15°左右，左膝稍有弯曲；锉至2/3时，右肘向前推进锉刀，身体倾斜到18°左右；锉到最后1/3行程时，右肘继续推进，身体随锉削时的反作用力退回到15°左右；锉削行程结束后，手和身体恢复到原来的姿势，同时将锉刀水平退回，如图1-6-7所示。

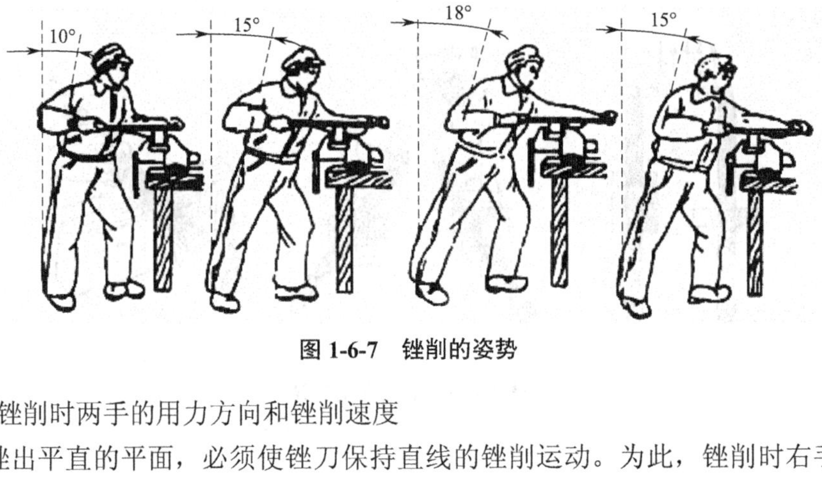

图 1-6-7　锉削的姿势

4）锉削时两手的用力方向和锉削速度

要锉出平直的平面，必须使锉刀保持直线的锉削运动。为此，锉削时右手的压力要随锉刀推动而逐渐增加，左手的压力要随锉刀推动而逐渐减小，回程时不加压力，以减小锉齿的磨损。锉削速度一般应在 40 次/min 左右，推出时稍慢，回程时稍快，动作要自然协调，如图 1-6-8 所示。

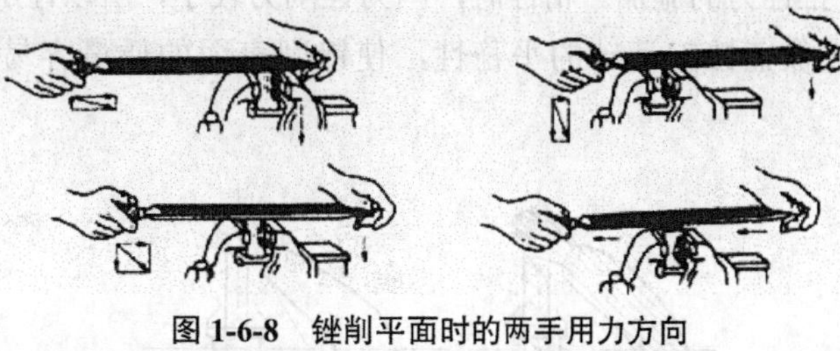

图 1-6-8　锉削平面时的两手用力方向

3. 锉削方法

1）顺锉法

顺锉法是最普通的锉削方法。锉刀运动方向与工件夹持方向始终一致，面积不大的平面常采用这种方法。顺锉法可得到整齐一致的锉痕，比较美观，精锉时常常采用，如图 1-6-9（a）所示。

2）交叉锉法

交叉锉法中锉刀运动方向与工件夹持方向成 30°~40°，且第一遍锉削与第二遍锉削交叉进行，由于锉痕是交叉的，容易判断锉削表面的不平程度，也容易把表面锉平。交叉锉法去屑较快，适用于粗锉平面，如图 1-6-9（b）所示。

3）推锉法

推锉法中锉刀的运动方向与锉身垂直。因为推锉刀的平衡易于掌握且锉削量很小，所以便于获得较平整的加工表面和较好的表面结构。推锉法一般适用于加工狭长面，由于推锉法不能发挥手臂的力量，因此锉削效率较低。如图 1-6-9（c）所示。

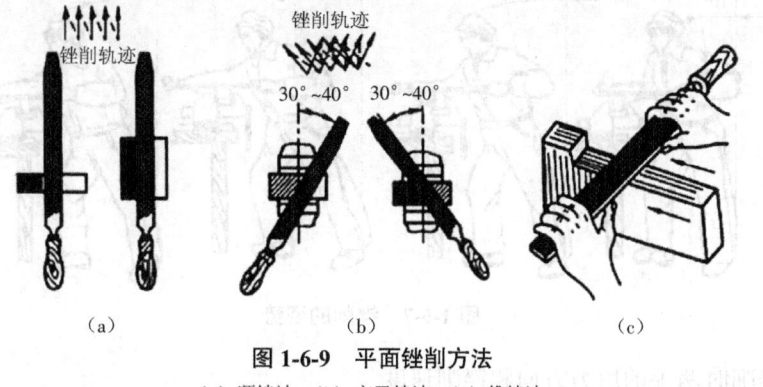

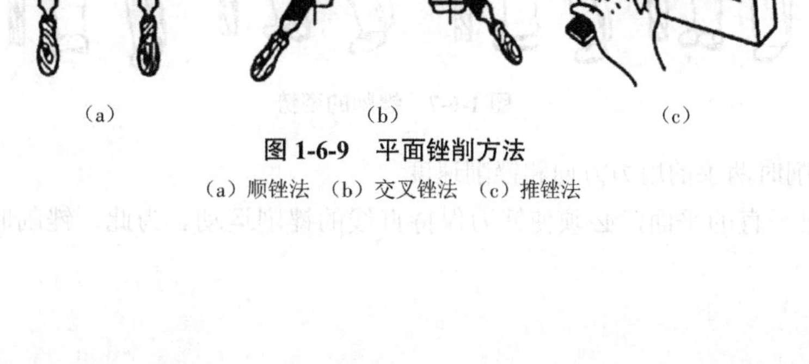

图 1-6-9　平面锉削方法

（a）顺锉法　（b）交叉锉法　（c）推锉法

4. 锉削注意事项

（1）锉刀必须装柄使用，以免刺伤手腕。松动的锉刀柄应装紧后再用，锉身或锉柄已经开裂或没有锉刀箍的锉刀不可使用。

（2）不准用嘴吹锉屑，也不要用手清除锉屑。当锉刀堵塞后，应用钢丝刷顺着锉纹方向刷去锉屑。

（3）对铸件上的硬皮或粘砂、锻件上的飞边或毛刺等，应先用砂轮磨去，然后锉削。

（4）锉屑时不准用手摸锉过的表面，因手有油污，再锉时易打滑。

（5）锉刀不能敲击工件，防止锉刀折断伤人。

（6）放置锉刀时，不要使其露出工作台面，以防锉刀跌落伤脚；也不能把锉刀与锉刀叠放或把锉刀与量具叠放。

思考与练习

（1）锉刀的种类有哪些？各适用于什么场合？

（2）锉削加工时如何合理地选用锉刀？

（3）简述锉削加工的规范姿势。

（4）平面锉削时常用的方法有哪几种？各种方法适用于哪种场合？

（5）锉削回程时为何不施加压力？

（6）锉削的速度大约为多少？

任务七　鸭嘴榔头的加工

一、图样与技术要求

如图 1-7-1 所示，工件毛坯尺寸为 120 mm×25 mm×25 mm，材料为 45 钢。

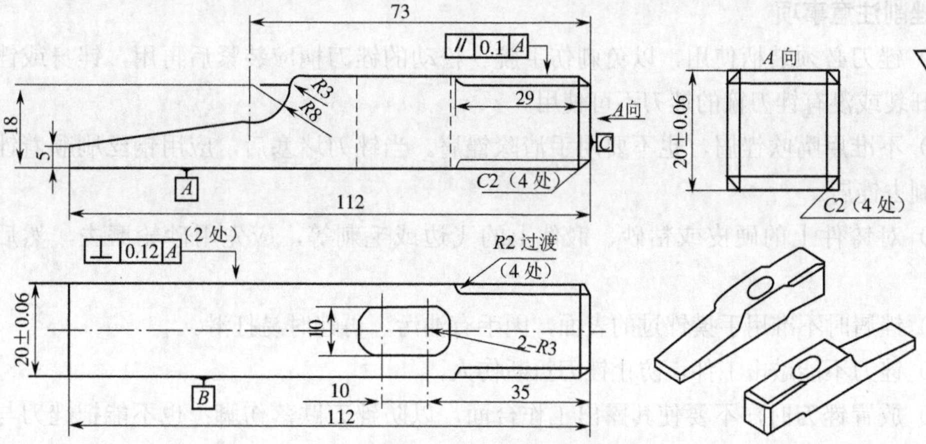

图 1-7-1　鸭嘴榔头零件图

表 1-7-1　评分表

技术要求	配分	实测结果	得分
工件长度 112 mm	8		
20±0.06 mm（2 处）	10		
腰孔宽度 10 mm	10		
平行度公差 0.1 mm（2 处）	12		
垂直度公差 0.12 mm（2 处）	12		
*R*3 mm、*R*8 mm 斜平面圆滑连接	10		
斜面平直度为 0.05 mm	10		
*R*3 mm 圆弧连接圆滑（4 处）	8		
倒角尺寸 *C*2（8 处）	8		
表面结构参数 *Ra*3.2 μm	2		
安全文明生产	10		

二、加工工艺分析

（1）检查来料尺寸是否符合图纸要求。

（2）按图纸要求锉出 112 mm×20 mm×20 mm 的长方体。

（3）以长面为基准锉一端面，达到基本垂直，表面结构参数为 *Ra*3.2 μm。

（4）按图纸在 *R*8 mm 处钻 $\phi 5$ mm 孔，用手锯按图纸加工线锯去多余的部分，并留锉削余量。

（5）用半圆锉按线粗锉 $R8$ mm 内圆弧，用板锉粗锉斜面与 $R3$ mm 圆弧至划线线条。后用细板锉细锉斜面，用半圆锉细锉 $R8$ mm 内圆弧面，再用细板锉细锉 $R3$ mm 外圆弧面。最后用细板锉及半圆锉推锉修整，达到各表面连接圆滑、光洁，纹理齐整。

（6）锉 5 mm 扁头部分，并保证工件长度 112 mm。

（7）按图纸划出控制加工线及钻孔的检查线，并用 $\phi9.8$ mm 的钻头钻孔。

（8）用圆锉锉通两孔，然后按图纸要求锉好腰孔。

（9）以长面及端面为基准划出加工线，要求两面同时划好，并按图纸尺寸划出 4 处 $C2$ 倒角加工线。

（10）锉 4 处 $C2$ 倒角达到要求。方法：先用圆锉粗锉出 $R2$ mm 圆弧，然后分别用粗、细板锉，粗、细锉 $C2$ 倒角，再用圆锉细加工 $R2$ mm 圆弧，最后用推锉法将锉纹推直，并用砂布打光。

（11）四角端部棱边倒角 $C2$。

（12）用砂布将各加工面打光，交件待验。

（13）尺寸检验合格后，再将腰孔加工成弧形喇叭口，20 mm×20 mm 的端面锉成微凸弧形面，最后将工件两端淬硬。

三、相关知识与技能

钻孔是用钻头在实心材料上加工出孔的方法。在钻床上钻孔时，钻头的旋转是主运动，钻头沿轴向移动是进给运动。

1. 麻花钻

标准麻花钻是钻孔的常用工具，一般用高速钢制成。

1）钻头的结构

麻花钻由柄部、颈部和工作部分组成。切削部分在钻孔时起主要切削作用。导向部分是指切削部分与颈部之间的部分，钻孔时起导向作用，同时也起着排屑和修光孔壁的作用。麻花钻柄部形式有直柄和锥柄两种：一般直径小于 13 mm 的钻头做成直柄，直径大于 13 mm 的钻头做成锥柄。钻头的规格、材料和商标等刻印在颈部。麻花钻的结构如图 1-7-2 所示。

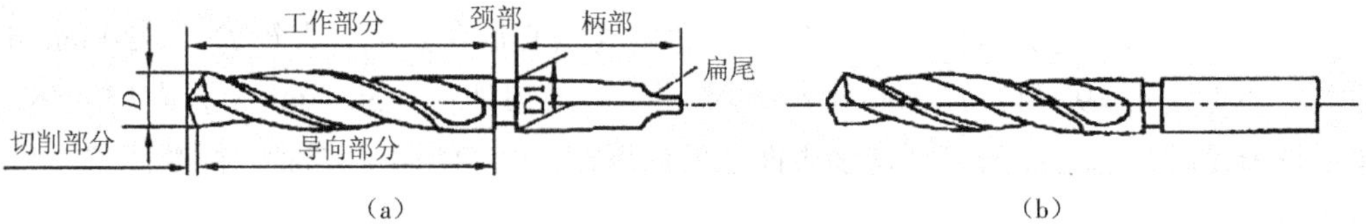

图 1-7-2　麻花钻

（a）锥柄　（b）直柄

2）麻花钻的切削角度（见图 1-7-3）

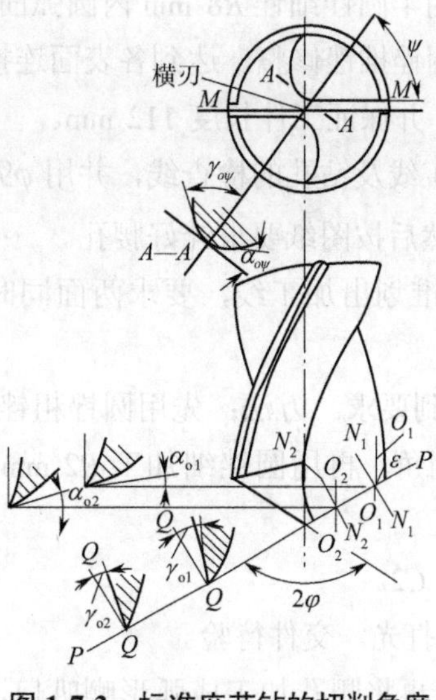

图 1-7-3　标准麻花钻的切削角度

（1）顶角 2φ。麻花钻的顶角又称锋角或钻尖角，它是两主切削刃在其平行平面 MM 上的投影之间的夹角。顶角的大小可根据加工条件在刃磨钻头时决定。标准麻花钻顶角 $2\varphi=118°\pm2°$，这时两主切削刃呈直线形。若 $2\varphi>118°$ 时，则两主切削刃呈内凹形；若 $2\varphi<118°$ 时，则两主切削刃呈内凸形。顶角的大小影响主切削刃上轴向力的大小。顶角愈小，则轴向力愈小，这时外缘处的刀尖角 ε 增大，有利于散热和提高钻头耐用度。但顶角减小后，在相同条件下，钻头所受的扭矩增大，切屑变形加剧，排屑困难，会妨碍冷却液的进入。

（2）前角 γ_o 在主截面（图 1-7-3 中 N_1–N_1 或 N_2–N_2）内，前刀面与基面之间的夹角称为前角。由于麻花钻的前刀面是一个螺旋面，沿主切削刃各点倾斜方向不同，所以主切削刃各点前角的大小是不相等的：近外缘处前角最大 $\gamma_{o1}=30°$；自外缘向中心逐渐减小，在钻心处至 $D/3$ 范围内为负值；接近横刃处的前角 $\gamma_{o2}=-30°$；横刃处 $\gamma_{o\psi}=-54°\sim-60°$。前角大小决定着切除的难易程度和切屑在前刀面上的摩擦阻力大小。前角愈大，切削愈省力。

（3）后角 α_o 在主截面内，后刀面与切削平面之间的夹角，称为后角。主切削刃上各点的后角刃磨不等，外缘处后角较小，愈接近钻心后角愈大。直径 D=15~30 mm 的钻头，外缘处 $\alpha_{o1}=9°\sim12°$；钻心处 $\alpha_{o2}=20°\sim26°$；横刃处 $\alpha_{o\psi}=30°\sim60°$。麻花钻后角的作用是为了减少后刀面与工件切削表面之间的摩擦。后角愈小，二者之间摩擦愈严重，但切削刃强度较高。

（4）横刃斜角 ψ。横刃斜角是横刃与主切削刃在钻头端面内的投影之间的夹角。

它是刃磨钻头时自然形成的，其大小与后角、顶角大小有关。后角刃磨正确的标准麻花钻，ψ=50°~55°。当后角磨得偏大时，横刃斜角就会减小，而横刃的长度会增加。

3）标准麻花钻的修磨

由于标准麻花钻存在较多的缺点，通常用修磨的方法改善其切削性能。一般是按钻孔的具体要求，有选择地对钻头进行修磨。麻花钻一般只刃磨两个后刀面，并同时磨出顶角、后角及横刃斜角，刃磨技术要求较高。其操作步骤如下。

（1）先检查砂轮表面是否平整，如有不平或跳动现象，需对砂轮进行修整。

（2）用右手握住钻头前端作为支点，左手紧握钻头柄部，将钻头的主切削刃放平，并置于砂轮中心平面以上，使钻头的轴线与砂轮圆柱母线所成角度为顶角的 1/2 左右，同时钻尾向下倾斜，如图 1-7-4 所示。

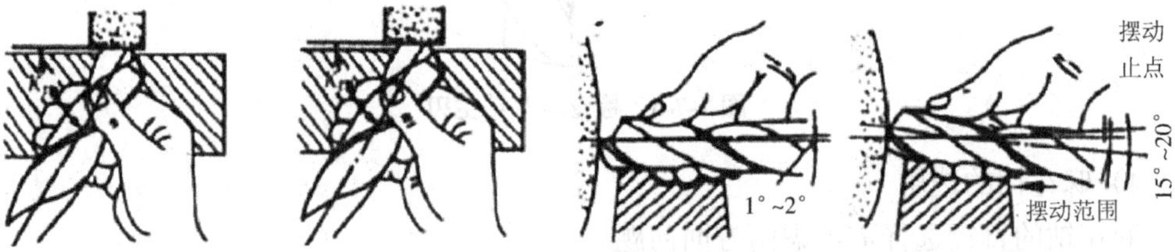

图 1-7-4　麻花钻的刃磨

（3）刃磨时，以钻头前端支点为圆心，左手握钻柄缓慢上下摆动并略做转动，同时磨出主切削刃和后刀面。摆动时向上不得高出水平线，向下不得过多，以防磨掉另一条主刀刃，如图 1-7-5 所示。

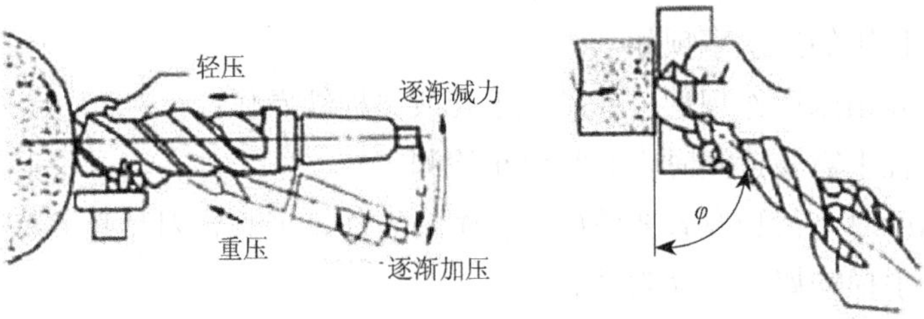

图 1-7-5　刃磨麻花钻主刃及后刀面

（4）将钻头转过 180°，用相同的方法刃磨另一条主切削刃和后刀面。两切削刃应经常交替刃磨，直到达到要求为止。

（5）按需要修磨横刃，将横刃磨短，钻心处前角磨大。

（6）麻花钻刃磨好后，常采用目测法检查。将钻头垂直竖立在与眼等高的位置，在明亮的背景下用肉眼看两刃；将钻头转过 180° 再观察，反复对比，直到觉得两刃基本对称。使用时，如发现仍有偏差，则需再次修磨。

（7）刃磨好的麻花钻可用角度尺检查。将角度尺的一边贴靠在麻花钻的棱边上，另一边放在麻花钻的刃口上，测量其刃长和角度；再将麻花钻转过 180°，用同样方法检查另一条主切削刃，如图 1-7-6 所示。

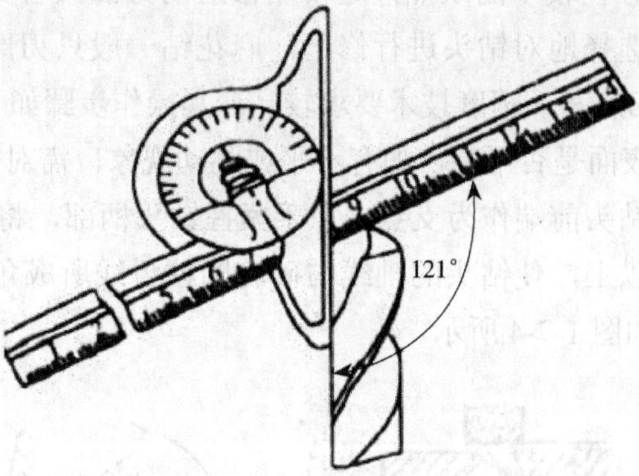

图 1-7-6　检查麻花钻主刃

4）麻花钻的选用

麻花钻的直径选择需考虑两方面问题。

（1）精度要求不高的孔加工，可用麻花钻直接钻出（直径等于孔径）。

（2）精度要求较高的孔加工，必须留下足够的扩、铰孔余量。麻花钻的长度一般为导向部分略长于孔深。

2. 常用钻床

（1）常用钻床有台式钻床、立式钻床和摇臂钻床。

（2）钻床使用的注意事项如下：

①严格遵守钻床操作规程，严禁戴手套或垫棉纱操作；

②钻孔过程中需要检测时，必须停车检测；

③操作钻床时，留长发者要戴好工作帽，工件、夹具、刀具必须装夹牢固、可靠；

④严禁用手摸铁屑、用嘴吹铁屑；

⑤钻深孔或铸铁工件时，要经常退屑，钻通孔时，要在工件底部垫上垫板或垫等高垫铁，以免钻坏工作台或平口钳。

3. 钻削用量与切削液的选择

1）钻削用量的选择

钻削用量是切削深度、进给量和钻削速度的总称。合理选择钻削用量，可提高钻孔精度、生产效率，并能防止机床过载或损坏。

（1）切削深度的选择。直径小于 30 mm 的孔可一次钻出；直径为 30~80 mm 的孔可分为两次钻削，先用（0.5~0.7）D（D 为孔径）的钻头钻底孔，然后用直径为 D 的钻头将孔扩大。这样可以减小切削深度及轴向力，保护机床，同时提高钻孔质量。

（2）进给量的选择。高速钢标准麻花钻的进给量可参考表 1-7-2 选取。孔的精度要求较高和表面结构参数值要求较小时，应取较小的进给量；钻孔较深、钻头较长、刚度和强度较差时，也应取较小的进给量。

表 1-7-2　高速钢标准麻花钻的进给量

钻头直径 /mm	<3	>3~6	>6~12	>12~25	>25
进给量 /（mm/r）	0.025~0.05	>0.05~0.10	>0.10~0.18	>0.18~0.38	>0.38~0.62

（3）钻削速度的选择。当钻头的直径和进给量确定后，钻削速度应按钻头的寿命选取合理的数值，一般根据经验选取，可参考表 1-7-3。孔深较大时，应取较小的切削速度。

表 1-7-3　高速钢钻头切削速度

工件材料	切削速度 /（m/min）
铸铁	14~22
碳钢	16~24
黄铜或青铜	30~60

2）切削液的选择

为了便于钻头散热冷却，减少钻削时钻头与工件、切屑之间的摩擦，消除黏附在钻头和工件表面上的积屑瘤，从而降低切削抗力、提高钻头寿命和改善加工孔表面的质量，钻孔时要加注足够的切削液。钻钢件上的孔时，可用 3%~5% 的乳化液；钻铸铁上的孔时，一般可不加或连续加注 5%~8% 的乳化液。钻各种材料选用的切削液见表 1-7-4。

表 1-7-4　钻各种材料选用的切削液

工件材料	切削液
各类结构钢	3%~5% 乳化液、7% 硫化乳化液
不锈钢、耐热钢	3% 肥皂加 2% 亚麻油水溶液、硫化切削油
纯铜、黄铜、青铜	5%~8% 乳化液
铸铁	可不用、5%~8% 乳化液、煤油
铝合金	可不用、5%~8% 乳化液、煤油、煤油与菜油的混合油
有机玻璃	5%~8% 乳化液、煤油

4. 钻孔的方法

1）划线

按照钻孔的位置、尺寸要求，划出孔的中心线，并打上样冲眼。按照孔的大小划出孔的圆周线。对于较大的孔，还应划出几个大小不等的检查圆、检查方框，以便钻孔时检查和找正钻孔位置，如图 1-7-7 所示。

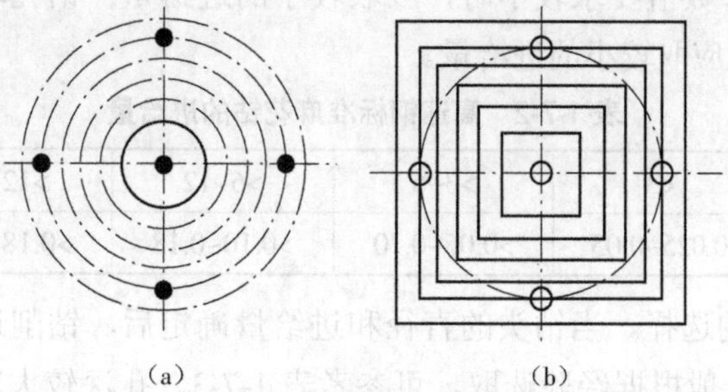
（a） （b）
图 1-7-7 检查麻花钻主刃
（a）检查圆 （b）检查方框

2）装夹钻头

（1）直柄钻头的装夹。直柄钻头用钻夹头夹持。先将钻头柄部伸入钻夹头的三爪内，其伸入长度不小于 15 mm，然后用钻钥匙旋转外套，使环形螺母带动三爪移动，做夹紧和放松动作，如图 1-7-8（a）所示。

（2）锥柄钻头的装夹。锥柄钻头的柄部与钻床主轴锥孔直接连接，要将锥孔和锥柄擦干净，并且要求钻头柄部的扁尾与主轴上的腰形孔对正，利用冲击力一次装夹，如图 1-7-8（b）所示。当钻头锥柄小于主轴锥孔时，需要过渡套连接，如图 1-7-8（c）所示。拆卸锥套和钻床主轴上的钻头时用斜铁，如图 1-7-8（d）所示。

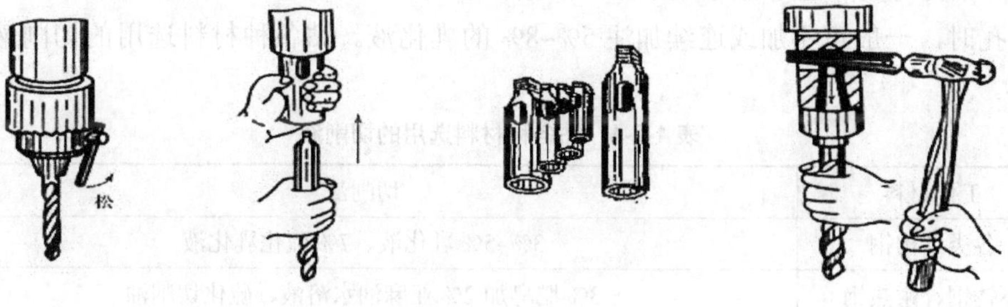

（a） （b） （c） （d）
图 1-7-8 钻头的装夹
（a）在钻夹头上安装钻头 （b）用锥套装夹 （c）锥套 （d）用斜铁拆钻头

3）起钻

开始钻孔时，先将钻头对准孔的中心钻出一浅坑，观察定心是否准确，并要不断校正，目的是使起钻浅坑与检查圆同心，防止出现偏心现象。

4）手动进给操作

当起钻达到钻孔的位置要求后，即可进行钻孔。手动进给时注意力度，不能使钻头发生弯曲变形，以免孔轴线歪斜。钻小孔或深孔时进给量要小，并且要经常退屑，以免切屑阻塞而折断钻头。钻孔将要穿透时，进给用力必须减小，以免进给突然过大，造成钻头折断或使工件随钻头一起转动造成事故。

5. 钻孔注意事项

（1）钻削不通孔时，应按钻孔深度调整好钻床上的挡块、深度标尺或采用其他控制措施，以免钻得过深或过浅，并注意退屑。

（2）钻削通孔时，当孔快要钻穿时，应减小进给力，以免发生“啃刀”，影响加工质量、折断麻花钻或使工件随着麻花钻转动造成事故。

（3）钻削深孔时，钻削深度达到麻花钻直径 3 倍时，就应退出排屑，并注意冷却润滑。

（4）钻 1 mm 以下小孔时，切削速度可选在 2 000~3 000 r/min 及以上，进给力小且平稳，不宜过大过快，防止麻花钻弯曲和滑移；且应经常退出麻花钻排屑，并加注切削液。

（5）钻 30 mm 以上的大孔，一般分成两次进行：第一次用 0.6~0.8 孔径的麻花钻；第二次用所需直径的麻花钻钻削。

（6）在斜面上钻孔时，可采用中心钻先钻底孔，或用铣刀在钻孔处铣削出小平面，也可用钻套导向等方法进行。

6. 其他加工环节注意事项及加工方法

在掌握钻孔方法的基础上，还要注意以下加工环节。

1）锉长方体

锉削基准选择原则如下：

（1）选择已加工的最大平整面作为锉削基准；

（2）选择质量较好的面作为锉削基准；

（3）选择划线基准、测量基准作为锉削基准；

（4）选择加工精度最高的面作为锉削基准。

长方体工件各表面的锉削必须按照一定的顺序进行，才能快速、准确地达到尺寸和相对位置精度要求，其一般原则如下：

（1）选择最大且表面质量相对较好的平面作为基准面进行锉削加工，达到规定的平面度要求；

（2）先锉大平面后锉小平面，以大平面控制小平面，此外，大平面锉削时更好把握，小平面锉削的难度比大平面大；

（3）先锉平行面后锉垂直面，即在达到规定的平行度要求后，再保证相关面的垂直度，一方面便于控制尺寸，另一方面在保证垂直度时可以进行平行度、垂直度两项误差的测量比较，减少累积误差。

2）立体划线

（1）该工件具有比较明显的立体划线特征。

（2）立体划线是在零件的不同表面（通常是相互垂直的表面）上划线。

（3）立体划线的前提条件有两个，一是要有三个或者三个以上的划线基准。二是立体划线的各面大多是互相垂直的。

3）加工腰孔

划线时，先用高度游标卡尺划出圆心的中心线，并打样冲眼。

钻孔时，先使用 $\phi2.5$ mm 钻头对准孔的中心钻出一浅坑，观察定心是否准确，并要校正，目的是使起钻浅坑与检查圆同心，再用 $\phi10$ mm 钻头对准引导孔钻出一浅坑，观察定心是否准确，并要校正，当起钻达到钻孔的位置要求后，即可扳动手柄完成钻孔。

钻孔后进行锉削加工，先用圆锉将两个相切的圆锉通，锉削时要保护好两个半圆，防止圆锉锉伤已钻削好的两个半圆，再用方锉将凸起锉掉，并保留腰孔前后八个切点，保证圆弧与平面圆滑连接。这样即可完成腰孔的加工。

半圆锉的选择原则与平锉相似，根据圆弧大小选择合适的半圆锉。

4）倒角

（1）划线。用高度游标卡尺在四个大面上划出 29 mm 倒角的加工界限。

（2）锉 $R2$ mm 圆弧。加工长方体上较长的棱，装夹相对的两条棱，装夹角度为 45°，用圆锉锉削加工 $R2$ mm 圆弧，圆弧起点为 29 mm 的加工界限。

（3）锉平面。在长方体大面的四条棱上，用平锉沿垂直棱的方向锉削加工窄平面的棱，该平面与上面加工好的 $R2$ mm 光滑连接。四条棱的倒角采用相同的加工工艺。

（4）锉削加工基准面 C 的四条棱。将工件装夹到台虎钳上，装夹时要保证锉削棱的位置为 45° 夹角。用平锉沿垂直棱的方向锉削加工窄平面，平面的宽度与之前锉好的四条棱等宽。

（5）锉削四个顶点。调整工件装夹位置，将工件基准面 C 四个顶点的锉削平面装夹到要锉削的角度，用平锉锉削加工四个正三角形。

思考与练习

（1）简述麻花钻的结构。

（2）如何选择钻孔时的切削用量？

（3）钻孔时，工件的常见装夹方式有哪些？

项目 2

车工

【项目描述】

普通车削加工是机械切削加工方法中应用最为广泛的方法之一，是加工轴类、盘类、套类零件的主要方法。采用车削加工的方法，可以加工各种回转体的内外表面，如内外圆柱面、圆锥面、切槽、车削螺纹等。

【学习目标】

（1）了解车削加工的基本知识，了解车削加工的工艺特点及加工范围。

（2）初步了解车床的型号、结构及传动系统，熟悉车床组成部分及其作用，掌握卧式车床的主要操作方法并能正确操作。

（3）熟悉车床常用附件的大致结构、特点、装夹方法和用途，掌握轴类零件装夹方法的特点。

（4）掌握普通车刀的组成、安装及刃磨，了解车刀的主要角度及作用，并能正确使用常用的刀具和量具。

（5）掌握车削端面、外圆、锥面、螺纹、切槽的车削方法及测量方法，熟悉车削所能达到的尺寸精度、表面结构参数值范围，能独立加工一般的零件，具有一定的操作技能和车削工艺知识。

任务一　车工一般知识

一、车工的作用及主要工作内容

车削是机械加工中最基本的一种切削方法，可完成的加工工作很多。一般公差等级在 IT8 级以下、表面结构参数 Ra 值在 1.6 μm 以上的旋转表面都可以在车床上完成加工。

车床运动包括工件旋转和刀具移动两种。工件的旋转运动为主运动，刀具的移动称为进给运动。在普通车床上主要工作内容包括：车外圆、车端面、切槽和切断、钻中心孔、钻孔、镗孔、铰孔、车内外圆锥面、车各种螺纹、车成形面、滚花和绕弹簧等（如图 2-1-1 所示）。

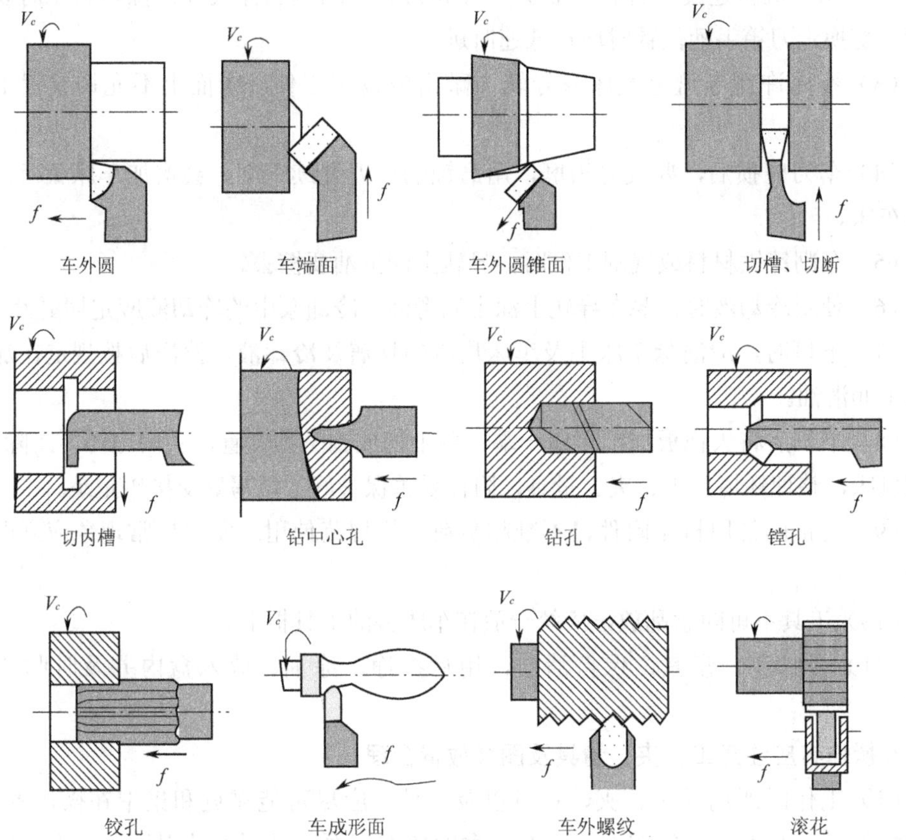

图 2-1-1　车削加工的基本内容

二、车工文明生产和安全操作技术

1. 文明生产

文明生产是工厂管理的一项十分重要的内容，它直接影响产品质量的好坏，影响设备和工、夹、量具的使用寿命。所以作为职业院校的学生，从开始学习基本操作技能时，就要养成文明生产的良好习惯。因此，要求学生实践操作时必须按照以下要求进行。

（1）开车前，应检查车床各部分机构是否完好，各传动手柄、变速手柄、限位、挡铁等位置是否正确，以防开车时因突然撞击而损坏机床，另外还应检查安全防护装置是否齐全。启动后，应使主轴低速空运转 5 min，使润滑油散布到车床需润滑的地方（这一步在冬天尤其重要），等车床运转正常后才能工作。

（2）调整机床速度、行程、装夹工件和刀具，测量工件尺寸及擦拭机床时要停车进行。变换走刀箱手柄位置时应在低速时进行。

（3）不允许在卡盘上及床身导轨上敲击或校正工件，床面上不允许放置工具或工件。

（4）车刀磨损后，要及时刃磨，用磨钝的车刀继续车削，会增加车床负荷，甚至损坏车床。

（5）车削铸铁材料或气割工件时，导轨上润滑油要擦去。

（6）使用冷却液时，要在导轨上涂上润滑油。冷却泵中的冷却液应定期更换。

（7）下班时，应清除车床上及车床周围的切屑及冷却液，擦净后按规定在加油部位加上润滑油。

（8）下班后将大拖板摇至尾座一端，各手柄放到空挡位置，关闭电源，清除铁屑，擦拭机床，整理环境，工、夹、量具，附件妥善保管好，填写好交接班记录。

（9）要正确使用机床附件，不准超负荷、超规范使用。发现异常现象应立即停车检查。

（10）工具不可随意乱放，应统一放在车床旁的工具柜上。

（11）爱护工、量具，保持清洁，用后擦净、涂油、放入盒内并及时归还到工具室。

2. 操作者应注意工、夹、量具及图样放置合理

（1）工作时所用的工、夹、量具以及工件，应尽可能靠近和集中在操作者周围。布置物件时，右手拿的放在右面，左手拿的放在左面，常用的放得近些，不常用的放得远些。物件放置应有固定位置，使用后要放回原处。

（2）工具箱的布置要分类，并保持清洁、整齐。

（3）图样、操作卡片应放在便于阅读的部位，并注意保持清洁和完整。

（4）毛坯、半成品和成品应分开，并按次序整齐排列，以便安放或拿取。

（5）工作的环境应经常保持整齐、清洁。

3. 安全操作技术

（1）开车前，应认真检查车床各部分是否完好，各手柄位置是否正确。开动车床后应使主轴低速空运转 5 min，待运转正常后才能工作。

（2）工作中主轴需要变速时，必须先停车再变速。

（3）工作时应穿工作服、戴套袖。女学生应戴工作帽。

（4）工作时不得戴戒指或其他饰品。

（5）工作时头不应靠工件太近，高速切削时，必须戴防护镜。

（6）工作时不准戴手套。

（7）不准用手刹住转动着的卡盘。

（8）车床转动时，不准测量工件，不准用手去触摸工件的表面。

（9）应该用专用的钩子清除切屑，不允许用手直接清除。

（10）工件装夹完毕，应随手取下卡盘扳手。棒料伸出主轴后端过长时，应使用料架或挡板。

（11）每个工作班结束后，应关闭机床总电源。

4. 事故应急处理

（1）立即停车，关闭电源。

（2）保护好现场。

（3）及时向教师及有关人员汇报，以便分析原因，总结经验教训。

思考与练习

（1）车工主要工作内容包括哪些？

（2）车工的安全操作技术有哪些？

（3）车工文明生产必须做到哪些方面？

任务二　车工操作

一、车床的结构及其传动关系

下面主要以实习中常用的 CA6140 卧式车床为例进行介绍。

1. 车床的主要结构

在所有车床中，普通车床的加工范围较广，可适用于机修车间或单件小批量生产。以 CA6140 型卧式车床为例，介绍车床的基本构成。

CA6140 普通卧式车床是可加工最大回转直径为 400 mm 的基础型车床，具有性能良好、操作简便、外形整齐美观等特点。该车床的形式及主要部分如图 2-2-1 所示。

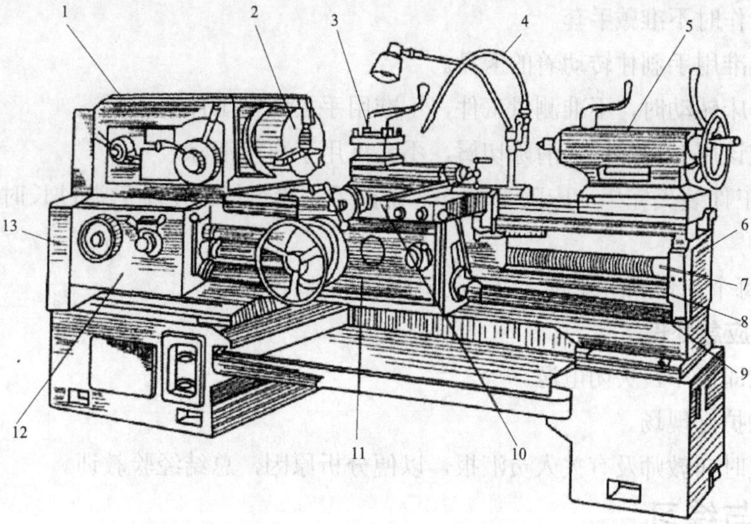

图 2-2-1　CA6140 卧式车床

1—主轴箱；2—卡盘；3—刀架；4—切削液管；5—尾座；6—床身；7—长丝杠；8—光杠；9—操纵杆；10—滑板；11—溜板箱；12—进给箱；13—交换齿轮箱

（1）主轴箱：是安装主轴和主轴变速机构的装置。电动机的运动经 V 形带传动给主轴箱，通过变速机构使主轴得到不同的转速。主轴又通过传动齿轮带动挂轮旋转，将运动传给进给箱。

（2）进给箱：是实现进给运动的变速机构，可按需要的进给设置或由螺距调整其变速机构，改变进给速度。

（3）溜板箱：是车床进给运动的操纵箱，可以将光杠传来的旋转运动变为车刀需要的纵向或横向直线运动，也可操纵开合螺母使丝杠带动刀架车削螺纹。

（4）滑板：由床鞍、中滑板和小滑板组成。床鞍在纵向车削时使用，中滑板在横向车削和控制切削深度时使用，小滑板在纵向车削较短的工件或车圆锥时使用。

（5）刀架：是用来夹持车刀使其做纵向、横向或斜向进给运动的。

（6）尾座：安装在床身导轨上的部件。在尾座的套筒内装上顶尖可以支撑工件，也可装上钻头、铰刀等在工件上钻孔、铰孔。

（7）床身：是车床的基础部件，可以用来连接各主要部件并保证各个部件之间有正确的相对位置。床身上的导轨，用以引导刀架和尾架相对于主轴箱进行正确的移动。

（8）丝杠：丝杠用来车螺纹，通过托板使车刀按要求的传动比做精确的直线移动。

（9）光杠：其将进给箱的运动传给溜板箱，使车刀按要求的速度做直线进给运动。自动走刀时也需要用光杠。

2. 车床的传动系统

CA6140 卧式车床的传动示意图如图 2-2-2 所示。主运动是通过电动机 1 驱动带 2，把运动输入给主轴 5。通过变速机构 4 变速，使主轴得到不同的转速，再经过卡盘 6（或夹具）带动工件旋转。而进给运动则是由主轴箱把旋转运动输出到交换齿轮箱 3，再通过进给箱 13 变速后由丝杠 11 或光杠 12 驱动溜板箱 9、床鞍 10、滑板 8、刀架 7，从而控制车刀的运动轨迹完成车削各种表面的工作。

CA6140 卧式车床主要有三种运动，也就有三条传动链，即主运动传动链、车削螺纹传动链、纵横向进给运动传动链。CA6140 卧式车床各主要部件之间的传动关系如图 2-2-3 所示。

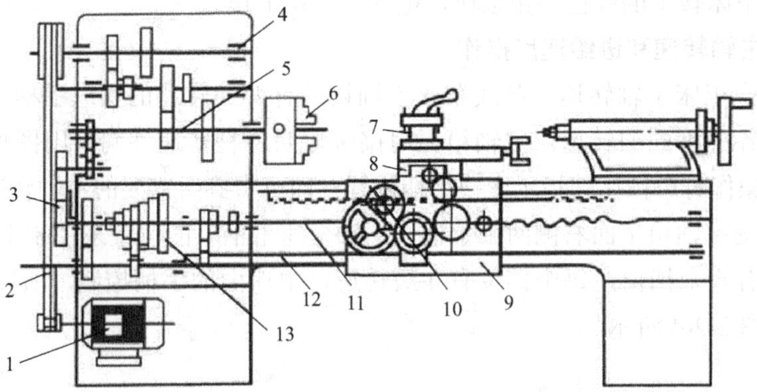

图 2-2-2　CA6140 卧式车床传动示意图

1—电动机；2—驱动带；3—齿轮箱；4—变速机构；5—主轴；6—卡盘；7—刀架；8—滑板；9—溜板箱；10—床鞍；11—丝杠；12—光杠；13—进给箱

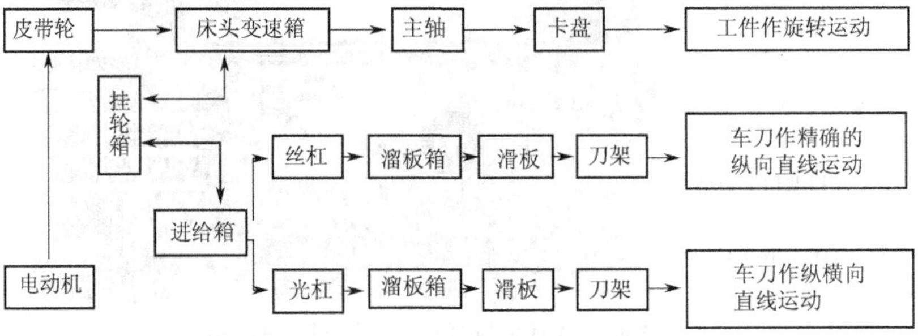

图 2-2-3　CA6140 卧式车床传动图

二、车床的基本操作

1. 车床的启动操作

（1）检查车床各变速手柄是否处于空挡位置，离合器是否处于正确位置，操纵杆是否处于停止状态。确认无误后，合上车床电源总开关。

（2）按下床鞍上的绿色启动按钮，电动机启动。

（3）向上提起溜板箱右侧的操纵杆手柄，主轴正转；操纵杆手柄回到中间位置，主轴停止转动；操纵杆向下压，主轴反转。

（4）主轴正反转的转换要在主轴停止转动后进行。

（5）按下床鞍上的红色停止按钮，电动机停止工作。

2. 变换主轴转速和进给速度操作

（1）变换车床主轴转速。卧式车床主轴箱外有变换转速的操纵手柄，改变手柄位置即可得到各种不同的转速。主轴箱上用铭牌注明各种转速并同时用图形表示出各手柄的位置，操作时可按铭牌指示变换手柄位置，即可得到所需要的主轴转速。

通过改变主轴箱正面右侧两个手柄（主轴变速手柄）的位置来控制主轴箱的变速。前面的手柄有 6 个挡位，每个挡位有 4 级转速，由后面的手柄控制，所以主轴共有 24 级转速，如图 2-2-4 所示。

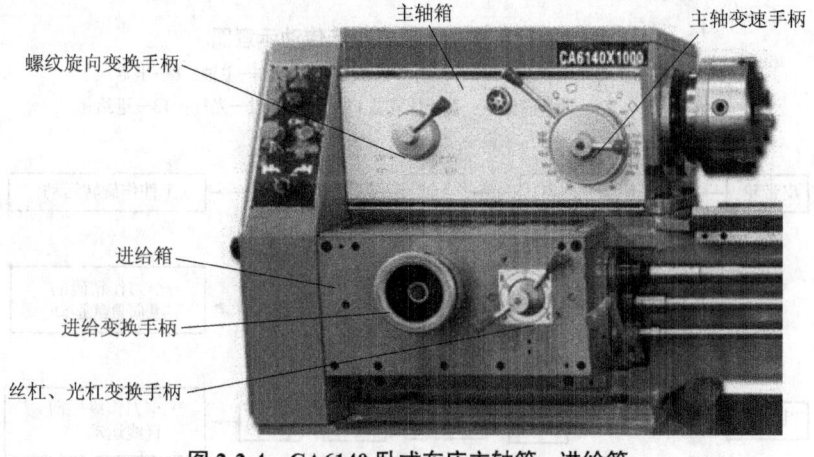

图 2-2-4　CA6140 卧式车床主轴箱、进给箱

主轴箱正面左侧的手柄（螺纹旋向变换手柄）用于螺纹的左右旋向变换和加大螺距，共有 4 个挡位，即右旋螺纹、左旋螺纹、右旋加大螺距螺纹和左旋加大螺距螺纹。

为保证安全，应在机床电动机已经关闭的前提下，进行变换主轴变速手柄的操作。变换主轴转速时，转动手柄的力不可过大。若发现手柄转不动或转不到位，主要是主轴箱内齿轮不能啮合，可用手转动卡盘，使齿轮的圆周位置改变，手柄即能扳动。

（2）变换车床进给速度。CA6140型车床的进给箱正面左侧有一个手柄（进给变换手柄），手柄向外拉出后可转动360°，圆周上有8个均布的数字，应根据铭牌指示调整，手轮转到需要的位置后，重新推入即可改变进给量或螺距大小。

右侧有前后两个手柄，前面的手柄是丝杠、光杠变换手柄。后面的手柄有4个挡位，与进给变换手柄配合，用以调整螺距或进给量。当调整螺距或进给量时，可通过查找进给箱油盖上的调配表来确定手轮和手柄的具体位置。

为保证安全，应在机床电动机已经关闭的前提下，进行变换进给箱手柄位置的操作，若发现手柄转不动，可用手转动卡盘。注意：转动卡盘时主轴转速应调整在高速位置，因为低速位置一般用手很难转动，待调整完毕后，再将主轴转速调整回初始状态，以免再次启动主轴产生误操作。

3. 滑板手动操作

溜板箱外操纵手柄用途及工作位置一般都用标牌标明，变换各手柄位置可使滑板做纵向或横向运动。

滑板又分为床鞍（旧称大拖板）、中滑板（旧称中拖板）和小滑板（旧称小拖板）。刀架在小滑板上面，可同时装夹4把车刀。CA6140型车床的溜板箱操纵手柄位置如图2-2-5所示。

图 2-2-5　CA6140 型车床溜板箱操纵手柄位置

（1）床鞍的横向移动由床鞍手柄控制。摇动床鞍手柄可使床鞍横向移动，手轮上刻度盘表示床鞍移动距离，刻度每转动1小格，移动距离为1 mm。旋紧压花螺钉，可将刻度圈锁紧；如松开螺钉，则可用手转动刻度圈调整零位。

（2）中滑板的横向移动由中滑板手柄控制。顺时针方向转动手柄时，滑板向前运动（即横向进给）；逆时针方向转动手轮时，向操作者运动（即横向退刀）。手轮轴上的刻度盘圆周等分为100格，手轮转过1格，横向移动0.05 mm。

（3）小滑板在小滑板手柄控制下可作短距离的纵向移动。小滑板手柄顺时针方向转动时，小滑板向左运动，逆时针方向转动时，小滑板向右运动。小滑板手轮轴上的刻度盘圆周等分为 100 格，手轮每转过 1 格，纵向或斜向移动 0.05 mm。

小滑板的分度盘在刀架需斜向进给车削短圆锥体时，可顺时针或逆时针在 90° 范围内偏转所需角度，调整时，先松开锁紧螺母，转动小滑板至所需角度位置后，再拧紧锁紧螺母将小滑板固定。

4. 溜板机动进给操作

（1）CA6140 型车床的纵、横向机动进给和快速移动采用单手柄操纵。自动进给手柄在溜板箱右侧，可沿十字槽纵、横扳动，手柄扳动方向与刀架运动方向一致，操作简单、方便。手柄在十字槽中央位置时，停止进给运动。在自动进给手柄顶部有一快进按钮，按下此钮，电动机快速工作，床鞍或中滑板手柄扳动方向作纵向或横向快速移动；松开按钮，电动机快速停止转动，快速移动中止。

（2）溜板箱正面右侧有一开合螺母操作手柄，用于控制溜板箱与丝杆之间的运动联系。

车削非螺纹表面时，开合螺母手柄位于上方。车削螺纹时，顺时针方向扳下开合螺母手柄，使开合螺母闭合并与丝杠啮合，将丝杆的运动传递给溜板箱，使溜板箱和床鞍按预定的螺距作纵向进给。加工螺纹完毕后，应立即将开合螺母手柄扳回到原位。

5. 尾座操作

车床尾座操作主要包括移动尾座和尾座套筒，车床尾座结构如图 2-2-6 所示。尾座可以沿着床身导轨前后移动，以支撑不同长度的工作。尾座套筒锥孔可供安装顶尖和钻头，套筒可以前后移动。其操作方法如下。

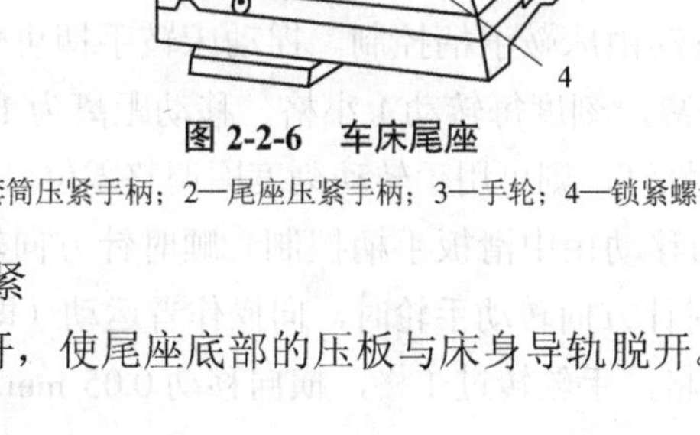

图 2-2-6　车床尾座

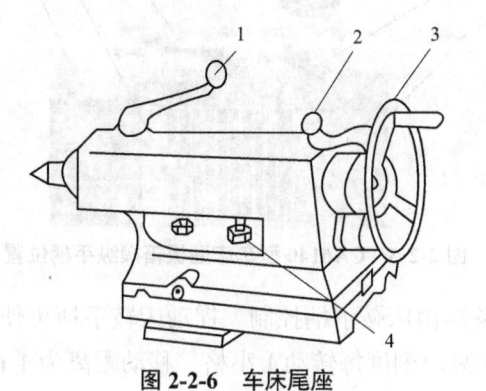

1—套筒压紧手柄；2—尾座压紧手柄；3—手轮；4—锁紧螺母

1）尾座的移动和锁紧

将尾座压紧手柄松开，使尾座底部的压板与床身导轨脱开。用手推动尾座，使尾座沿着床身导轨移动。将尾座压紧手柄扳紧，使尾座底部的压板紧压在床身导轨上，锁紧尾座。如果将车床尾座固定在导轨某一位置上，则可以直接用扳手将锁紧螺母扳紧。

2）尾座套筒的移动和锁紧

摇动手轮，使套筒前、后移动。注意，套筒不要伸出过长，以免影响刚度和防止套筒伸出到极限而使套筒内的丝杠与螺母脱开。

当需要固定尾座套筒时，直接扳紧套筒压紧手柄，锁紧尾座套筒。

三、车床的润滑和维护保养

为了保证车床的正常运转，减少磨损，延长使用寿命，应对车床的所有摩擦部位进行润滑，并注意日常维护保养。

1. 车床的润滑方式

（1）浇油润滑，通常用于外露的滑动表面，如床身导轨面和滑板导轨面等，用棉纱擦干净后用油壶浇油润滑，如图 2-2-7 所示。

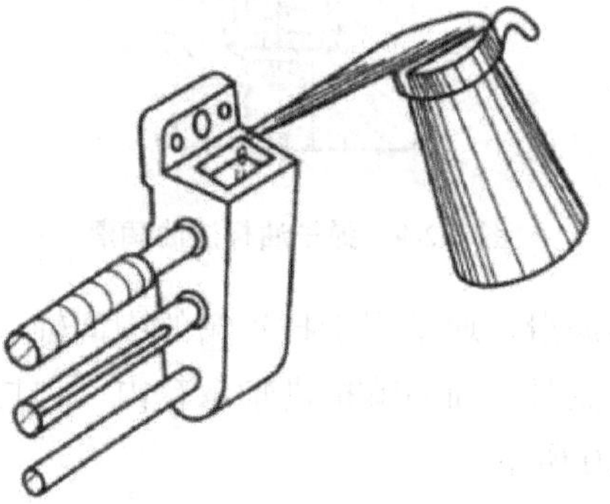

图 2-2-7　浇油润滑

（2）溅油润滑，通常用于密封的箱体中，如车床的主轴箱，它利用齿轮的转动把润滑油甩溅到油槽中，然后通过铜管输送到各个轴承进行润滑。

（3）油绳导油润滑，通常用于车床进给箱和溜板箱的油池中，它利用毛线吸油和渗油的能力，把机油慢慢地引到需要的润滑处，如图 2-2-8 所示。

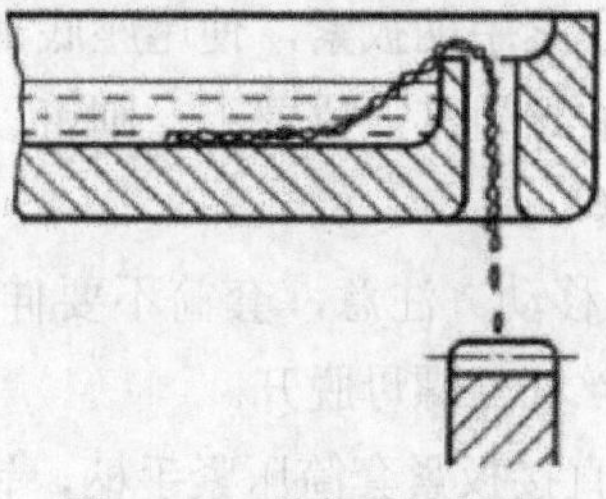

图 2-2-8　油绳导油润滑

（4）弹子油杯注油润滑，通常用于尾座和拖板摇手柄转动的轴承处。注油时，以油嘴把弹子压下，注入润滑油，使用弹子油杯的好处是可以防尘防屑，如图 2-2-9 所示。

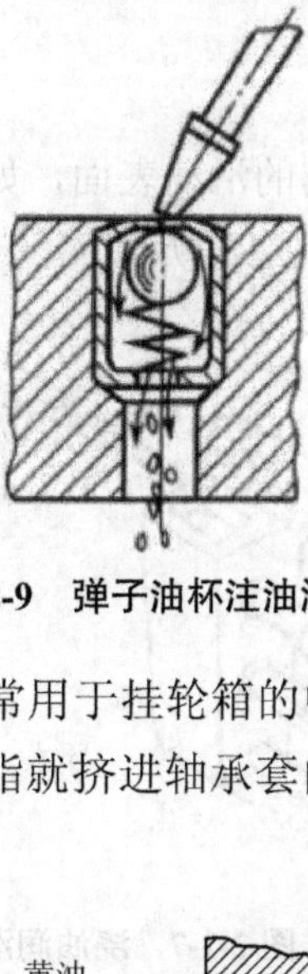

图 2-2-9　弹子油杯注油润滑

（5）黄油（油脂）杯润滑，通常用于挂轮箱的中间轴。使用时，先在黄油杯中装满工业油脂，当拧紧油杯盖时，油脂就挤进轴承套内。它具有存油时间长、不需要每天加油的特点，如图 2-2-10 所示。

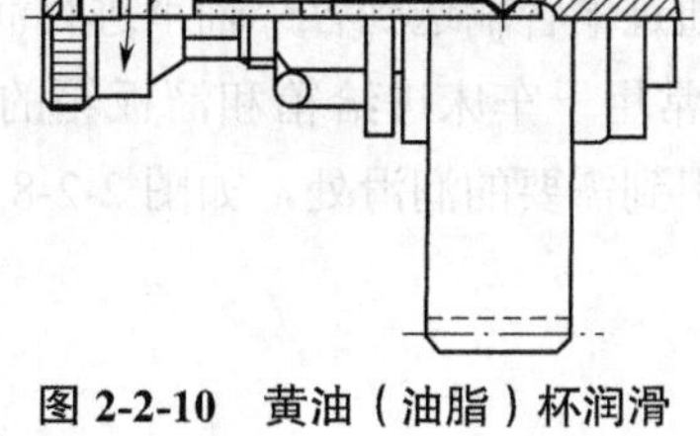

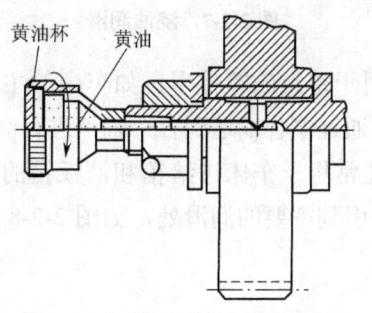

图 2-2-10　黄油（油脂）杯润滑

（6）油泵输油润滑，利用车床内的油泵来提供润滑油，通常用于转速高、润滑油需要量大的机构中。

2. 车床的维护保养

（1）工作结束后，应把铁屑清扫干净，擦净车床导轨面，要求无油污、无铁屑，并浇油润滑，使车床外表清洁，场地整齐。

（2）每周要求车床床身导轨面和中、小滑板导轨面及传动部位清洁、润滑、油眼畅通，油标油窗清晰，清洗机床油毛毡，并保持车床外表清洁和场地整齐等。

（3）维护保养完毕后，应将尾座、溜板箱移动至车床尾部。

思考与练习

（1）简述车床的主要结构有哪些？

（2）简述车床滑板手动操作的基本过程。

（3）车床的润滑方式有哪些？

（4）车床日常维护保养包括哪些内容？

任务三 车削刀具及刃磨

一、车床上常用刀具的种类

1. 按用途分类

车刀按用途分为端面车刀、外圆车刀、切槽刀、螺纹车刀、镗孔刀和成型刀等，按基本角度（主偏角）分为：45°、90°、75° 车刀。图 2-3-1 所示为常用车刀，其用途如下。

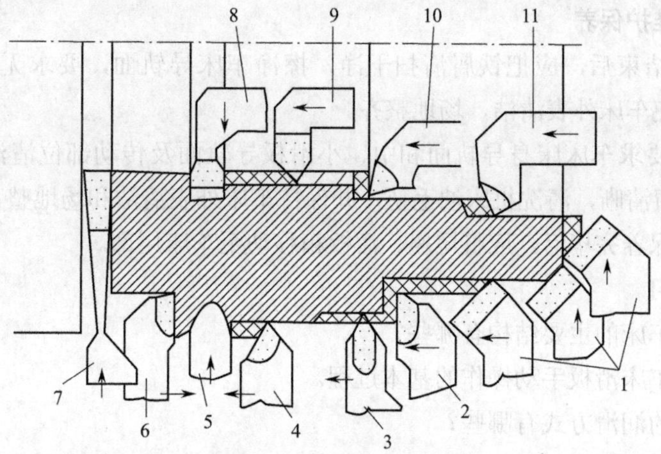

图 2-3-1　常用车刀种类

1—45° 弯头车刀；2—90° 外圆车刀（右偏刀）；3—外螺纹车刀；4—75° 外圆车刀；5—成形车刀；6—90° 外圆车刀（左偏刀）；7—切断刀；8—内沟槽车刀；9—内螺纹车刀；10—盲孔车刀；11—通孔车刀

（1）75° 外圆车刀主要用于车削工件外圆，和 90° 车刀相比有更好的抗冲击能力，一般用于粗车。

（2）45° 弯头车刀主要用于车削工件端面、外圆和倒角。

（3）90° 外圆车刀（右偏刀）主要用于车削工件的外圆、台阶和端面。

（4）切断刀主要用于切断工件或在工件上切出沟槽。

（5）成形车刀用于车削台阶处的圆角、圆槽，或车削各种特殊型面工件。

（6）外螺纹车刀主要用于车削各种螺纹。

（7）内孔车刀用于车削工件的内孔。

2. 按刀具切削部分的材料分类

按刀具切削部分的材料分为硬质合金、高速钢、陶瓷、金刚石、涂层刀具等。目前常用的车刀材料有高速钢和硬质合金两大类。

1）高速钢车刀

常用的高速钢牌号有 W18Cr4V、W6Mo5Cr4V2（每个化学元素后的数字，是指材料中含该元素的百分比，如 W18 表示含有 18% 的钨）。它是一种含钨、铬、钒等合金元素较多的工具钢。高速钢刀具制造简单，刃磨方便，磨出的刀具刃口锋利且韧性好，能承受较大的冲击力，因此常用于冲击力较大的工件加工，也常用于精车和各种螺纹加工的粗、精车。但高速钢刀具的耐热性较差，因此不能用于高速切削。

2）硬质合金车刀

硬质合金是用钨和钛的碳化物粉末加钴作为黏结剂，高压压制成型后再用高温烧结而成的粉末冶金制品。硬质合金的红硬性很好，在 1 000 ℃左右的高温下仍能保持良好的切削性能，它的硬度较高，耐磨性也很好。因此可选用比高速钢刀具高几倍甚至几十倍的切削速度，并且能切削难加工材料和淬硬材料。

硬质合金材料与高速钢比较，有很多优点，但也存在缺点，主要是韧性较差，怕冲击，刃口不如高速钢锋利。但是这些不足可以通过选择合理的刀具角度来改善，所以硬质合金刀具材料应用最为广泛。

二、车刀的几何角度和作用

1. 车刀的组成

车刀是由刀体和刀柄两部分组成的。刀体部分担负切削工作，又称切削部分，刀柄用来装夹车刀。图 2-3-2 所示为不同式样车刀的组成。

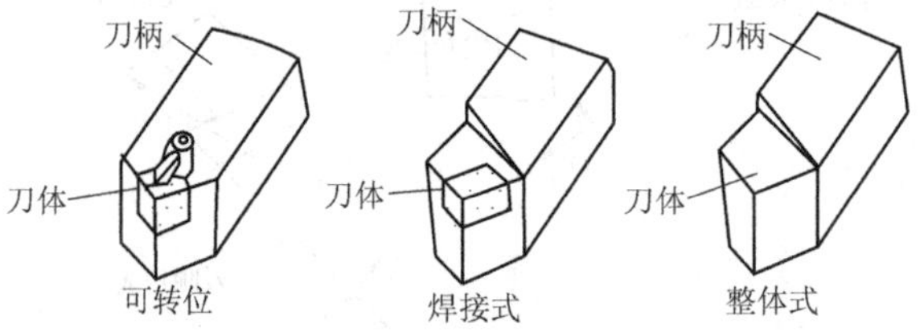

图 2-3-2 车刀的组成

刀体是一个楔形的几何体，由刀面和刀刃组成，如图 2-3-3 所示。

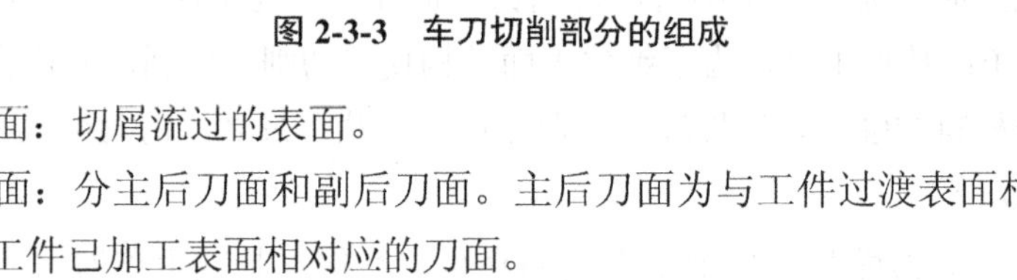

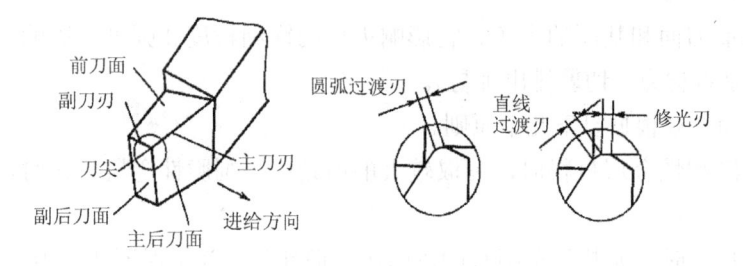

图 2-3-3 车刀切削部分的组成

（1）前刀面：切屑流过的表面。

（2）后刀面：分主后刀面和副后刀面。主后刀面为与工件过渡表面相对应的刀面。副后刀面为与工件已加工表面相对应的刀面。

（3）主切削刃：前刀面和主后刀面的交线，它担负着主要的切削工作。

（4）副切削刃：前刀面和副后刀面的交线，它担负着次要的切削工作。

（5）刀尖：主切削刃和副切削刃交会的一小段切削刃称为刀尖。实际上，刀尖刃磨时不是很尖，为增加其强度总是磨成圆弧形或直线形过渡刃。

（6）修光刃：副切削刃近刀尖处一小段平直的切削刃称为修光刃。装刀时必须使修光刃与进给方向平行，且修光刃长度必须大于进给量，才能起修光作用。

2. 刀具的切削角度及其作用

车刀切削部分共有 6 个独立的基本角度，它们是前角、主后角、副后角、主偏角、副偏角和刃倾角；还有两个派生角度，即刀尖角和楔角，如图 2-3-4 所示。

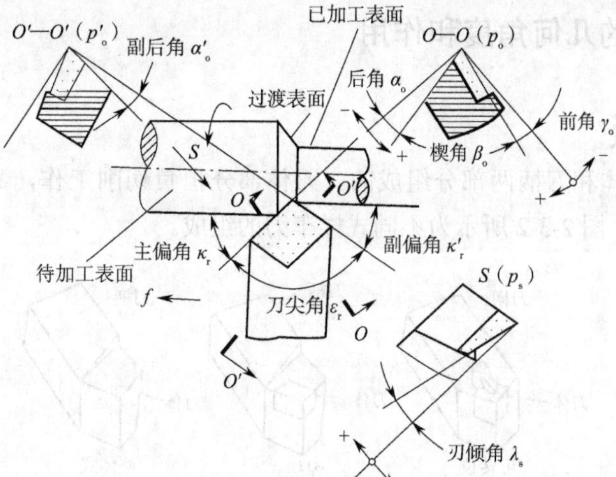

图 2-3-4　车刀切削部分主要几何角度

1）前角（(γ_0)）

前角为前刀面和基面的夹角。它影响刃口的锋利程度和强度，影响切削变形和切削力（切削是否省力，切屑排出难易）。

选择前角主要根据以下几个原则。

（1）车削塑性金属材料时，可取较大的前角；车削脆性金属材料时，应取较小的前角。

（2）粗加工时，尤其是车削有硬皮的铸、锻件时，为了保证切削刃有足够的强度，应取较小的前角；精加工时，为了细化表面结构使得切削刃锋利，前角应取较大些。

（3）车刀材料的强度、韧性较差，前角应取小些；反之，前角可取得大些。

2）后角(α_0)

主后角：主后刀面与主切削平面间的夹角，其作用是减少车刀主后刀面和工件过渡表面间的摩擦。

副后角：副后刀面与副切削平面间的夹角，其作用是减少车刀副后面和工件已加工表面间的摩擦。

选择后角主要根据以下几个原则。

（1）粗加工时，应取较小的后角（硬质合金车刀，α_0 取 5°~7°；高速钢车刀，α_0 取 6°~8°）。

（2）精加工时，应取较大的后角（硬质合金车刀，α_0 取 8°~10°；高速钢车刀，α_0 取 8°~12°）。

工件材料较硬时，后角应取小些；工件材料较软时，后角可取大些。

3）主编角（κ_r）

主偏角是主切削刃在基面上的投影与进给方向间的夹角。它改变主切削刃的受力及导热能力，影响切削的深度。选择主偏角首先应考虑工件的形状；加工台阶轴类工件，主偏角必须等于或略大于 90°；加工中间切入的工件，主偏角一般选用 45°~60°；加工刚性差的轴类工件主偏角要选大些，以便减小切削过程中的径向力。

4）副偏角（κ_r'）

副偏角是副切削刃在基面上的投影与背离进给方向间的夹角。它减少副切削刃与工件已加工表面间的摩擦。减小副偏角可以减小切削残余面积，因此可以减小工件的表面结构参数值；相反，副偏角太大时，刀尖角 ε_r 就减小，影响刀头强度。副偏角一般选择 6°~8°。

5）刃倾角（λ_s）

刃倾角是主切削刃与基面之间的夹角。刃倾角的主要作用是可以控制切屑的排出方向。当刃倾角为负值时，可增强刀头强度，在受冲击时，还可以保护刀尖。

选择刃倾角主要根据以下几个原则。

（1）精加工时刃倾角取正值，粗加工时刃倾角取负值。

（2）冲击负荷较大的断续切削，应取较大负值的刃倾角。

（3）加工高硬度材料时，应取负的刃倾角，以提高刀具强度。

（4）成形车削一般选择刃倾角为零，粗加工时可以选择为负角，精车时选择为正角。

6）刀尖角（ε_r）

主、副切削刃在基面上的投影间的夹角称为刀尖角，其影响刀尖强度和散热性能。

3. 车刀的刃磨

1）砂轮的选用

常用的砂轮有白色的氧化铝砂轮和灰绿色的碳化硅砂轮两种。

（1）氧化铝砂轮：适用于刃磨高速钢、碳素工具钢等刀具以及硬质合金车刀的刀柄部分。

（2）碳化硅砂轮：适用于刃磨硬质合金的刀片部分。

2）车刀的刃磨步骤和方法

车刀的刃磨一般有如下步骤，但是也要根据不同的车刀适当改变。

（1）粗磨主后面，同时磨出主偏角及主后角，见图 2-3-5（a）。

（2）粗磨副后面，同时磨出副偏角及副后角，见图 2-3-5（b）。

（3）粗磨前面，同时磨出前角及刃倾角，见图 2-3-5（c）。

（4）修磨前刀面。

（5）修磨主后刀面和副后刀面。

（6）修磨刀尖圆弧，见图 2-3-5（d）。

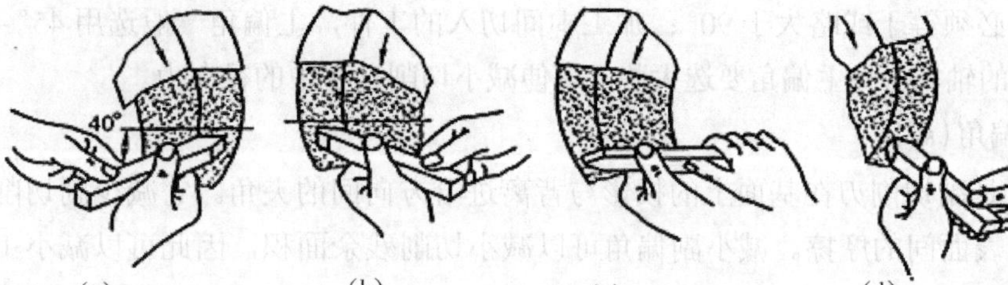

图 2-3-5　车刀的刃磨

（a）粗磨主后面　（b）粗磨副后面　（c）粗磨前面　（d）修磨刀尖圆弧

4. 车刀的角度检测

（1）目测法：观察车刀角度是否合乎要求，刀刃是否锋利，表面是否有裂痕和其他缺陷。

（2）用样板测量。

（3）用量角器测量。

当检测角度要求准确的车刀时，用车刀量角器进行测量。

5. 刃磨车刀的安全注意事项

（1）磨刀前，要对砂轮机的防护设施进行认真检查。

（2）磨刀时要带防护眼镜。

（3）磨刀时应站在砂轮的侧面，以防砂轮碎裂飞出伤人。

（4）磨刀时两手握紧车刀，不能用力过大，以防打滑伤手。

（5）刃磨硬质合金刀时，不准把刀头放入水中冷却，以防刀片碎裂。刃磨高速钢车刀时，应随时用水冷却，以防车刀过热退火降低硬度。

（6）砂轮磨削表面须经常修整，使砂轮没有明显的跳动。

（7）磨刀后随手关闭砂轮机电源。

思考与练习

（1）车刀按用途可以分为哪些种类？

（2）简述车刀的几何角度和作用。

（3）刃磨车刀时砂轮的选用方法是什么？

（4）简述车刀刃磨的方法和步骤。

（5）车刀角度检测的方法有哪些？

任务四　手动进给车削台阶轴

一、图样与技术要求

如图 2-4-1 所示，工件毛坯尺寸为 $\phi45\times80$ mm，材料为 45 钢，工件加工不做切断要求。评分表见表 2-4-1。

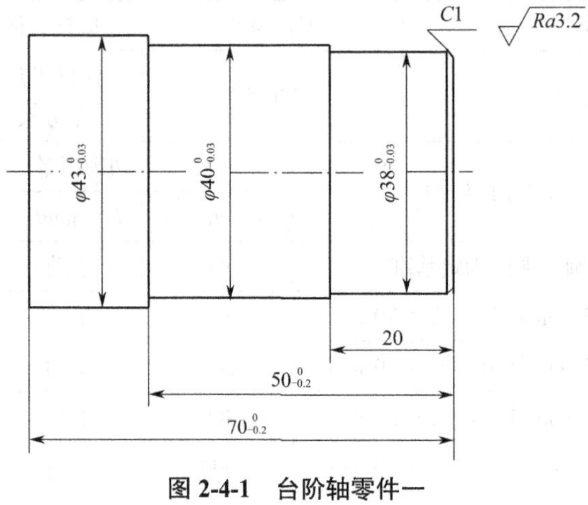

图 2-4-1　台阶轴零件一

表 2-4-1　评分表

技术要求	配分	实测结果	得分
$\phi\ 38_{-0.03}^{\ 0}$ mm 外圆	15		
$\phi\ 40_{-0.03}^{\ 0}$ mm 外圆	15		

续表

技术要求	配分	实测结果	得分
$\phi\ 43_{-0.03}^{\ 0}$ mm 外圆	15		
长度 20 mm	10		
长度 $50_{-0.2}^{\ 0}$ mm	10		
长度 $70_{-0.2}^{\ 0}$ mm	10		
倒角 *C*1（1 处）	5		
表面结构参数 *Ra*3.2 μm	15		
安全文明操作	5		

二、加工工艺分析

零件各个轮廓加工工艺安排见表 2-4-2。

表 2-4-2 普车加工工艺卡

<table>
<tr><td>零件图号</td><td>图 2-4-1</td><td colspan="2" rowspan="2">普通车床加工工艺卡</td><td colspan="2">机床型号</td><td>CA6140</td></tr>
<tr><td>零件名称</td><td>台阶轴</td><td colspan="2">机床编号</td><td></td></tr>
<tr><td colspan="4">刀具表</td><td colspan="3">量具表</td></tr>
<tr><td colspan="2">刀具名称</td><td colspan="2">刀具参数</td><td colspan="2">量具名称</td><td>规格 /mm</td></tr>
<tr><td colspan="2">90° 外圆车刀</td><td colspan="2">YT15</td><td colspan="2">游标卡尺
千分尺</td><td>0~150/0.02
25~50/0.01</td></tr>
<tr><td rowspan="2">工序</td><td rowspan="2">工艺内容</td><td colspan="4">切削用量</td><td rowspan="2">加工性质</td></tr>
<tr><td>S/（r/min）</td><td colspan="2">F/（mm/r）</td><td>a_p/mm</td></tr>
<tr><td>普车</td><td>车外圆、端面确定基准</td><td>400</td><td colspan="2">手动</td><td>手动</td><td>手动</td></tr>
<tr><td>1</td><td>粗加工 ϕ43 mm 外圆，长度 70 mm</td><td>400</td><td colspan="2">手动</td><td>手动</td><td>手动</td></tr>
<tr><td>2</td><td>精加工 ϕ43 mm 外圆，长度 70 mm</td><td>600</td><td colspan="2">手动</td><td>手动</td><td>手动</td></tr>
<tr><td>3</td><td>粗加工 ϕ40 mm 外圆，长度 50 mm</td><td>400</td><td colspan="2">手动</td><td>手动</td><td>手动</td></tr>
<tr><td>4</td><td>精加工 ϕ40 mm 外圆，长度 50 mm</td><td>600</td><td colspan="2">手动</td><td>手动</td><td>手动</td></tr>
<tr><td>5</td><td>粗加工 ϕ38 mm 外圆，长度 20 mm</td><td>400</td><td colspan="2">手动</td><td>手动</td><td>手动</td></tr>
<tr><td>6</td><td>精加工 ϕ38 mm 外圆，长度 20 mm</td><td>600</td><td colspan="2">手动</td><td>手动</td><td>手动</td></tr>
<tr><td>7</td><td>倒角 C1</td><td>600</td><td colspan="2">手动</td><td>手动</td><td>手动</td></tr>
</table>

三、相关知识与技能

1. 车刀的装夹方法和要求

（1）装夹车刀时，刀尖应对准工件中心线，如图 2-4-2 所示。

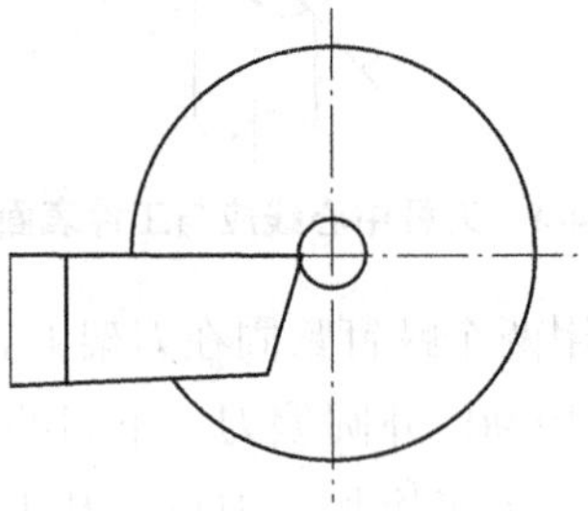

图 2-4-2 刀尖应对准工件中心线

①按车床中心高的数值，用钢直尺对中心。

②刀尖与顶尖找平，对中心，如图 2-4-3 所示。

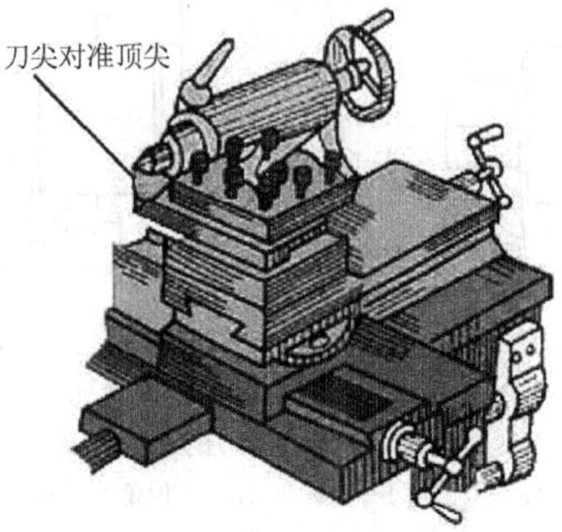

图 2-4-3 刀尖应对准顶尖中心线

（2）车刀伸出部分长度应尽可能短些，若伸出过长，刚性不足，容易引起振动。伸出长度约为刀杆厚度的 1~1.5 倍。

（3）刀杆中心线应与工件表面垂直，否则会引起主偏角和副偏角的数值发生变化，如图 2-4-4 所示。

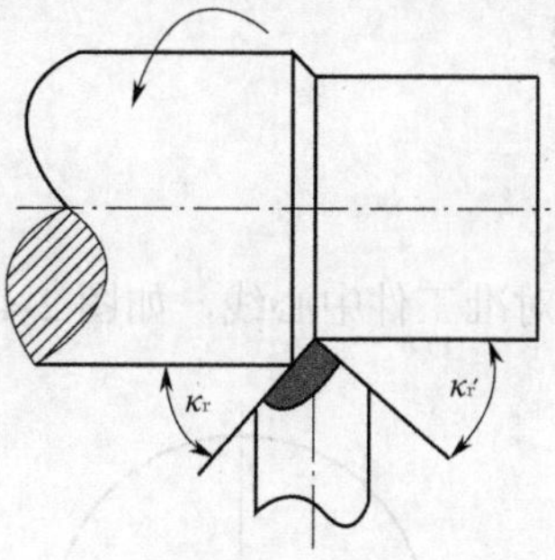

图 2-4-4　刀杆中心线应与工件表面垂直

（4）车刀要夹紧，至少要用两个螺钉紧固在刀架上。

（5）车台阶工件，通常使用 90° 外圆偏刀。车刀的装夹应根据粗、精车和余量的多少来区别。如粗车时余量多，为了增加吃刀量，减少刀尖压力，车刀装夹可取主偏角小于 90° 为宜（一般为 85° ~90° ），见图 2-4-5（a）。精车时为了保证台阶平面和轴心线垂直，应取主偏角大于 90°（一般为 93° 左右），见图 2-4-5（b）。

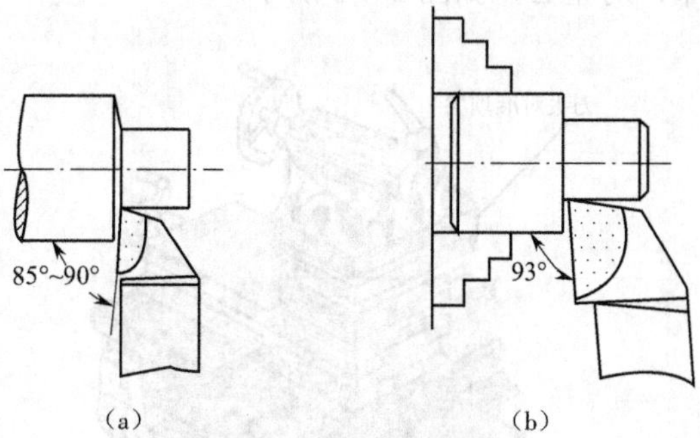

图 2-4-5　车刀的装夹

（a）主偏角小于 90°　（b）主偏角大于 90°

2. 工件的装夹

车削之前，工件先要在车床上进行安装，根据轴类工件的形状、大小、加工精度和数量的不同，其装夹方法也有所不同，以保证加工的质量和生产效率。

1）三爪自定心卡盘装夹

三爪自定心卡盘如图 2-4-6（a）所示，结构如图 2-4-6（b）所示。当用卡盘扳手转动小锥齿轮时，大锥齿轮也随之转动，在大锥齿轮背面平面螺纹的作用下，使 3 个爪同时向心移动或退出，以夹紧或松开工件。它的特点是对中性好，自动定心精度可达到 0.05~0.15 mm；可以装夹直径较小的工件，如图 2-4-6（c）所示。当装夹直径较大的外圆工件时可用 3 个反爪进行，如图 2-4-6（d）所示。但三爪自定心卡盘因为夹紧力不大，所以一般只适用于质量较小的工件。当对质量较大的工件进行装夹时，宜用四爪单动卡盘或其他专用夹具。

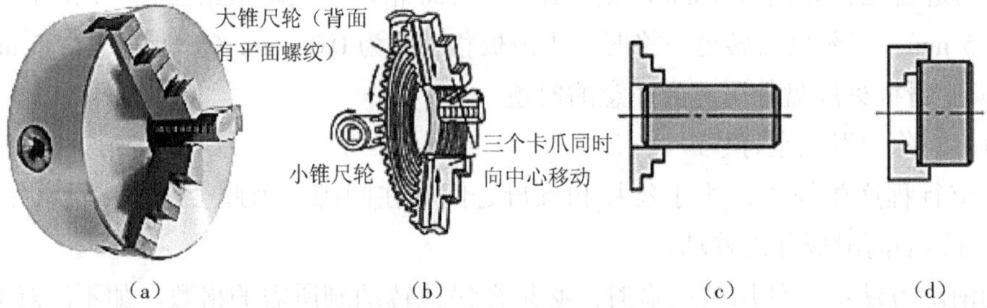

图 2-4-6　三爪卡盘

（a）三爪卡盘　（b）结构　（c）正卡爪夹持棒料　（d）反爪夹持大棒料

2）工件找正

所谓工件找正，就是把被加工的工件装夹在卡盘上，使工件的中心与车床主轴旋转中心一致，这一过程称为找正。找正方法：先用划针尖靠近工件外圆表面，用手转动卡盘，观察工件表面与划针尖之间间隙的大小，然后根据间隙大小调整卡爪或用铜棒敲击工件。经过几次反复调整，直到工件旋转一周，各处针尖与工件表面距离均等时为止。

（1）轴类零件找正。轴类零件通常找正工件外圆 A、B 两点，如图 2-4-7（a）所示。其方法是先找正 A 点外圆，后找正 B 点外圆。找正 A 点外圆时，应调整卡爪；找正 B 点外圆时，用铜棒敲击。

（2）盘类零件找正。盘类零件通常既要找正外圆，又要找正平面，如图 2-4-7（b）所示。找正 A 点外圆时，通过移动卡爪进行调整；找正 B 点平面时，用铜棒敲击。

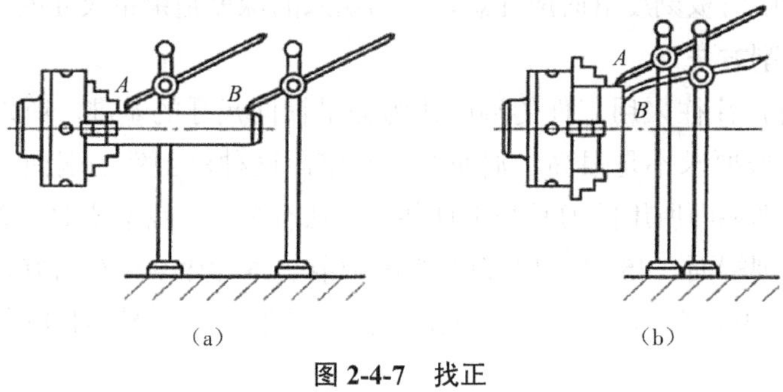

图 2-4-7　找正

（a）轴类零件　（b）盘类零件

3. 刻度盘的原理和应用

1）刻度盘的原理

以中滑板刻度盘为例：刻度盘装在中滑板丝杠上，当摇动手柄带动刻度盘转一周时，丝杆也转了一周，这时固定在中滑板上的螺母就带动中滑板、车刀移动一个导程。如果横向进给丝杠导程为 5 mm，刻度盘分为 100 格，当摇动进给丝杠一周时，中滑板就移动 5 mm。当刻度盘转过一格时，中滑板移动量为 0.05 mm（5÷100 = 0.05 mm）。

2）中滑板刻度盘使用时应注意的问题

（1）消除中滑板空行程。

①空行程产生原因：由于丝杠和螺母之间存在间隙，因此会产生中滑板空行程，即刻度盘转动而滑板并没移动。

②消除方法：当使用刻度盘时，必须慢慢地转动到所需的格数，如不注意多转了几格（或吃刀量过大时，绝不能简单地退回几格，因为有间隙存在，若只退回几格实际上中滑板并没有退回），必须向相反方向退回全部空行程，然后再转到所需要的格数，如图 2-4-8 所示。

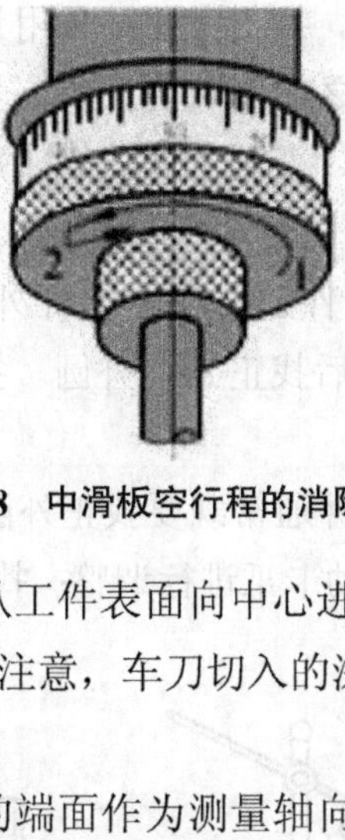

图 2-4-8　中滑板空行程的消除方法

（2）由于工件的旋转，车刀从工件表面向中心进刀后，切下的部分是切削深度的两倍。因此使用中滑板刻度盘时应注意，车刀切入的深度是余量尺寸的一半。

4. 端面的车削方法

车削工件时，往往采用工件的端面作为测量轴向尺寸的基准，所以必须先进行加工，表面结构一般要求不是很高，端面要求平整，这种对工件的端面进行车削的方法叫车削端面。车削端面所用的刀具与工件装夹方法有关。卡盘装夹时，常用 45° 弯头车刀和 90° 车刀车削端面。45° 车刀的刀头强度和散热条件比 90° 车刀好，常用于车削工件的端面和倒角。90° 车刀刀尖强度较差，常用于精加工。如图 2-4-9 所示。

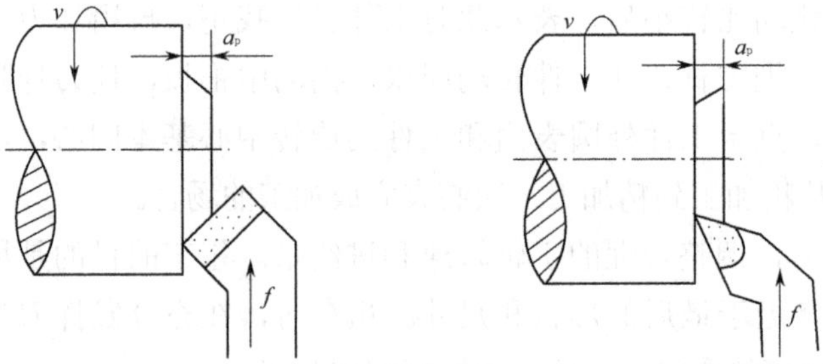

（a）　　　　（b）

图 2-4-9 车端面

（a）45° 车刀车削端面 （b）90° 车刀车削端面

车削端面的步骤如下：

（1）在卡盘上装夹工件，找正外圆、端面并夹紧；

（2）按要求装夹车刀，调整合理的转速和进给量；

（3）摇动床鞍和中滑板，使刀尖离开待加工端面 2~3 mm；

（4）将床鞍在床身上锁紧；

（5）摇动中滑板和小滑板使刀尖接触工件端面，退回中滑板（小滑板不动）；

（6）将小滑板刻度对零，或记住小滑板刻度；

（7）用小滑板刻度按需求调整切削深度；

（8）用双手摇动中滑板手轮，保持速度均匀，横向车削至端面中心后退刀；

（9）依次车削，直到车平端面或符合图样要求为止。

注意：如果装刀时不对准工件旋转中心，在车端面至中心时会留有凸头或造成刀尖碎裂，如图 2-4-10 所示。

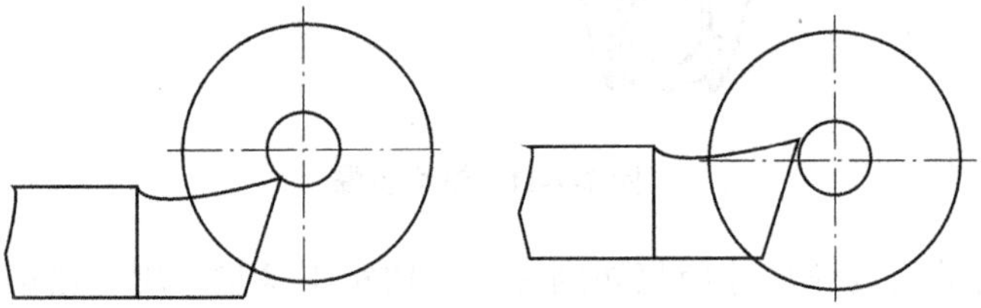

（a）　　　　（b）

图 2-4-10 装刀高度

（a）刀尖过低易被压断 （b）刀尖过高不易切削

5. 外圆的车削方法

1）粗车的步骤

（1）在卡盘上装夹、找正后夹紧。按照车削长度，将工件装夹在卡盘上，并伸出适当的长度，用手稍微将工件夹紧，将工件转动起来，用铁棒敲击工件，观察工件旋转时的跳动，待跳动比较小后，表示此时工件已经找正，再用加力导筒夹紧工件。或者在刀架上装夹一把刀杆，让工件转动起来，摇动中拖板，使刀杆和工件接触后，再缓慢摇动中拖板，直至工件外圆表面和工件的旋转中心基本同心，最后夹紧工件。该方法适合于工件从粗加工到精加工一次装夹完成加工的场合。

（2）装夹车刀，调整合理的主轴转速和进给量。粗车的目的是尽快地切去多余的金属层，使工件接近于最后的形状和尺寸。粗车时，在充分发挥刀具、车床性能的情况下，吃刀量应尽可能取得大一些，以缩短切削时间。

（3）将车刀摇至工件的尾端，距离工件端面 3~5 mm 处，开动车床。

（4）试切削，用中滑板刻度控制切削深度 1~2 mm，纵向进给 3 mm 左右，纵向退回车刀（横向不动），停车，测量工件直径尺寸，如图 2-4-11 所示。

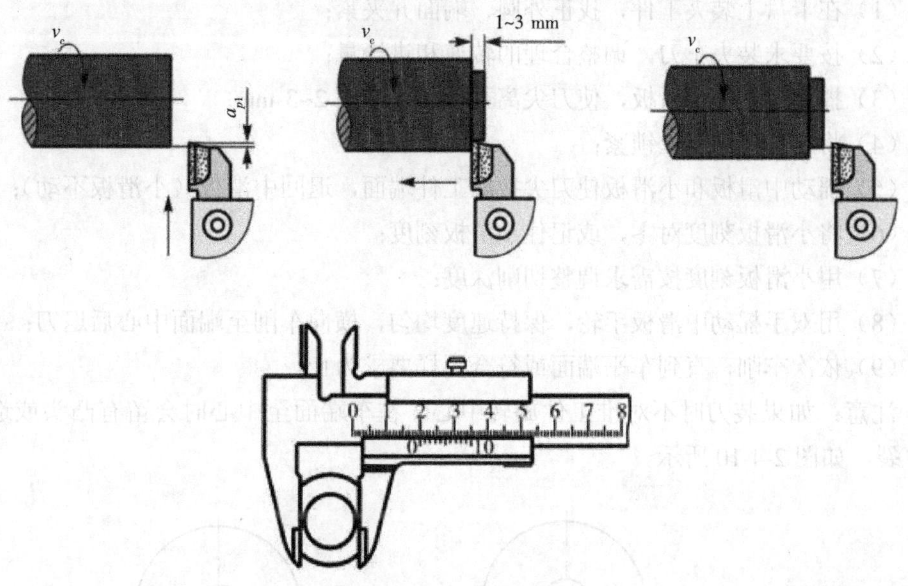

图 2-4-11　试切步骤

（5）根据测量的直径尺寸调整切削深度，留精车余量。用高速钢车刀低速精车留余量 0.1~0.2 mm；用硬质合金车刀精车留余量 0.4~1.0 mm。调整好背吃刀量，再纵向进给，用双手摇动床鞍手轮，尽可能匀速，进给到规定长度时，摇动中滑板手柄，退出车刀，床鞍快速移动回原位，如图 2-4-12 所示。

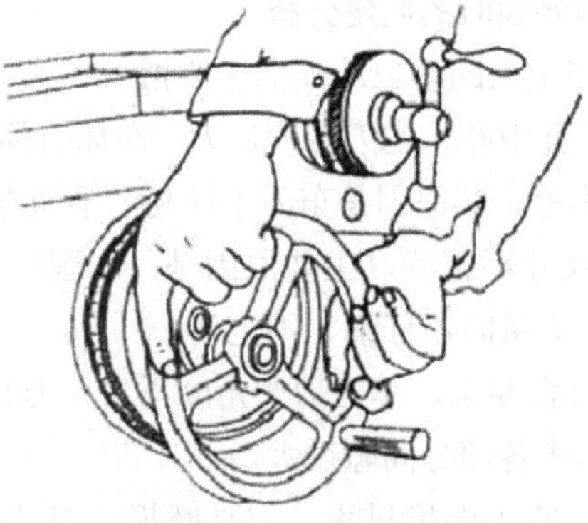

图 2-4-12　双手摇轮

2）精车的步骤

（1）按要求装夹精车刀，调整合理的转数和进给量。

（2）用刀尖在工件的末端外圆处对刀，以中滑板刻度盘控制切削深度 0.1~0.2 mm，试切削 3 mm 长度。

（3）停车，测量工件外径尺寸。

（4）根据测量的外径尺寸与图样尺寸要求，调整切削深度。

（5）用双手摇动床鞍手轮，保持速度均匀，纵向车削至长度要求。

（6）停车，检验尺寸，符合图样要求后卸下工件。

3）外圆长度尺寸的控制

根据外圆长度要求用钢直尺、卡钳或样板，量好刀尖至工件端面的距离，用刀尖车出一条线痕（称刻线痕法）。然后根据线痕进行车削，当车削完毕后，再进行测量直至符合要求。如图 2-4-13 所示。

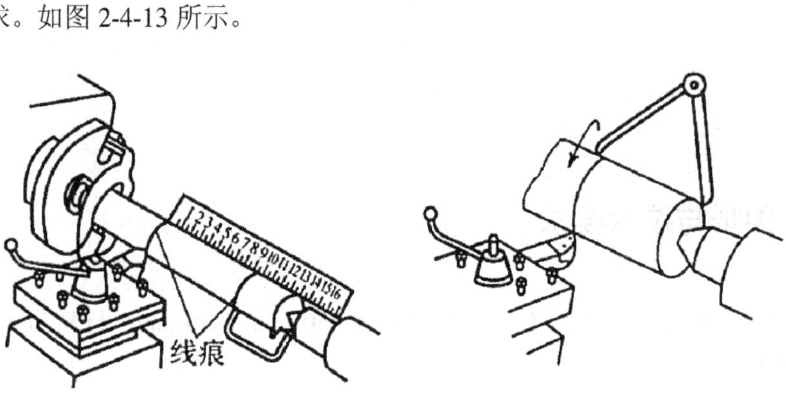

（a）　（b）

图 2-4-13　刻线痕确定车削长度

（a）用钢直尺和样板刻线痕　（b）用内卡钳在工件上刻线痕

6. 车削轴类零件端面和外圆时的质量分析

（1）工件端面不平，产生凸凹现象或端面中心留“小头”。其原因是车刀刃磨或安装不正确，刀尖没有对准工件中心，吃刀深度过大，车床有间隙，拖板移动。

（2）工件端面表面结构差。其原因是车刀不锋利，手动走刀摇动不均匀或太快。

（3）工件直径或长度尺寸不正确，主要原因是：车削时粗心大意，看错尺寸；刻度盘计算错误或操纵失误；丈量时不仔细、不正确。

（4）工件表面结构不符合要求，车刀刃磨角度不对，刀具安装不正确或刀具磨损，以及切削用量选择不当，车床各部分间隙过大。

（5）工件外径有锥度。吃刀深度过大，刀具磨损，刀具或滑板松动。用小滑板车削时转盘下基准线未对准“0”线。精车时加工余量不足等情况。

思考与练习

（1）简述手动进给车削端面的方法。

（2）简述刀具安装和对刀的方法。

（3）车削外圆时用刻度盘控制加工长度的方法是什么？

（4）简述外圆车削的基本过程和方法。

（5）车削端面和外圆时应注意哪些问题？

任务五　机动进给车削台阶轴

一、图样与技术要求

如图 2-5-1 所示，台阶轴工件毛坯尺寸为 $\phi 50$ mm×100 mm，材料为 45 钢。评分表见表 2-5-1。

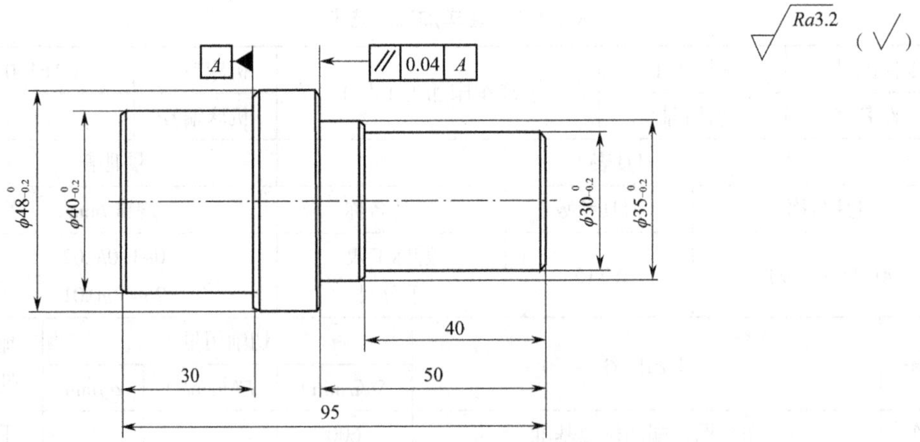

图 2-5-1　台阶轴零件二

表 2-5-1　评分表

技术要求	配分	实测结果	得分
$\phi\ 30_{-0.2}^{0}$ mm 外圆	10		
$\phi\ 35_{-0.2}^{0}$ mm 外圆	10		
$\phi\ 40_{-0.2}^{0}$ mm 外圆	10		
$\phi\ 48_{-0.2}^{0}$ mm 外圆	10		
长度 40 mm	5		
长度 50 mm	5		
长度 30 mm	5		
长度 95 mm	10		
平行度公差 0.04 mm	10		
倒角 C1（5 处）	3		
表面结构参数 Ra3.2 μm（4 处）	12		
安全文明操作	10		

二、加工工艺分析

零件各个轮廓加工工艺安排如表 2-5-2 所示。

表 2-5-2　普车加工工艺卡

<table>
<tr><td>零件图号</td><td>图 2-5-1</td><td rowspan="2" colspan="2">普通车床加工工艺卡</td><td>机床型号</td><td colspan="2">CA6140</td></tr>
<tr><td>零件名称</td><td>台阶轴</td><td>机床编号</td><td colspan="2"></td></tr>
<tr><td colspan="3">刀具表</td><td colspan="4">量具表</td></tr>
<tr><td colspan="2">刀具名称</td><td>刀具参数</td><td colspan="2">量具名称</td><td colspan="2">规格 /mm</td></tr>
<tr><td colspan="2">90° 外圆车刀</td><td>YT15</td><td colspan="2">游标卡尺
千分尺</td><td colspan="2">0~150/0.02
25~50/0.01</td></tr>
</table>

<table>
<tr><td rowspan="2">工序</td><td rowspan="2">工艺内容</td><td colspan="3">切削用量</td><td rowspan="2">加工性质</td></tr>
<tr><td>S/（r/min）</td><td>F/（mm/r）</td><td>a_p/mm</td></tr>
<tr><td>普车</td><td>车外圆、端面确定基准</td><td>600</td><td>—</td><td>1</td><td>手动</td></tr>
<tr><td>1</td><td>粗加工 ϕ48 mm 外圆，长度至尺寸要求</td><td>600</td><td>0.2</td><td>1~2</td><td>自动</td></tr>
<tr><td>2</td><td>粗加工 ϕ40 mm 外圆，长度至尺寸要求</td><td>600</td><td>0.2</td><td>1~2</td><td>自动</td></tr>
<tr><td>3</td><td>精加工 ϕ48 mm、ϕ40 mm 外圆及倒角 C1</td><td>1 000</td><td>0.1</td><td>0.5~1</td><td>自动</td></tr>
<tr><td>4</td><td>调头夹 ϕ40 mm 外圆，平端面，保证总长</td><td>600</td><td>—</td><td>1</td><td>手动</td></tr>
<tr><td>5</td><td>粗加工 ϕ35 mm 外圆，长度至尺寸要求</td><td>600</td><td>0.2</td><td>1~2</td><td>自动</td></tr>
<tr><td>6</td><td>粗加工 ϕ30 mm 外圆，长度至尺寸要求</td><td>600</td><td>0.2</td><td>1~2</td><td>自动</td></tr>
<tr><td>7</td><td>精加工 ϕ35 mm、ϕ30 mm 外圆及倒角 C1</td><td>1 000</td><td>0.1</td><td>0.5~1</td><td>自动</td></tr>
</table>

三、相关知识与技能

1. 机动进给车削的优点

操作省力、进给均匀，加工后表面结构参数值小等。

2. 机动进给车削工件的操作方法

机动进给时，扳动机动进给手柄（进给手柄可十字扳动，如图 2-5-2 所示），刀具移动方向与手柄扳动方向一致；车削到规定位置时，横向退刀，测量工件，多次进给后达到工件规定尺寸。机动进给手柄的顶部有一个按钮，可接通快速进、退刀。

图 2-5-2　机动进给

特别提示：横向机动进给应注意中滑板向前移动时刀架前面不要超过卡盘中心，以防止中滑板丝杠与螺母脱开。如反向进给，则应防止中滑板后退时与刻度盘相撞而损坏。

3. 工件的找正方法

每当接刀工件装夹时，找正必须从严要求，否则会造成接刀偏差，影响加工质量。一般使用百分表按照工件已精加工的表面找正，如图 2-5-3 所示。

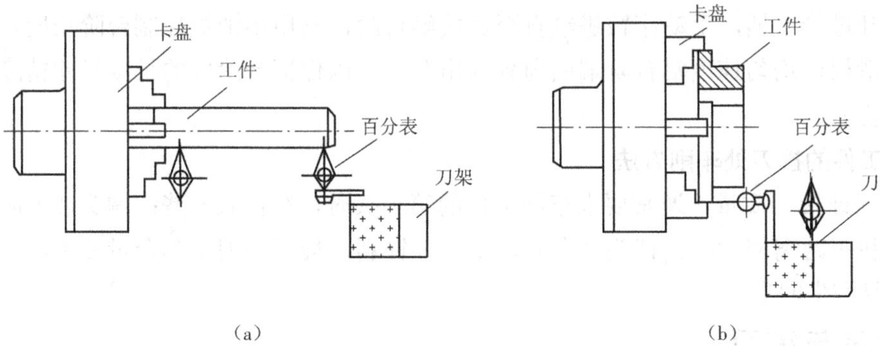

图 2-5-3　百分表找正

（a）找正外圆　（b）找正端面

将百分表固定在刀架上，百分表指针压在工件外圆或端面，并使其转过一圈，转动工件一周，如果工件的回转中心与卡盘中心重合，则指针不动；如果工件的回转中心与卡盘中心不重合，工件表面靠近指针的，则指针向右偏转，工件表面远离指针的，则指针向左偏转，调整卡爪或工件位置直至符合要求。百分表找正的定心精度可达0.02~0.01 mm。

4. 机动车削外圆的基本步骤

（1）选择主轴转速和进给量，调整相关手柄的位置。

（2）对刀，移动刀架，使车刀刀尖接触工件表面，对零点时必须启动车床。

（3）对完刀后，用刻度盘调整切削深度。在用刻度盘调整切前深度时，应了解中滑板刻度盘的刻度值，也就是每转过一小格时车刀的横向切削深度值。然后根据切前深度计算出需要转过的格数。

（4）试切，检查切削深度是否准确，横向进刀。为了保证加工的尺寸精度，应采用试切法车削。试切尺寸不合格会有两种情况：如果尺寸偏大，则应再次横向进刀定背吃刀量 a_p；如果尺寸偏小，则将车刀横向退出一定的距离，再行试切，重复上述步骤（1）~（3），直到尺寸合格为止。各次所定的背吃刀量均应小于各次直径余量的一半。

如果刻度盘手柄摇过了，或试切后发现尺寸太小而必须退刀时，为了消除丝杆和螺母之间的间隙，应反转半周左右，再转至所需的刻度值上。

（5）纵向手动进给前车外圆。启动车床，工件旋转→试切削→机动进给→纵向车削外圆→车至接近需要长度时停止进给→改用手动进给→车至长度尺寸退刀，停车。

（6）测量外圆尺寸。

（7）倒角。选择 45° 车刀，然后移动床鞍至工件外圆和端面相交处进行倒角。

5. 机动精车外圆时注意事项

采用机动进给精加工台阶轴，刀具进行纵向车削时，当接近台阶 0.7 mm 至 1 mm 时，断开进给手柄，手动床鞍进给直至其接触台阶，利用小滑板控制台阶长度。手动进给中滑板以均匀速度沿台阶端面向外摇出车刀，以保证台阶轴的长度尺寸精度和表面结构要求。

6. 工件的接刀处车削方法

为保证接刀质量，通常要求车削工件的第一头时，车得长一些；调头装夹时，两点间的找正距离应大些；调头精车时，车刀要锋利，最后一刀精车余量要小，否则工件上容易产生凹痕。

思考与练习

（1）简述机动车削外圆的加工方法。

（2）简述工件的找正方法。

（3）简述接刀工件的车削过程。

任务六　车削外圆沟槽及切断

一、图样与技术要求

如图 2-6-1 所示，台阶轴工件毛坯尺寸为 $\phi30$ mm×80 mm，材料为 45 钢。评分表见表 2-6-1。

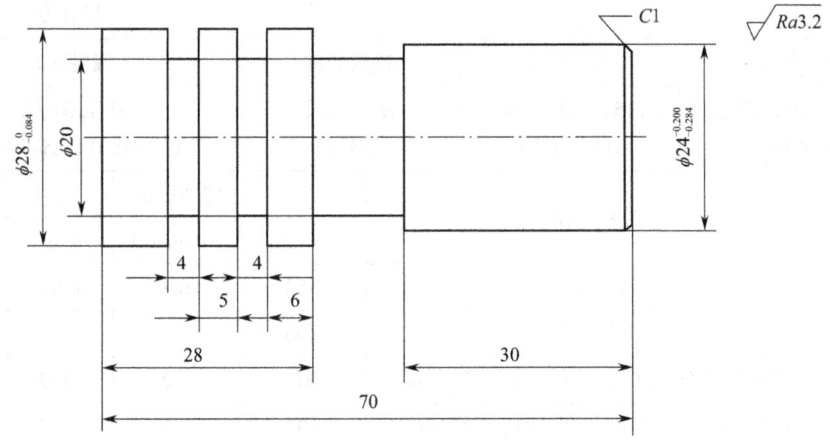

图 2-6-1　台阶轴零件三

表 2-6-1　评分表

技术要求	配分	实测结果	得分
$\phi\ 24_{-0.284}^{-0.200}$ mm 外圆	12		
$\phi\ 28_{-0.084}^{\ 0}$ mm 外圆	12		
$\phi20$ mm 退刀槽外圆	10		
$\phi20$ mm 外沟槽外圆（2 处）	20		
长度未注公差（7 处）	7		
钻中心孔	10		

续表

技术要求	配分	实测结果	得分
倒角 $C1$	3		
表面结构参数 $Ra3.2$ μm	16		
安全文明操作	10		

二、加工工艺分析

零件各个轮廓加工工艺安排如表 2-6-2 所示。

表 2-6-2　普车加工工艺卡

<table>
<tr><td>零件图号</td><td>图 2-6-1</td><td colspan="2" rowspan="2">普通车床加工工艺卡</td><td colspan="2">机床型号</td><td>CA6140</td></tr>
<tr><td>零件名称</td><td>台阶轴</td><td colspan="2">机床编号</td><td></td></tr>
<tr><td colspan="4">刀具表</td><td colspan="3">量具表</td></tr>
<tr><td>刀具名称</td><td>刀具参数</td><td colspan="2">量具名称</td><td colspan="3">规格 /mm</td></tr>
<tr><td>90° 外圆车刀、切槽刀
中心钻</td><td>YT15、高速钢车刀
B3 中心钻</td><td colspan="2">游标卡尺
千分尺</td><td colspan="3">0~150/0.02
0~25/0.01、25~50/0.01</td></tr>
<tr><td rowspan="2">工序</td><td colspan="2" rowspan="2">工艺内容</td><td colspan="3">切削用量</td><td rowspan="2">加工性质</td></tr>
<tr><td>S/（r/min）</td><td>F/（mm/r）</td><td>a_p/mm</td></tr>
<tr><td>1</td><td colspan="2">车端面</td><td>350</td><td>0.2</td><td>0.5~1</td><td>自动</td></tr>
<tr><td>2</td><td colspan="2">钻中心孔</td><td>700</td><td>—</td><td>—</td><td>手动</td></tr>
<tr><td>3</td><td colspan="2">粗车外圆至 $\phi29$ mm，保证长度 75 mm</td><td>600</td><td>0.2</td><td>1~2</td><td>自动</td></tr>
<tr><td>4</td><td colspan="2">粗车外圆至 $\phi25$ mm，保证长度 42 mm</td><td>600</td><td>0.2</td><td>1~2</td><td>自动</td></tr>
<tr><td>5</td><td colspan="2">精车 $\phi24$ mm 外圆尺寸，保证长度 42 mm</td><td>800</td><td>0.1</td><td>0.5</td><td>自动</td></tr>
<tr><td>6</td><td colspan="2">精车 $\phi28$ mm 外圆至尺寸，保证长度 28 mm</td><td>800</td><td>0.1</td><td>0.5</td><td>自动</td></tr>
<tr><td>7</td><td colspan="2">切 $\phi20$ mm 退刀槽，保证尺寸</td><td>300</td><td>—</td><td>—</td><td>手动</td></tr>
<tr><td>8</td><td colspan="2">切密封槽保证尺寸</td><td>300</td><td>—</td><td>—</td><td>手动</td></tr>
<tr><td>9</td><td colspan="2">倒角倒钝</td><td>300</td><td>—</td><td>—</td><td>手动</td></tr>
<tr><td>10</td><td colspan="2">调头，夹 $\phi24$ mm 外圆，找正 $\phi28$ mm 外圆，车左端面，保证长度 28 mm</td><td>350</td><td>0.2</td><td>0.5~1</td><td>自动</td></tr>
</table>

三、相关知识与技能

1. 槽的种类和作用

在零件上设置沟槽的目的，一是为了便于后序的加工，如车削螺纹的退刀槽、磨削加工的越程槽等；二是为了保证零件在装配时轴向定位的准确性，如轴肩槽；三是为了随意移动或紧固工作，如 T 形槽和燕尾槽等；四是在相互移动的配合面上，设置不同形式的润滑槽，以及设置起密封或防尘作用的密封槽和防尘槽等。在工件表面上车沟槽的方法叫车槽。

沟槽的结构形式有矩形槽、成形槽、斜沟槽、端面槽等。根据部位不同，沟槽可分为外沟槽和内沟槽。如图 2-6-2 所示，总的来说，沟槽的尺寸要求不高。

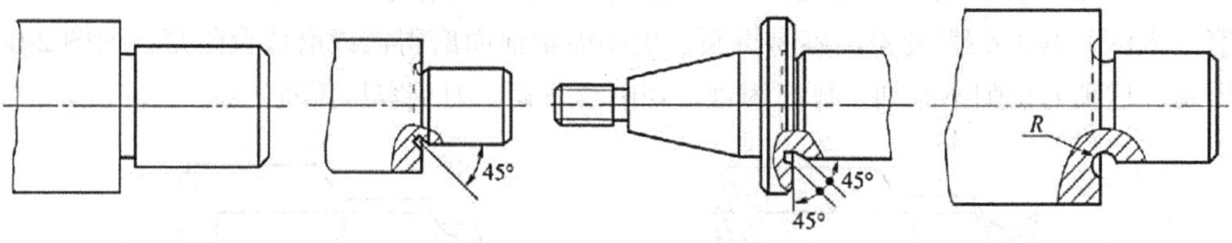

图 2-6-2　沟槽的种类

2. 切断刀（切槽刀）的选择

1）高速钢切断刀

高速钢切断刀如图 2-6-3 所示。

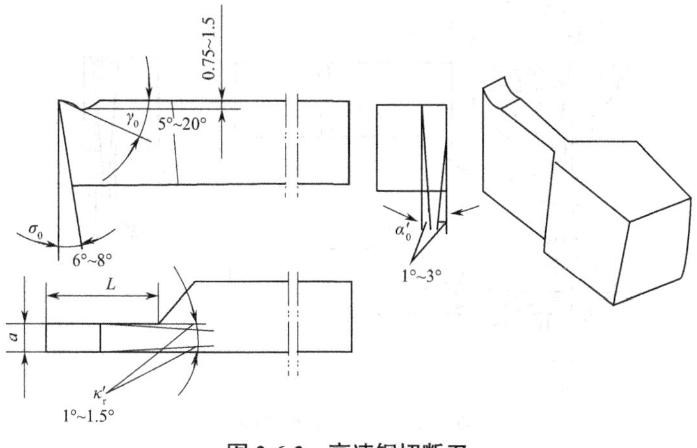

图 2-6-3　高速钢切断刀

（1）前角（γ_o）：切断中碳钢材料时 γ_o =20°~30°，切断铸铁材料时 γ_o =0°~10°。

（2）后角（α_o）：切断塑性材料时后角取大些，切断脆性材料时后角取小些，一般

取 α_o =6°~8°。

（3）副后角（ α_o' ）：切断刀有两个对称的副后角，其作用是减少副后刀面与工件已加工表面的摩擦，一般取 α_o' =1°~3°。

（4）主偏角（ κ_r ）：切断刀以横向进给为主，因此 κ_r =90°。为防止切断时在工件端面中心外留有小凸台以及切断空心工件时不留飞边，可以把主切削刃略磨斜些。

（5）副偏角（ κ_r' ）：切断刀的两个副偏角必须对称，否则会因两边所受切削抗力不均而影响平面度和断面对轴线的垂直度。为了不削弱刀头强度，一般取 κ_r' =1°~1.5°。

（6）主切削刃宽度（a）：主切削刃太宽会因切削力太大而振动，同时浪费材料，太窄又会削弱刀体强度，因此主切削刃选择合适的宽度为宜。

（7）刀体长度（L）：刀体太长也容易引起振动和使刀体折断。

（8）卷屑槽：切断刀的卷屑槽不宜磨得太深，一般为 0.75~1.50 mm，如果卷屑槽磨得太深，其刀头强度差，容易折断；更不能把前面磨得低或磨成台阶形，如图 2-6-4 所示，这种刀切削不顺利，排屑困难，切削负荷大，刀头容易折断。

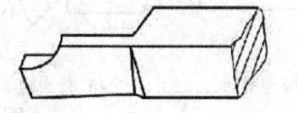

图 2-6-4　刃磨不正确的卷屑槽

2）硬质合金切断刀

用硬质合金切断刀高速切断工件时，切屑和工件槽宽相等，容易将切屑堵塞在槽内。为了排屑顺利，可把主切削刃两边倒角或磨成人字形，如图 2-6-5 所示。

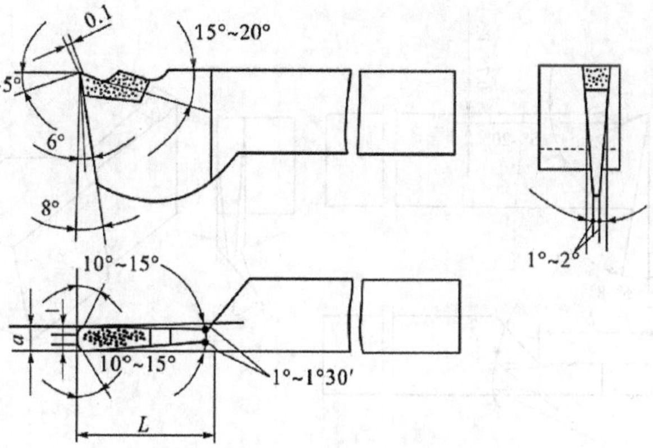

图 2-6-5　硬质合金切断刀

高速切断时，会产生很大的热量。为防止刀片脱焊，在开始切断时应浇注充分的切削液。为增加刀体的强度，常将切断刀体下部做成凸圆弧形。

3）反切刀

切削直径较大的工件时，由于刀头较长，刚性较差，容易引起振动，这时可采用反向切断法，即工件反转，用反切刀来切断，如图 2-6-6 所示。这样切断时，切削力的方向与重力方向一致，不容易引起振动。另外，反向切断时切屑从下面排出，不容易堵在工件槽内。

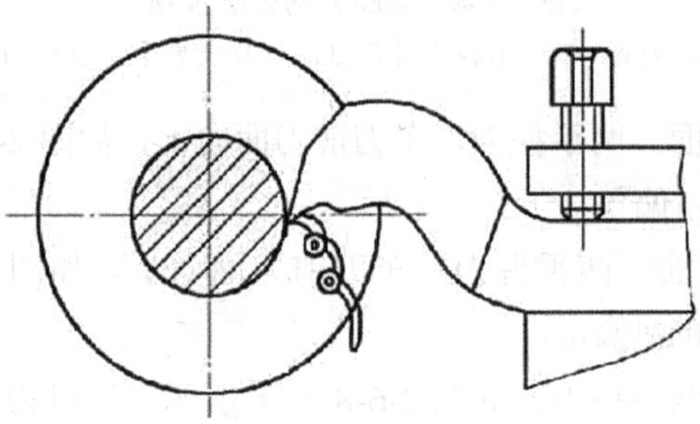

图 2-6-6 反切断法和反切刀

使用反向切断时，卡盘与主轴连接部分必须装有保险装置。此时刀架受力是向上的，故刀架应有足够的刚性。

4）弹性切断刀

用高速钢做成的片状刀体，装夹在弹性刀柄上；切削用量过大时，弹性刀柄因受力而产生变形；因刀柄弯曲中心位于刀柄之上，刀头会自动让刀，可避免因扎刀而折断刀具，如图 2-6-7 所示。

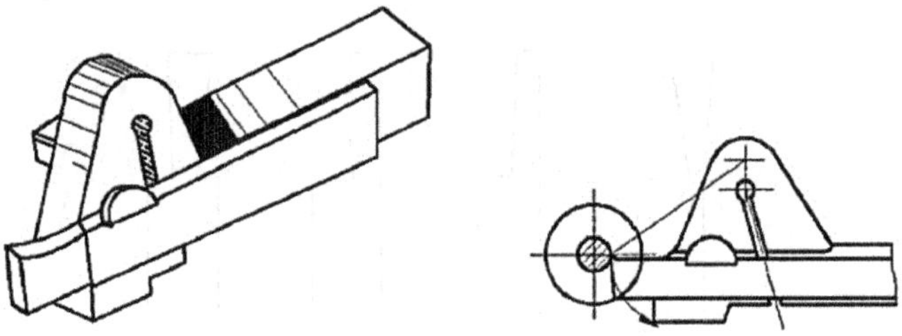

图 2-6-7 弹性切断刀

5）车槽刀

外沟槽车刀的角度和形状与切断刀基本相同。在车较窄的外沟槽时，车槽刀的主切削刃宽度应与槽宽相等，刀体长度要略大于槽深。

3. 切断刀（切槽刀）的刃磨

切断刀与切槽刀极为相似，刃磨与切槽刀基本相同。

（1）切断刀的刃磨主要包括四个基本步骤，如图 2-6-8 所示。

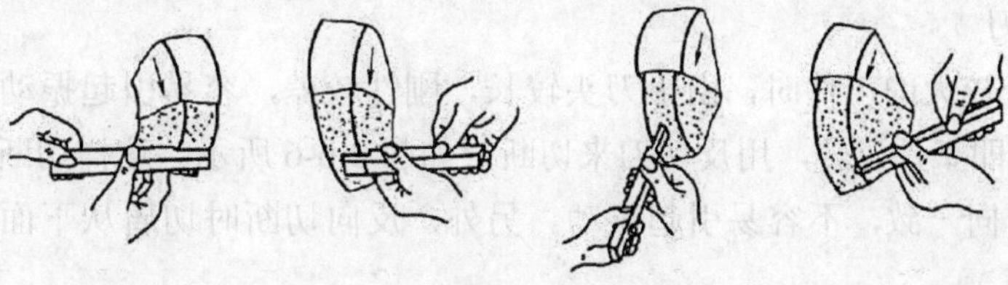

图 2-6-8 切断刀的刃磨步骤

（a）刃磨左侧副后刀面 （b）刃磨右侧副后刀面 （c）刃磨主后刀面 （d）刃磨前刀面

①刃磨左侧副后刀面。两手握刀，车刀前刀面向上，如图 2-6-8（a）所示，同时刃磨出车刀左侧副后角和副偏角。

②刃磨右侧副后刀面。两手握刀，车刀前刀面向上，如图 2-6-8（b）所示，同时刃磨出车刀右侧副后角和副偏角。

③刃磨主后刀面。两手握刀，如图 2-6-8（c）所示，同时刃磨出车刀主后角。

④刃磨前刀面。两手握刀，如图 2-6-8（d）所示，同时刃磨出车刀前角。

⑤修磨刀尖。两手握刀，分别在刀尖处刃磨直线形或圆弧形过渡刃。

（2）在刀具刃磨过程中，还要注意以下几点。

①刀具在刃磨时，如果两侧副偏角太大，那么切断刀刀头强度变差，容易造成折断，如图 2-6-9（a）所示；也不能磨成负值，这样就不能用直进法切断，切槽时，容易出现夹刀现象，会造成槽的两个侧面与工件中心不垂直，如图 2-6-9（b）所示；副切削刃不平直，会造成切削困难，如图 2-6-9（c）所示；切刀左侧面不能磨去太多，如图 2-6-9（d）所示，否则不能切割有高台阶的工件。

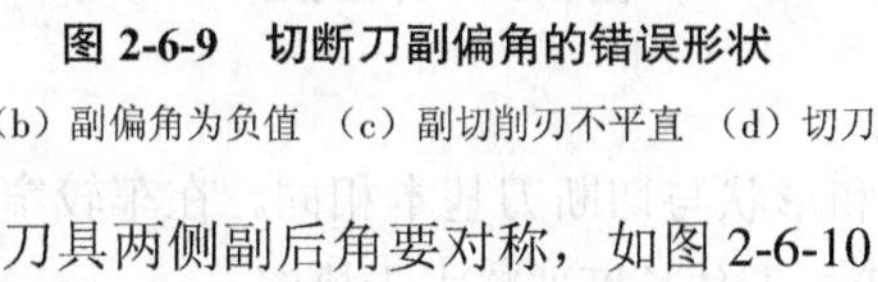

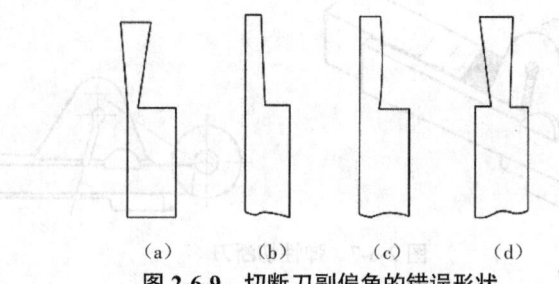

图 2-6-9 切断刀副偏角的错误形状

（a）副偏角太大 （b）副偏角为负值 （c）副切削刃不平直 （d）切刀左侧面磨去太多

②刃磨时，还要注意刀具两侧副后角要对称，如图 2-6-10（a）所示；一侧副后角为负值，如图 2-6-10（b）所示，切断时，该副后角会与工件侧面摩擦或使切断刀折断；两侧副后角也不可过大，否则刀头强度变差，易折断，如图 2-6-10（c）所示。

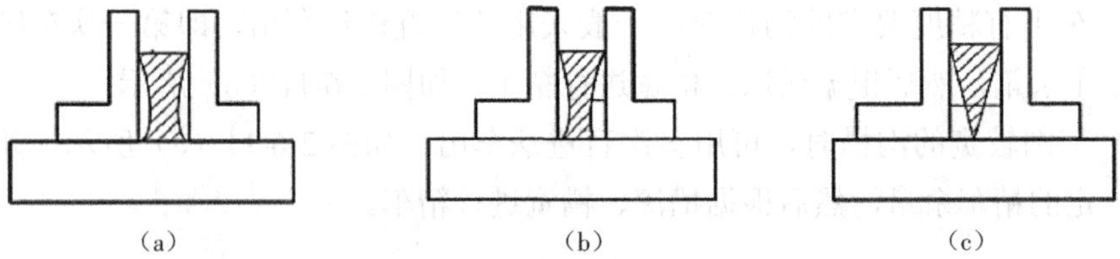

图 2-6-10　切断刀副后角的刃磨要求

（a）两侧副后角要对称　（b）一侧副后角为负值　（c）两侧副后角过大

（3）刃磨安全注意事项。

①磨刀时要戴防护眼镜。

②车刀刃磨时，不能用力过大，以免打滑伤手。

③车刀高低应控制在砂轮水平中心，刀尖略为向上翘。

④车刀刃磨时应做水平方向的左右移动，以免砂轮出现凹坑。

⑤应避免在砂轮侧面刃磨。

⑥砂轮磨削表面必须经常修整，砂轮应没有明显跳动。

⑦刃磨车刀时，操作者应站在砂轮的侧面。

⑧刃磨高速钢车刀时，应注意随时冷却，以防退火；硬质合金车刀在刃磨时，车刀不能放在水中冷却，以防刀片碎裂，同时在刃磨过程中不能用力过猛，否则车刀刀头的焊接处在高温下容易脱落。

⑨刃磨两侧副后角时，应以车刀的底面为基准，用金属直尺或 90° 角尺检查。

4. 切断刀（切槽刀）的安装

刀具的装夹是否正确，对加工质量有直接影响。

（1）为了增加切断刀和车槽刀的刚性，安装时车刀不宜伸出过长。

（2）车槽刀的主切削刃中心线必须垂直于工件轴线，确保两副后角对称，否则车出的槽壁不会平直。

（3）安装车槽刀时，刀具的主切削刃必须与车床主轴中心线平行，否则会造成车槽刀的损坏。

（4）用切断刀切断实心工件时，主切削刃必须与工件旋转中心等高，否则不能车到工件中心，而且容易崩刀，甚至折断车刀。

（5）切断刀的刀头更加窄长，刚性更差。切断时，刀头伸进工件内部，散热条件差，排屑困难，易引起振动，刀头容易折断。因此安装工件时，应尽量将切断处靠近卡盘，以增加工件的刚性。

5. 外沟槽的车削方法

（1）车削精度不高及宽度较窄的沟槽时，可用刀宽等于槽宽的车槽刀，采用一次直进法车出，如图 2-6-11（a）所示。

（2）车削有精度要求的沟槽时，一般采用两次直进法车出，即第一次车槽时槽壁两侧留精车余量，然后根据槽深、槽宽进行精车，如图 2-6-11（b）所示。

（3）车削较宽的沟槽时，可用多次直进法车出，如图 2-6-11（c）所示，并在槽壁两侧留一定的精车余量，然后根据槽深、槽宽进行精车。

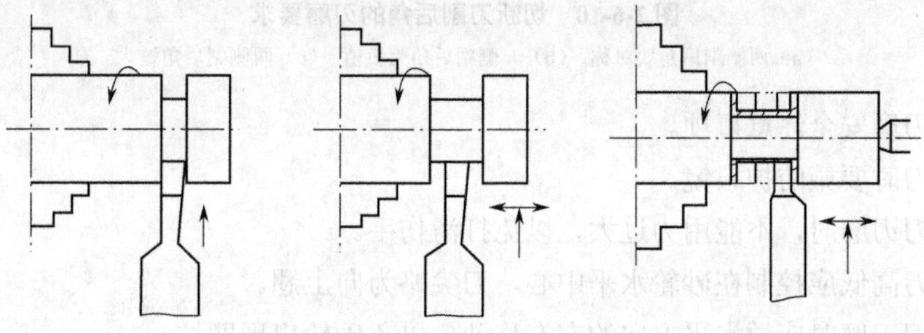

图 2-6-11　矩形沟槽的加工方法

（a）一次直进法车削　（b）两次直进法车削　（c）多次直进法车削

（4）车削较窄的梯形槽时，一般用成形刀一次完成，如图 2-6-12 所示。

（5）车削较窄的圆弧槽时，一般用成形刀一次车出。

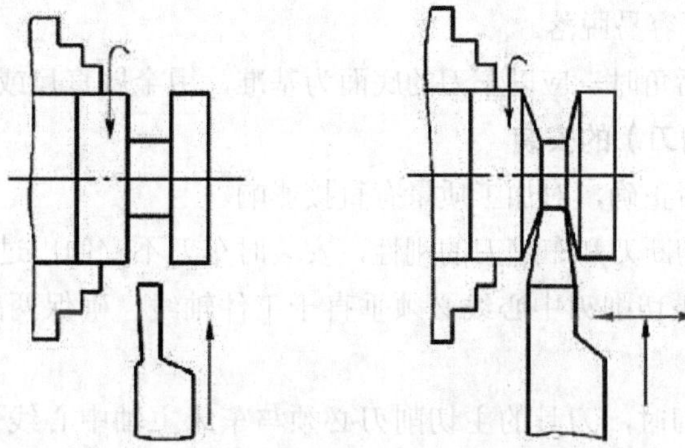

图 2-6-12　较窄的梯形槽的加工方法

6. 切断的方法

切断的方法有直进法、左右借刀法、反切法，如图 2-6-13 所示。直进法常用于切断铸铁等脆性材料及直径较小的棒料，左右借刀法常用于切断钢等塑性材料及直径较大的棒料。

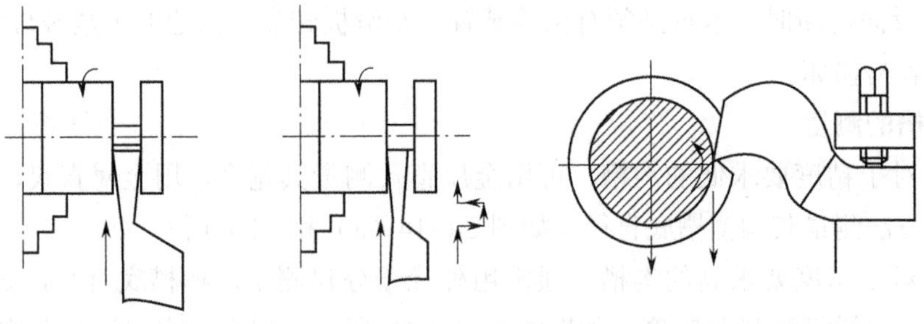

图 2-6-13　切断方法

(a) 直进法　(b) 左右借刀法　(c) 反切法

1）直进法

直进法是指垂直于工件轴线方向进给切断。这种方法效率高，但对车床、切断刀的刃磨和安装都有较高的要求，否则容易造成刀头折断。直进法操作简便，节省材料，应用较广。

2）左右借刀法

切断刀在工件轴线方向反复往返移动，同时两侧径向进给，直至工件被切断。这种方法用于工艺系统（车床、刀具、夹具和工件组成的系统）刚性不足的场合。

3）反切法

采用反切法时车床主轴反转，车刀反装，切削平稳，排屑顺利，但要求车床卡盘必须要有保险装置。

7. 切槽和切断操作的注意事项

切槽和切断虽然操作简单，但要掌握好很不容易，特别是切断，操作时如不注意，刀头就会折断。操作时应注意以下事项。

（1）工件和刀架一定要装夹牢固，刀架要锁紧，以防松动。切断时，切断处应靠近卡盘，以增加工件刚性，减少切削时的振动。

（2）安装切断刀时，刀尖一定要对准工件中心。车刀过低容易折断刀具，过高切断处将留下凸台。

（3）切削速度不宜过大或过小，手动进给切断时，进给要均匀，主轴和刀架各部分配合间隙要小。

（4）切削钢件时应使用切削液，促进切断过程的润滑和散热。切削铸铁工件时不能加切削液，但必要时应使用煤油进行冷却润滑。

（5）工件直径较大或工件一夹一顶安装时，不可直接切到工件中心，留下 2~3 mm，退刀后将工件扳断。

（6）切断刀前刀面不可磨得太低或将断屑槽磨得太深，以免排屑不畅使刀头容易折断。

（7）反向切断时，卡盘必须有保险装置，小滑板转盘上两边的压紧螺母也应锁紧，否则车床容易损坏。

8. 沟槽的测量

（1）对于精度要求低的沟槽，可用金属直尺测量其宽度，用金属直尺、外卡钳相互配合等方法测量其沟槽槽底直径，如图 2-6-14（a）和（b）所示。

（2）对于精度要求高的沟槽，通常用外径千分尺测量沟槽槽底直径，如图 2-6-14（c）所示；用样板测量其宽度，如图 2-6-14（d）所示；用游标卡尺测量其宽度，如图 2-6-14（e）所示。

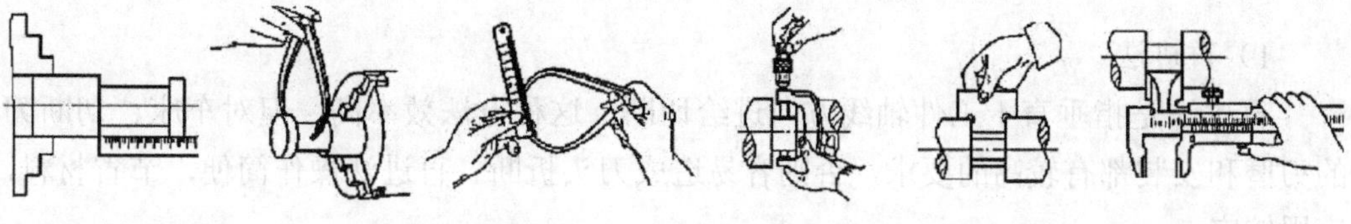

（a）（b）（c）（d）（e）

图 2-6-14 沟槽的检查和测量

9. 钻中心孔

在使用顶尖装夹时需在工件一端或两端先钻出中心孔。在装夹工件前应按照图纸要求定好工件长度，再打中心孔。中心孔的形状必须按照图纸要求，符合标准。

中心孔的类型按国家标准 GB/T145—2001 规定有 A 型（不带护锥）、B 型（带保护锥）、C 型（带螺孔）和 R 型（弧形）四种，如图 2-6-15 所示。中心孔的圆柱部分可用来储存油脂、保护顶尖，使顶尖与 60° 锥孔很好地配合。

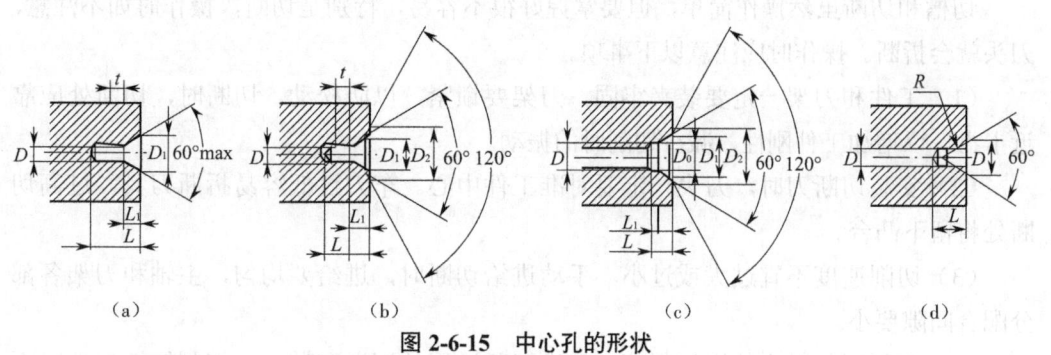

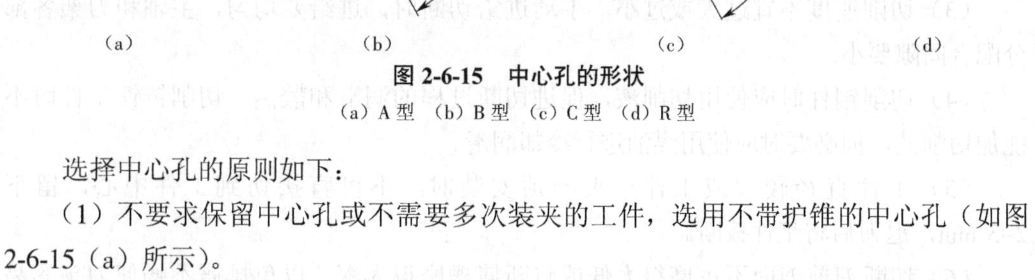

（a）（b）（c）（d）

图 2-6-15 中心孔的形状

（a）A 型 （b）B 型 （c）C 型 （d）R 型

选择中心孔的原则如下：

（1）不要求保留中心孔或不需要多次装夹的工件，选用不带护锥的中心孔（如图 2-6-15（a）所示）。

（2）中心孔的端部多一个 120° 的圆锥孔，可保护 60° 锥孔的边缘不被碰伤，多次装夹的零件或要求保留中心孔的工件，采用带护锥的中心孔（如图 2-6-15（b）所示）。

（3）中心孔中的螺纹用于工件之间的紧固连接，如果被加工零件需要在轴头上固定其他零件时，应选用带螺纹孔的中心孔（如图 2-6-15（c）所示）。

（4）R 型中心孔口的圆弧可减小与顶尖之间的摩擦，以提高定位精度（如图 2-6-15（d）所示）。

用中心钻钻中心孔之前，应当先将端面平整后再钻孔。钻孔时，主轴采用较高转速；进刀要缓慢，并应加注冷却液；为了防止钻头折断，须及时清除钻屑。

10. 工件装夹注意事项

（1）在使用顶尖装夹工件粗车时，由于加工余量较大，应选用活顶尖，能避免在切削力过大时损坏顶尖孔。精车时，主要考虑加工精度，切削余量小，切削力也小，可选用死顶尖并加黄油润滑，防止因过热发生烧损。

（2）工件中心孔与顶尖之间的配合松紧要适宜，不宜过松或过紧。若太松，则工件不能定心，在车削时往往产生振动。若过紧，在切削过程中，随切削温度上升，工件逐渐伸长使工件顶得更紧；如果是死顶尖，则摩擦加剧，会产生烧损现象；如果是活顶尖，则容易因压力过大而损坏顶尖内部结构。

（3）在使用顶尖安装工件之前，还要校正尾座顶尖，使尾座顶尖和前顶尖在同一轴线上，否则车削的外圆将成为锥面。校正时，将尾座移向卡盘处，在卡盘中心夹持已加工的圆锥，使顶尖接近圆锥尖部，目测是否对准，若不重合，需将尾座体做横向调节，使之符合要求。再装上工件，车一刀后测量工件两端的直径，根据直径的大小来调整尾座的横向位置。如果工件右端直径大、左端直径小，则尾座应向操作者方向偏移；反之，向相反方向偏移。

（4）尾座顶尖套筒应伸出短些，以增强尾座的刚性，减小切削时的振动，并要紧固好尾座螺钉。

思考与练习

（1）切断刀的种类有哪些？

（2）切断刀具安装的方法是什么？

（3）简述切断、切槽的方法。

（4）沟槽的测量方法有哪些？

（5）中心孔的选择原则是什么？

任务七　车削外圆锥

一、图样与技术要求

如图 2-7-1 所示，工件毛坯尺寸为 $\phi 50$ mm×70 mm，材料为 45 钢。评分表见表 2-7-1。

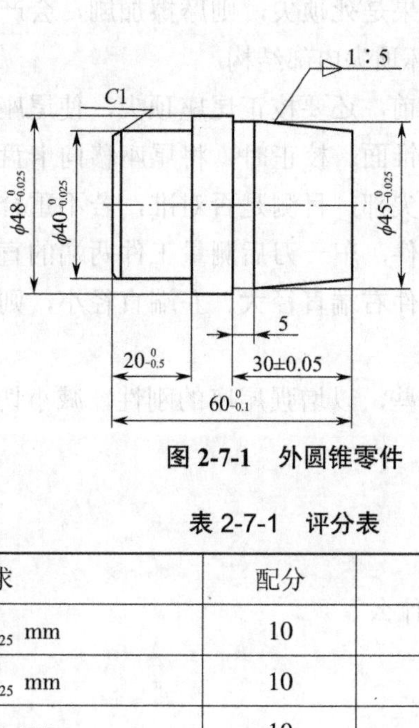

图 2-7-1　外圆锥零件

表 2-7-1　评分表

技术要求	配分	实测结果	得分
外圆 $\phi\ 48^{\ 0}_{-0.025}$ mm	10		
外圆 $\phi\ 40^{\ 0}_{-0.025}$ mm	10		
外圆 $\phi\ 45^{\ 0}_{-0.025}$ mm	10		
锥度 1 ：5（5° 42'±4'）	20		
长度 30±0.05 mm	6		
长度 5 mm	5		
长度 $60^{\ 0}_{-0.1}$ mm	5		

续表

技术要求	配分	实测结果	得分
长度 $20_{-0.05}^{0}$ mm	5		
倒角 $C1$	3		
表面结构参数 $Ra3.2$ μm	16		
安全文明操作	10		

二、加工工艺分析

零件各个轮廓加工工艺安排见表 2-7-2。

表 2-7-2　普车加工工艺卡

零件图号		图 2-7-1	普通车床加工工艺卡		机床型号	CA6140	
零件名称		外圆锥			机床编号		
刀具表					量具表		
刀具名称		刀具参数			量具名称	规格 /mm	
90° 外圆车刀		YT15、高速钢车刀			游标卡尺 千分尺	0~150/0.02 25~50/0.01	
工序	工艺内容		切削用量				加工性质
			S/（r/min）	F/（mm/r）		a_p/mm	
1	车端面		1 100	—		—	手动
2	粗车外圆至 ϕ46 mm，保证长度 29 mm		450	0.33		2	自动
3	掉头装夹已车好的台阶 46 mm×29 mm 处		—	—		—	手动
4	车端面		1 100	—		—	手动
5	粗车外圆 ϕ48 mm，保证长度 30 mm，留余量		450	0.33		2	自动
6	粗车外圆 ϕ40 mm，保证长度 20 mm，留余量		450	0.33		2	自动
7	精车 $\phi\ 40_{-0.025}^{0}$ mm 外圆至尺寸，保证长度 $20_{-0.05}^{0}$ mm		1 100	0.08		0.2	自动
8	精车 $\phi\ 48_{-0.025}^{0}$ mm 外圆至尺寸，保证长度 30 mm		1 100	0.08		0.2	自动
9	倒角倒钝		450	—		—	手动

续表

10	调头，精车端面，保证总长$60_{-0.1}^{0}$ mm	1 100	—	—	手动
11	精车$\phi45_{-0.025}^{0}$ mm外圆至尺寸，保证长度30 mm	1 100	0.08	0.2	自动
12	转动小滑板粗车锥度1∶5圆锥体，并逐步找正1∶5锥度，留精车余量0.3~0.5 mm	560	—	0.5	手动
13	精车圆锥体，保证长度25 mm	1 100	—	0.2	手动
14	倒钝	450	—	—	手动

三、相关知识与技能

1. 圆锥的应用

在机械工程中圆锥面配合应用广泛，如顶尖尾柄与尾座套筒的配合、锥销与锥孔的配合等。圆锥面配合紧密，装拆方便，经过多次拆卸后仍能保证有准确的定心作用，小锥度配合表面还能传递较大的转矩，因此，大直径的麻花钻都使用锥柄，如图2-7-2所示。

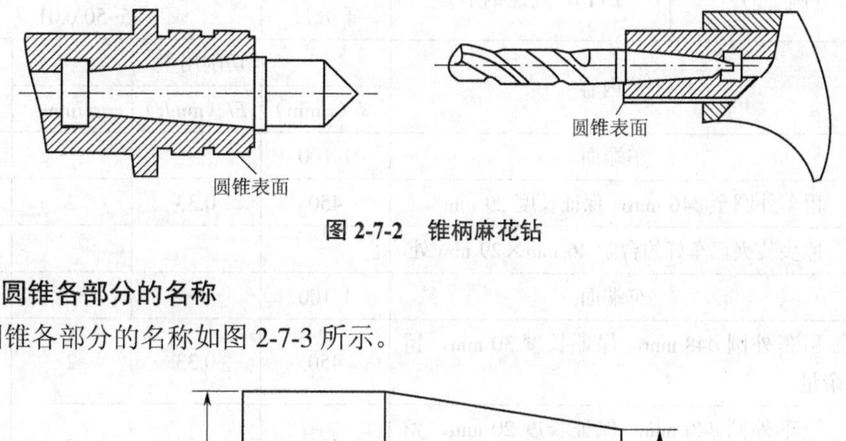

图 2-7-2　锥柄麻花钻

2. 圆锥各部分的名称

圆锥各部分的名称如图2-7-3所示。

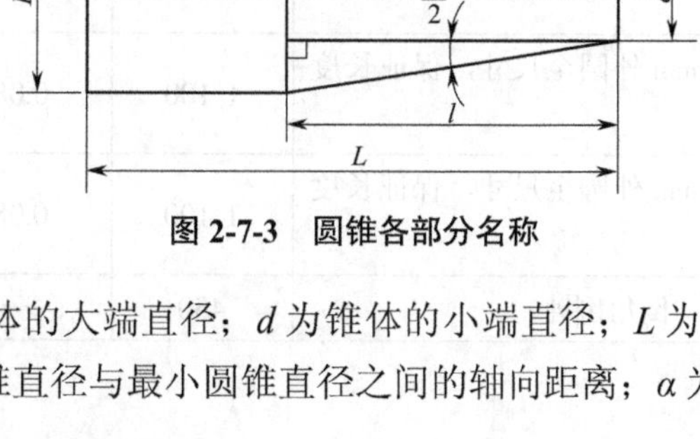

图 2-7-3　圆锥各部分名称

图中，D为圆锥体的大端直径；d为锥体的小端直径；L为零件的长度；l为圆锥体的长度，即最大圆锥直径与最小圆锥直径之间的轴向距离；α为圆锥角。

圆锥体的大小端直径之差与圆锥体长度之比称为锥度 C，即

$$C=\frac{D-d}{l}$$

$$\tan\frac{\alpha}{2}=\frac{D-d}{2l}$$

圆锥面的加工除了尺寸精度、形位精度和表面结构要求外，还有锥度的要求。

3. 车圆锥面刀具的选用与安装

车削圆锥面所用刀具除与车削外圆、内孔相同的刀具外，还可选用宽刃车刀、锥形铰刀。安装车刀时，车刀刀尖必须严格对准工件的旋转中心，否则车出的圆锥素线不直，影响加工质量。

4. 圆锥的车削方法

由于圆锥的素线与轴线相交成圆锥半角，因此在车削圆锥时，只有车刀的运动轨迹与圆锥的素线平行，才能车削出正确的圆锥面。

常见车削圆锥面的方法有转动小滑板法、偏移尾座法、仿形法和宽刃刀法等。

1）转动小滑板法

转动小滑板法是指将小滑板沿顺时针或逆时针方向，按工件的圆锥半角来转动一个角度，使车刀的运动轨迹与所要加工圆锥在水平轴平面内的素线平行，用双手配合，均匀、不间断地转动小滑板手柄，手动进给车削圆锥面的方法，如图 2-7-4 所示。

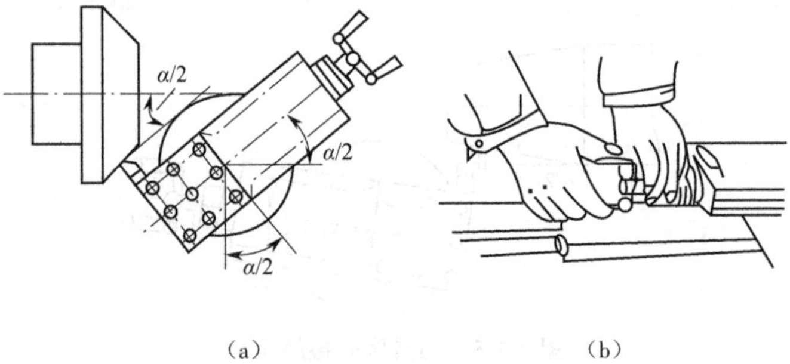

（a）　（b）

图 2-7-4　转动小滑板法

（a）小滑板转动角度　（b）双手交替转动小滑板手柄

（1）转动小滑板的方法。

①用扳手将小滑板下面转盘上的两个螺母松开。

②按工件上外圆锥面的情况确定小滑板的转动方向。

车削正外圆锥面（又称顺锥），即圆锥大端靠近主轴、小端靠近尾座方向时，小滑板应按逆时针方向转动；车削反外圆锥面（又称倒锥），即圆锥小端靠近主轴、大端靠近尾座方向时，小滑板应按顺时针方向转动。

③根据工件的圆锥半角，转动小滑板至所需位置，使小滑板基准零线与圆锥半角

刻线对齐，然后锁紧转盘上的螺母。

④如果工件圆锥半角不是整数，其小数部分可用目测的方法估计，大致对准后，再通过试车法逐步找正。

⑤还需检查和调整小滑板导轨与镶条间的配合间隙。

（2）转动小滑板法的特点。

①能车削圆锥角 α 较大的圆锥面。

②操作简单，可加工任意锥角的内外锥面，能车削整圆锥表面和圆锥孔，应用范围广；在同一工件上车削不同锥角的圆锥面时，调整角度方便。

③只能手动进给，对操作者要求高，劳动强度大，工件表面结构值较难控制，只适用于单件、小批量生产。

④加工长度受小滑板行程的限制，只能加工锥体长度较短的圆锥面。

（3）转动小滑板法刀具的安装。

车刀的刀尖必须严格对准工件的回转中心，否则车出的圆锥素线不是直线而是双曲线（车刀装夹方法及对准工件中心的方法与车端面时相同）。

（4）车削外圆锥面的步骤。

①按最大圆锥直径（增加 1 mm 余量）和圆锥长度将圆锥部分先车削成圆柱体。

②移动中、小滑板，使车刀刀尖与轴端外圆面刚好接触，如图 2-7-5 所示。然后将小滑板向后退出，将中滑板刻度调至零位，作为粗车外圆锥面的起始位置。

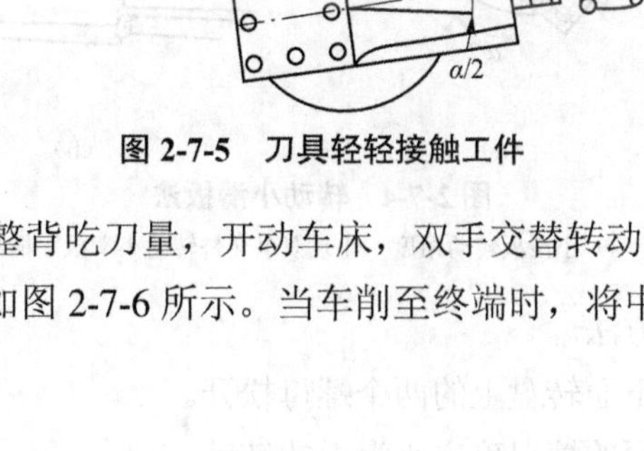

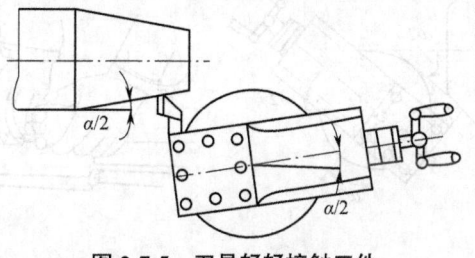

图 2-7-5　刀具轻轻接触工件

③移动中滑板，调整背吃刀量，开动车床，双手交替转动小滑板手柄，进给切削，速度应保持均匀一致，如图 2-7-6 所示。当车削至终端时，将中滑板退出，小滑板快速后退复位。

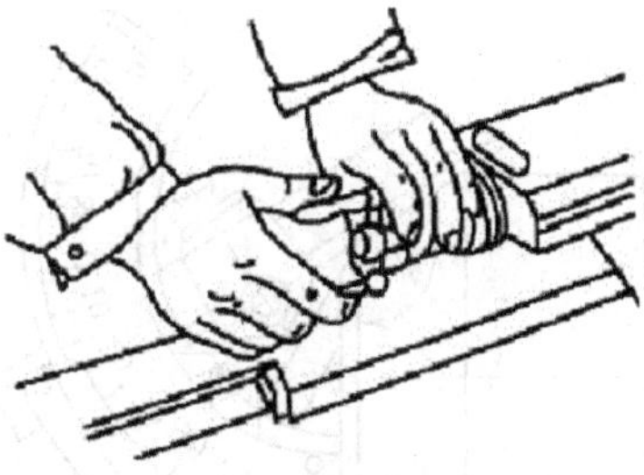

图 2-7-6　双手均匀手动进给

④重复步骤③，调整背吃刀量，手动进给粗车外圆锥面，直至工件能塞入套规约 1/2 为止。

⑤用套规、样板或游标万能角度尺检测圆锥角，找正小滑板转角。

⑥找正后，继续粗车外圆锥面，留精车余量 0.5~1.0 mm。

⑦把小滑板转角调整准确后，精车外圆锥面至要求尺寸。

（5）外圆锥面圆锥角的找正。

精车外圆锥面时，必须找正外圆锥面的圆锥角。因此，要求在粗车了一半外圆锥面时开始找正，找正圆锥角或锥度的主要方法如下。

①用角度样板透光检测：如图 2-7-7 所示，用角度样板检查时，主要通过透光的多少来找正小滑板的角度，反复多次，直到达到要求为止。

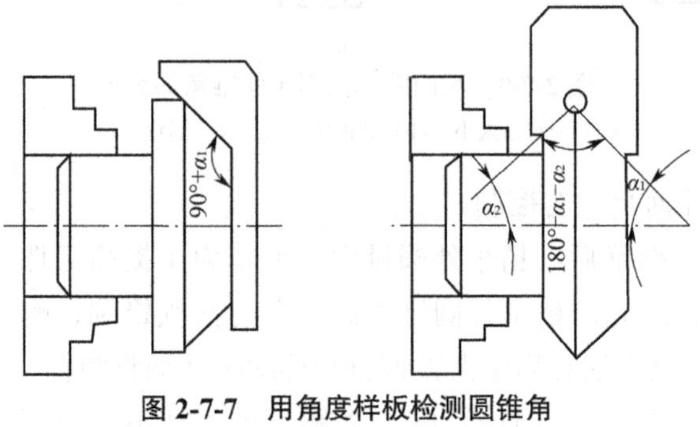

图 2-7-7　用角度样板检测圆锥角

②用游标万能角度尺检测：将游标万能角度尺调整到要测量的角度，使基尺通过工件中心靠在端面上，刀口形直尺靠在外圆锥面的素线上，测量出圆锥半角，根据要求进行调整，如图 2-7-8 所示。

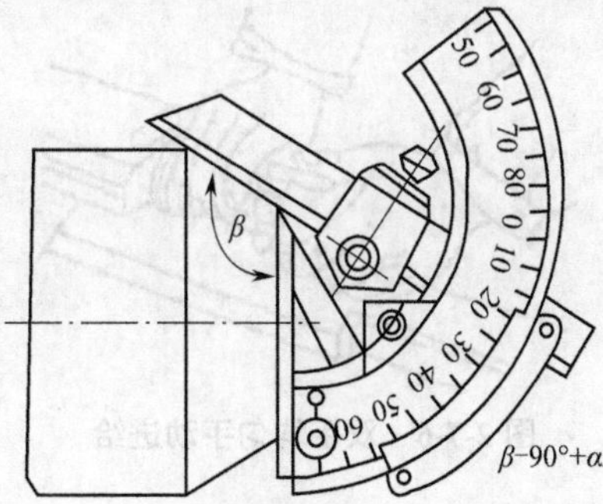

图 2-7-8　用游标万能角度尺检测圆锥半角

③用套规着色检测：将套规轻轻套在工件上，捏住套规左、右两端分别上下摆动，应均无间隙。若大端有间隙，说明圆锥角太小；若小端有间隙，说明圆锥角太大，如图 2-7-9 所示。这时可松开转盘螺母，按需要用铜锤轻轻敲动小滑板使其微量转动，然后拧紧螺母。试车后再检测，直至符合要求为止。

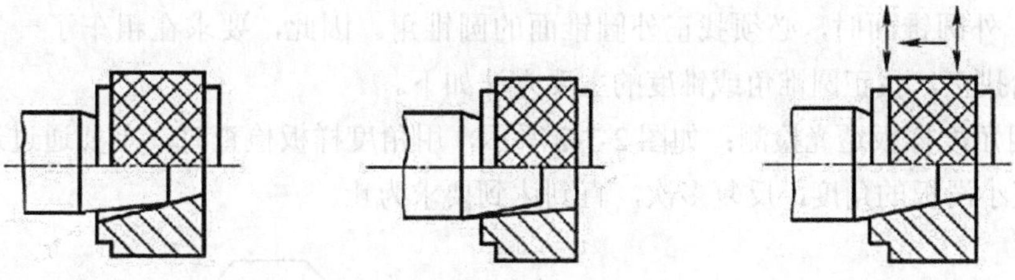
(a)　(b)　(c)

图 2-7-9　用间隙部位判定圆锥角的大小

(a) 圆锥角太小　(b) 圆锥角太大　(c) 圆锥角合适

（6）外圆锥面几何尺寸的控制。

小滑板转角调整准确后，精车外圆锥面主要是为了提高工件的表面质量和控制外圆锥面的尺寸精度。因此，精车外圆锥面时，车刀必须锋利、耐磨，进给要均匀、连续。其背吃刀量的控制方法有用中滑板调整和移动床鞍调整两种。

①中滑板调整背吃刀量（计算法）。先测量出工件小端面至套规过端面的距离 L（如图 2-7-10 所示），用下式计算出背吃刀量 a_p：

$$a_p = L\tan\frac{\alpha}{2} = L\frac{C}{2}$$

式中　a_p——界限量规或台阶中心距离工件端面 L 时的背吃刀量 /mm；

α——工件圆锥角 /（°）；

C——工件锥度；

L——套规台阶中心到工件小端面的距离 /mm。

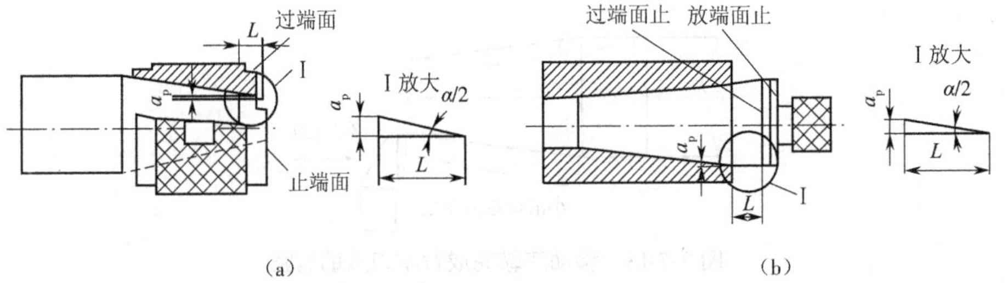

图 2-7-10 测量距离 *L*

（a）外锥尺寸检查 （b）锥孔尺寸检查

然后移动中、小滑板，使刀尖轻轻接触工件圆锥小端外圆表面后，退出小滑板。中滑板按背吃刀量 a_p 值进刀，小滑板手动进给，精车外圆锥面至尺寸，如图 2-7-11 所示。

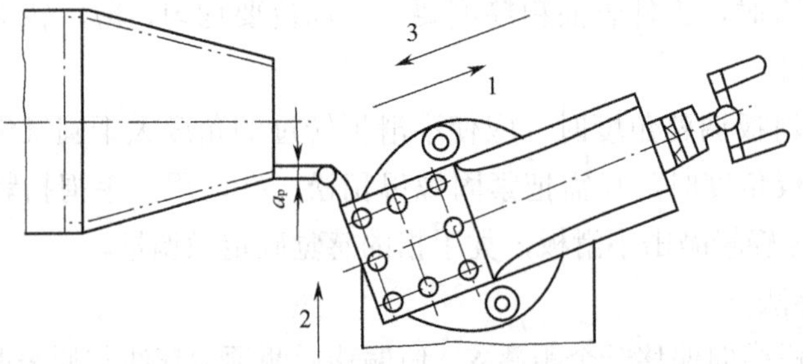

图 2-7-11 用中滑板调整精车背吃刀量 a_p

②移动床鞍调整背吃刀量。根据量出的长度 a，使车刀刀尖轻轻接触工件圆锥小端的外圆锥表面，向后退出小滑板，使车刀沿轴向离开工件端面一个 a 的距离（调整前应先消除小滑板丝杠间隙），如图 2-7-12 所示。然后移动床鞍，使车刀与工件端面接触，如图 2-7-13 所示。此时，虽然没有移动中滑板，但车刀已经切入一个需要的背吃刀量 a_p。

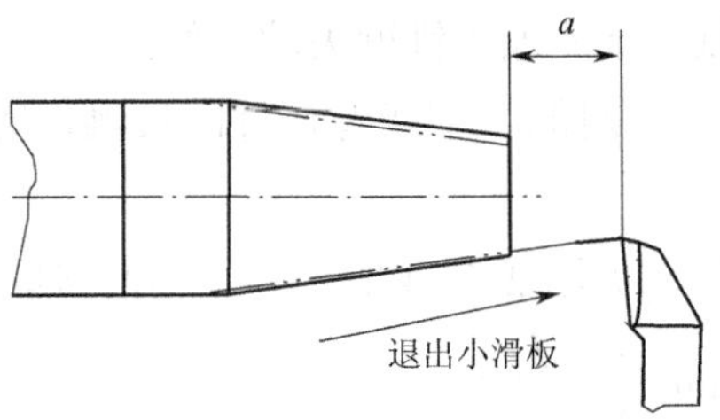

图 2-7-12 退出小滑板调整精车背吃刀量

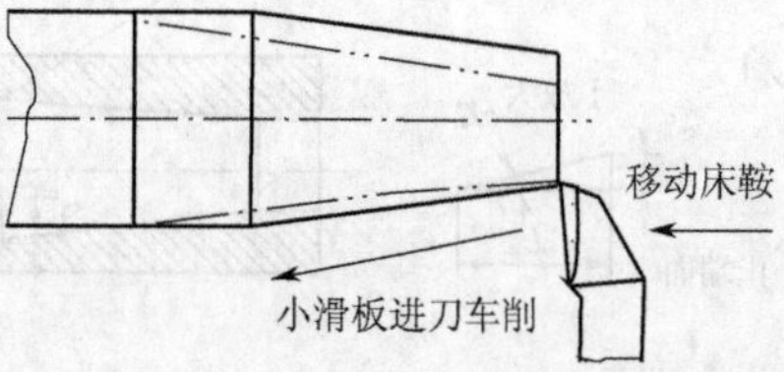

图 2-7-13　移动床鞍完成背吃刀量的调整

（7）滑板转位法车锥面时，要注意以下几点。

①车刀必须对准工件旋转中心，否则会产生双曲线误差。

②手动进给时，应双手握小滑板手柄，均匀移动小滑板，工件表面应一刀车出。

③粗车时，进刀量不宜太大，应先找正锥度，以防把工件车小而报废。

④用游标万能角度尺、游标量角器检查锥度时，测量边应通过工件中心。

⑤用套规检查时，工件表面粗糙度要小，涂色要均匀、薄，转动量一般在半圈之内，以免造成误判。

⑥在转动小滑板调整角度时，应使小滑板转过的角度大于圆锥半角，然后逐步找正。当微调小滑板角度时，只需把紧固螺母稍松一些，用左手拇指紧贴在小滑板与中滑底盘上，用铜棒轻轻敲击小滑板，凭手指的感觉确定微调量。

2）偏移尾座法

将尾座上滑板横向偏移一个距离 S，使偏移后两顶尖连线与原来两顶尖中心线相交角度为 $\alpha/2$，当床鞍带着车刀沿着平行于主轴方向移动切削时，工件就被车成圆锥体。这种车削圆锥面的方法称为偏移尾座法。尾座的偏移方向取决于工件大小头在两顶尖间的加工位置，其偏移量与工件的总长有关。

（1）偏移尾座法的特点：

①适用于加工锥度小、精度要求不高、锥形部分较长的工件；

②可采用纵向机动进给车削，加工表面刀纹均匀，表面结构参数值小；

③由于加工工件需要对顶安装，所以这种方法不能车削整个锥体，也不能车削圆锥孔，受尾座偏移量的限制，不能加工锥度大的工件。

④因顶尖在中心孔中是歪斜的，不能有良好的接触，所以顶尖和中心孔磨损都不均匀。

（2）尾座偏移量的计算。如图 2-7-14 所示，尾座偏移量 S 可按下列公式近似计算：

$$S=\frac{D-d}{2L_0}L=L\tan\frac{\alpha}{2}=\frac{L_0}{2}C$$

式中　S——尾座偏移量；

L_0——工件锥体部分长度；

L——工件总长度；

D、d——锥体大、小头直径；

C——工件锥度。

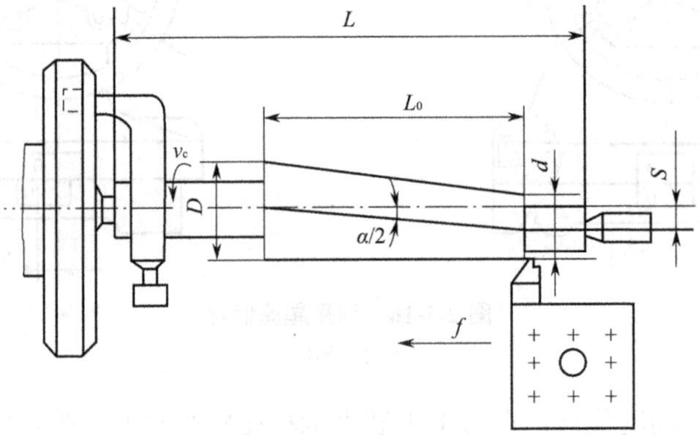

图 2-7-14　尾座偏移量

（3）工件安装步骤如下：

①调整尾座在车床导轨上的位置，使前后两顶尖的距离等于工件总长，且尾座套筒伸出长度小于套筒总长的 1/2；

②先将工件两端中心孔内加注黄油，然后在工件一端装夹鸡心夹头，最后把工件装夹在两顶尖之间，且松紧程度适中，如图 2-7-15 所示。

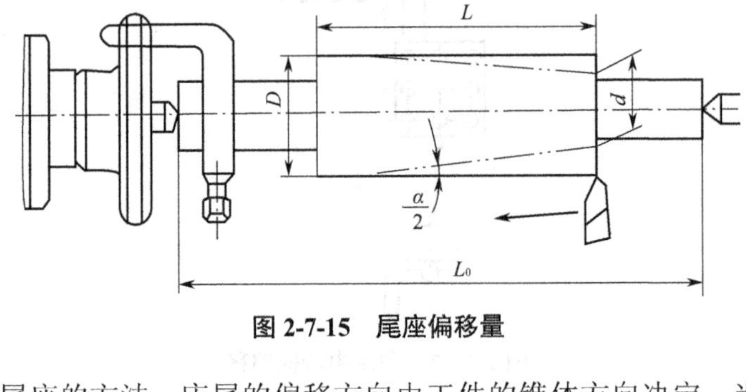

图 2-7-15　尾座偏移量

（4）偏移尾座的方法。床尾的偏移方向由工件的锥体方向决定。当工件的小端靠近床尾处时，床尾应向里移动；反之，床尾应向外移动。先将前后两顶尖对齐（尾座上下层零线对齐），然后根据所得偏移量 S 采用以下四种方法偏移尾座上层位置。

①利用尾座下层刻度偏移。松开尾座紧固螺母，用六角扳手转动尾座上层两侧的螺钉 1、2，使尾座上层向里（操作者方向）移动距离 S（如图 2-7-16 所示）。调准后，拧紧尾座紧固螺母，以防加工中偏移量 S 发生变化。该方法可在尾座有刻度的车床上采用。

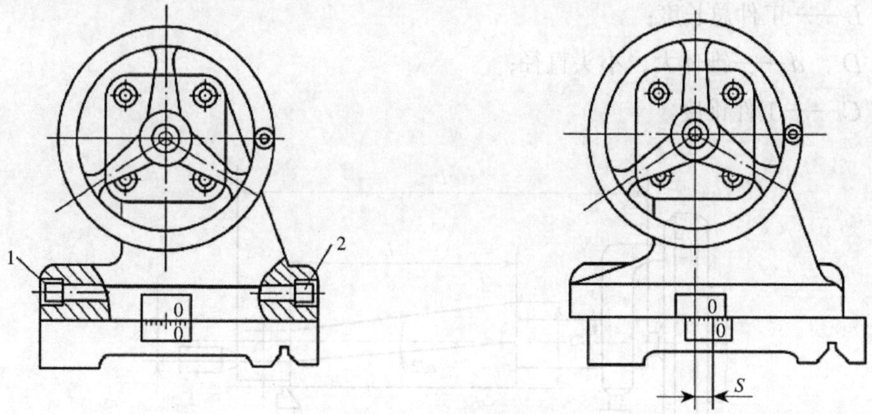

图 2-7-16　利用尾座偏移

1，2—螺钉

②利用中滑板刻度偏移。在刀架上装夹铜棒比较平整的一端，使铜棒平整的一端与尾座套筒轻轻接触，记下中滑板刻度数，再根据尾座偏移量将中滑板移动利用距离 S（如图 2-7-17 所示，横向移动尾座上层位置），使尾座套筒与铜棒端面接触。此时，尾座便在横向上偏移了距离 S。

移动中滑板时，要注意消除中滑板丝杠与螺母间的间隙。

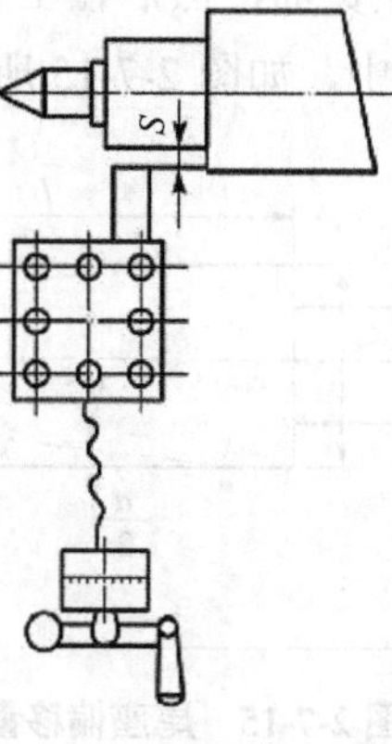

图 2-7-17　利用中滑板偏移

③利用百分表偏移。将百分表固定在刀架上，使百分表的测量头与尾座套筒垂直接触，并与车床中心等高。将百分表调至零位，然后偏移尾座，当百分表上指针转至偏移值 S 时，把尾座固定，如图 2-7-18 所示。该方法可准确地调整偏移量。

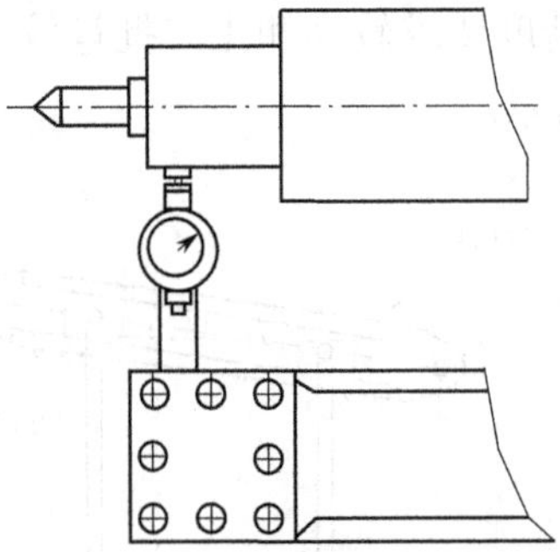

图 2-7-18　利用百分表偏移

④利用锥度量棒或标准样件偏移。将锥度量棒或标准样件安装于两顶尖之间，在刀架上固定一百分表，使百分表测量头与量棒锥面垂直接触，并与车床中心等高。偏移尾座时，纵向移动床鞍，使百分表在圆锥面两端的读数一致后，再将尾座固定，如图 2-7-19 所示。使用该方法偏移尾座，必须选用与工件等长的锥度量棒或标准样件，否则，加工出的锥度不正确。

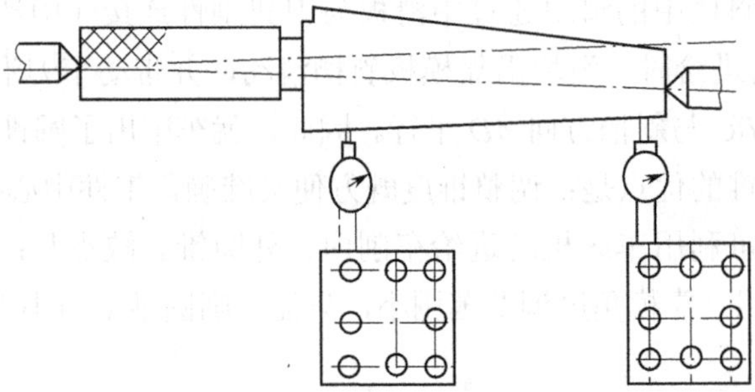

图 2-7-19　利用锥度量棒偏移

（5）锥度的控制。为确保圆锥面加工的精度，车削圆锥面时分为粗车、精车两阶段进行。

①粗车外圆锥面。在粗车外圆锥面到达其长度的 1/2 时，应先进行锥度检查。若锥度偏大，则应反向偏移尾座以减小偏移量 S；若锥度偏小，则应同向偏移尾座以增大偏移量 S。反复调整正确后才可粗车，并留半精车余量 0.5~1.0 mm。粗车外圆锥面时，可采用机动进给。

②精车外圆锥面时用计算法或移动床鞍法确定背吃刀量。用机动进给精车外圆锥面达图纸要求。当批量生产时，工件的总长和中心孔的大小、深浅必须保持一致，否则加工出的工件锥度将不一致。

3）仿形法

仿形法（靠模法）是刀具按仿形装置进给对工件进行车削加工的一种方法。这种方法适用于车削长度较长、精度要求较高和生产批量较大的内、外圆锥工件。仿形法车削圆锥的原理如图 2-7-20 所示。

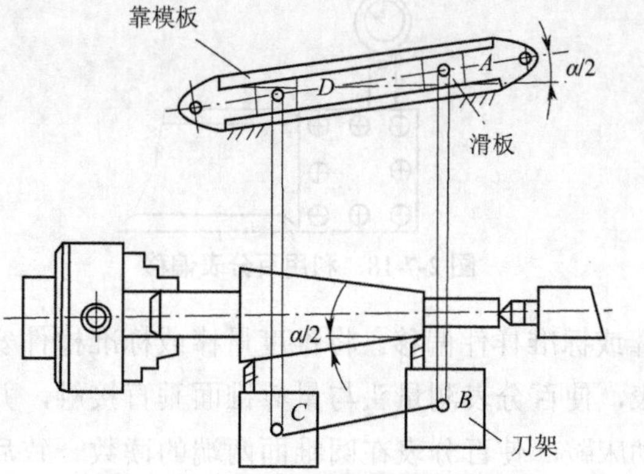

图 2-7-20　靠仿形法车削圆锥原理图

在车床的床身后装一个固定靠模板，靠模板上有斜槽，斜槽角度可按所车削圆锥的圆锥半角调整。斜槽中的滑块通过中滑板与刀架刚性连接（中滑板丝杠在车削时已抽去）。当床鞍纵向进给时，滑块沿靠模板斜槽滑动，并带动车刀沿平行于斜槽的方向移动，其运动轨迹 BC 与斜槽方向 AD 平行。因此，就车削出了圆锥。

仿形法车削圆锥的优点是：调整锥度既方便又准确，工件中心孔与顶尖接触良好，锥面加工质量高，可利用车床机动进给车削内、外圆锥。缺点是：只有在带有靠模附件的车床上才能使用，靠模角度调节范围小，只能车削圆锥半角小于 12° 以下的圆锥。

4）宽刀车削法

宽刀车削法指利用宽刃刀一次或数次将外圆锥面车削成形的方法。其工作原理实质上是属于成型法，所以要求切削刃必须平直，切削刃与主轴轴线的夹角应为 $\alpha/2$，即工件圆锥半角的一半。同时要求车床有较好的刚性，否则易引起振动。

（1）宽刀车削法的特点。

用宽刃刀车削外圆锥面时，容易产生振动。在不影响操作的情况下，可将小滑板间隙调小些，工件伸出长度应尽可能短些，切削用量的选择也要合理，当车削产生振动时，应适当减慢主轴转速。

（2）宽刀车削法的刀具选择与安装。

①宽刃刀选择。对于 30°、45°、60°、75° 的圆锥半角，可选用与主偏角对应的车刀，对于其他的圆锥半角，可选用与主偏角相近的车刀。切削刃长度要大于圆锥素线长度，否则应使用接刀成形。切削刃要平直，如图 2-7-21 所示，否则会使圆锥素线不直。

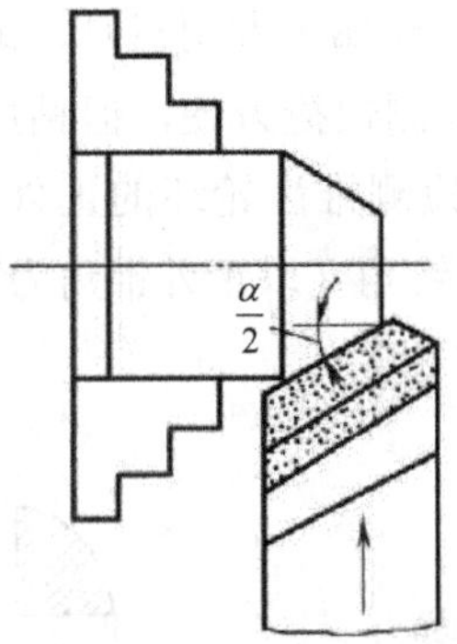

图 2-7-21　宽刃刀车削圆锥

②宽刃刀的安装。宽刃刀的装夹与 45° 端面车刀相似，必须与锥体的锥面角度相同，装夹时可用样板找正，也可用万能角度尺找正，如图 2-7-22 所示。

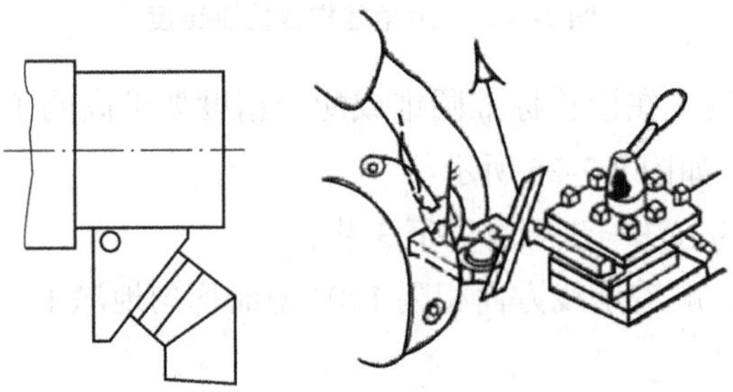

图 2-7-22　宽刃刀的安装找正

5. 外圆锥面的检测

外圆锥面的检测主要指圆锥角度检测和尺寸精度检测。

1）角度或锥度的检测

（1）用游标万能角度尺测量。游标角度尺测量范围为 0°~320°，用其测量精度不高，只适用于单件、小批量生产。用游标万能角度尺检测外圆锥角度时，应根据被测角度的大小，选择不同的测量方法，如图 2-7-23 所示。

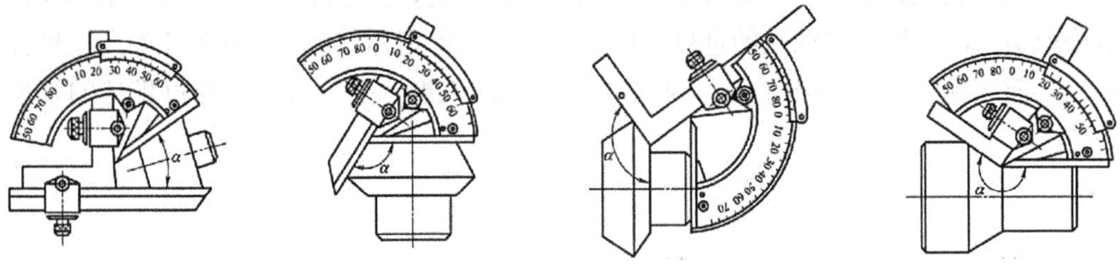

图 2-7-23　用游标万能角度尺测量工件锥度的方法

（a）0°~50°　（b）50°~140°　（c）140°~230°　（d）230°~320°

（2）用角度样板检测。角度样板属于专用量具，成批车削圆锥零件时，一般会事先做出专用的样板。用角度样板检测快捷方便，但精度较低，且不能测得实际的角度值。图 2-7-24（a）为用角度样板检测锥齿轮坯的正外锥角（以端面为基准），圆（b）为用角度样板检测锥齿轮坯的反外锥角（以正外锥角为基准）。

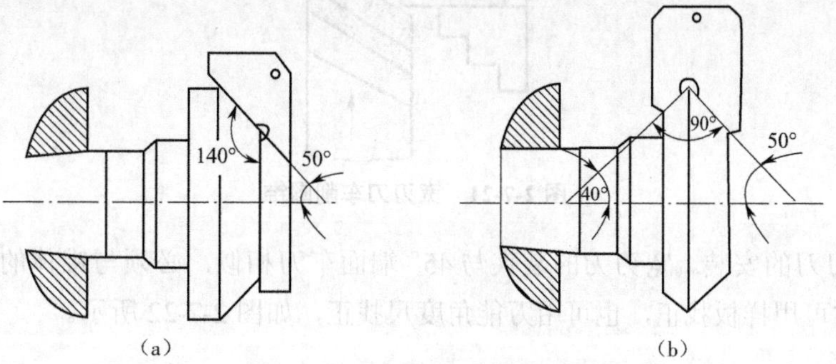

图 2-7-24　用角度样板检测锥度

（3）用涂色法检测。在检验标准圆锥或配合精度要求高的工件时，可用标准锥度塞规或锥度套规检验，如图 2-7-25 所示。

用套规检查圆锥的方法与步骤有以下 3 步。

①在工件表面上，顺着素线方向相隔 120° 薄而均匀地涂上三条显示剂，如图 2-7-26 所示。

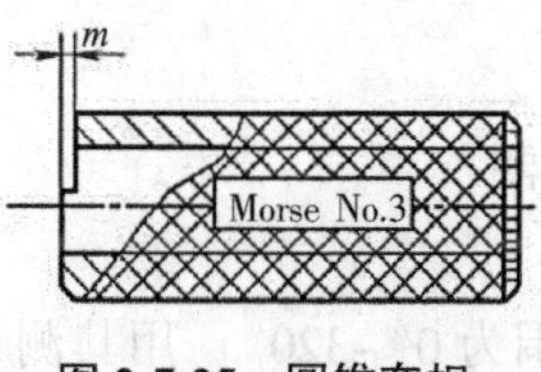

图 2-7-25　圆锥套规

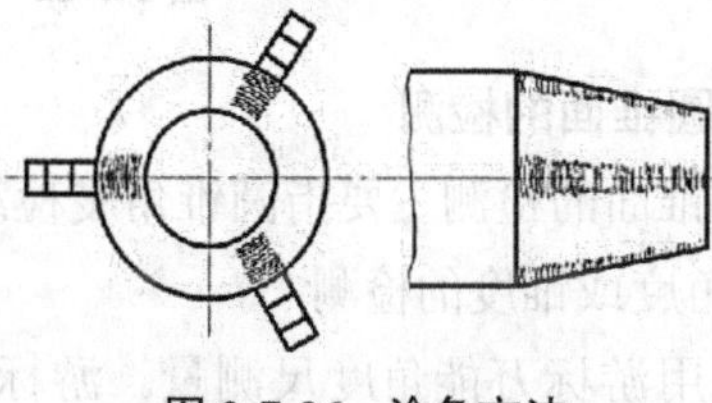

图 2-7-26　涂色方法

②把套规轻轻套在工件上转动不超过半圈，如图 2-7-27 所示。

③取下套规，观察工件锥面上显示剂的擦去情况，如果涂在工件上显示剂的摩擦痕迹很均匀，则说明圆锥孔的锥度正确，如图 2-7-28 所示。如果锥体的小端处有摩擦痕迹，而大端处没有摩擦痕迹，则说明圆锥体的锥度小了；反之说明圆锥体的锥度大了。

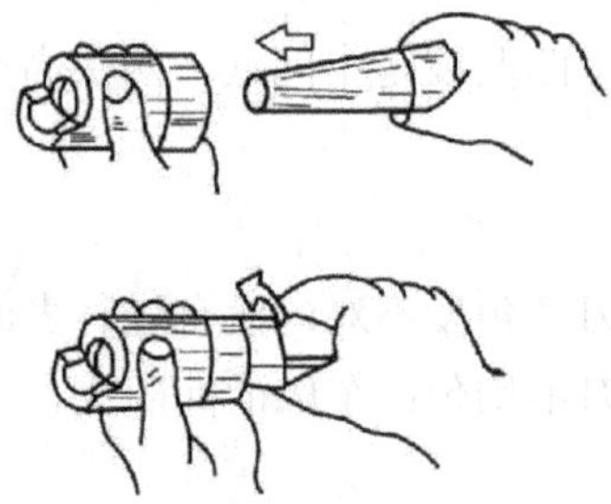

图 2-7-27　用套规检查圆锥

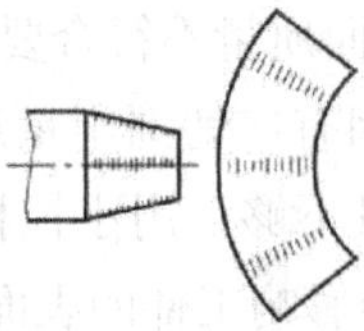

图 2-7-28　合格的圆锥展开图

2）圆锥尺寸测量

（1）用千分尺或游标卡尺测量。对精度要求较低的圆锥或在加工中粗测圆锥尺寸时，一般使用千分尺或游标卡尺测量。

（2）用圆锥套规检测。根据工件的直径和公差，在圆锥套规的小端处开有轴向距离为 m 的缺口，如图 2-7-25 所示，以表示通端和止端。检测时，如果锥体的小端面在缺口之间，则说明小端直径尺寸合格，如图 2-7-29（a）所示；若锥体小端面未能进入缺口，则说明小端直径大了，如图 2-7-29（b）所示；若锥体小端平面超过了止端，则说明小端直径小了，如图 2-7-29（c）所示。

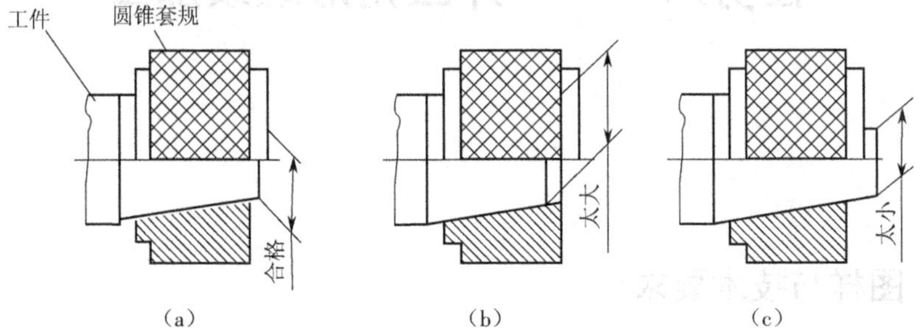

图 2-7-29　用圆锥套规检测外圆锥的尺寸

6. 车圆锥体的质量问题分析

1）锥度不准确

其原因有：计算上的误差；小拖板转动角度和床尾偏移量不精确；车刀、拖板、床尾没有固定好，在车削中移动；工件表面质量太差，量规或工件上有毛刺或没有擦干净，造成检验和测量误差。

2）锥度准确而尺寸不准确

其原因有：粗心大意，测量不及时、不仔细；进刀量控制不好，尤其是最后一刀没有掌握好进刀量而造成误差。

3）圆锥母线不直

圆锥母线不直是指锥面不是直线，锥面上产生凹凸或是中间低、两头高现象。主要原因是车刀安装没有对准中心。

4）表面质量不符合要求

其原因有：切削用量选择不当；车刀磨损或刃磨角度不对；没有进行表面抛光或者抛光余量不够；用小拖板车削锥面时，手动走刀不均匀；车床的间隙大，工件刚性差也可能会影响工件的表面质量。

思考与练习

（1）车削外圆锥面的方法有哪些？

（2）检测外圆锥面锥度的方法有哪些？

（3）检测外圆锥面尺寸的方法有哪些？

（4）简述利用转动小滑板法车削外圆锥面的方法。

（5）简述利用偏移尾座法车削外圆锥面的方法。

任务八　外三角形螺纹加工

一、图样与技术要求

如图 2-8-1 所示，工件毛坯尺寸为 $\phi60$ mm×70 mm，材料为 45 钢。评分表见表 2-8-1。

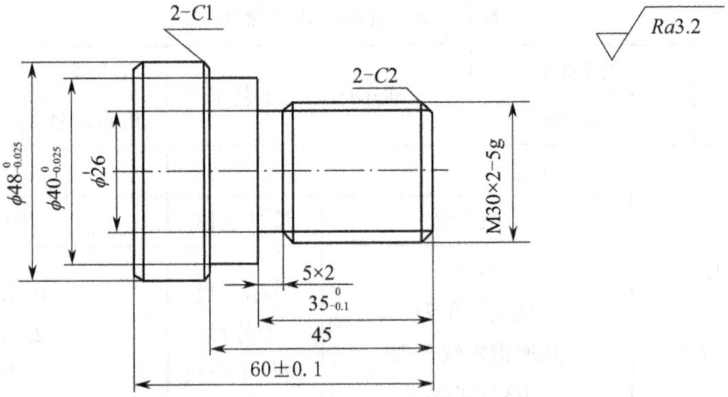

图 2-8-1 外三角螺纹加工零件

表 2-8-1 评分表

技术要求	配分	实测结果	得分
$\phi\ 48_{-0.025}^{\ 0}$ mm	10		
$\phi\ 40_{-0.025}^{\ 0}$ mm	10		
槽 5 mm×2 mm	5		
三角螺纹大径 ϕ30 mm	5		
三角螺纹中径 M30×2–5g	20		
三角螺纹牙型角 30°	10		
长度 $35_{-0.1}^{\ 0}$ mm	5		
长度 60±0.1 mm	10		
长度 45 mm	5		
倒角 *C*1	2		
倒角 *C*2	2		
表面结构参数 *Ra*3.2 μm	6		
安全文明操作	10		

二、加工工艺分析

零件各个轮廓加工工艺安排如表 2-8-2。

表 2-8-2　普车加工工艺卡

零件图号	图 2-8-1	普通车床加工工艺卡	机床型号	CA6140
零件名称	外三角螺纹		机床编号	
刀具表			量具表	
刀具名称	刀具参数		量具名称	规格 /mm
90° 外圆车刀 切槽（断）刀 三角螺纹车刀 中心钻	YT15 高速钢车刀 高速钢螺纹车刀 B3 中心钻		游标卡尺 千分尺 螺纹千分尺	0~150/0.02 25~50/0.01 25~50

工序	工艺内容	切削用量			加工性质
		S/（r/min）	F/（mm/r）	a_p/mm	
1	车端面	1 100	—	—	手动
2	粗车外圆至 ϕ45 mm，保证长度 30 mm	450	0.33	2	自动
3	掉头装夹已车好的台阶 ϕ45 mm×30 mm 处	—	—	—	手动
4	车端面，钻中心孔，用顶尖支撑	1 100	—	—	手动
5	粗车外圆 ϕ48 mm，保证长度 65 mm，留余量	450	0.33	2	自动
6	粗车外圆 ϕ40 mm，保证长度 45 mm，留余量	450	0.33	2	自动
7	粗车外圆 ϕ30 mm，保证长度 35 mm，留余量	450	0.33	2	自动
8	精车三角螺纹大径 ϕ30 mm 外圆至尺寸	1 100	0.08	0.15	自动
9	精车 $\phi\ 40_{-0.025}^{\ 0}$ mm 外圆至尺寸，保证长度 45 mm	1 100	0.08	0.15	自动
10	精车 $\phi\ 48_{-0.025}^{\ 0}$ mm 外圆至尺寸，保证长度 65 mm	1 100	0.08	0.15	自动
11	车 5 mm×2 mm 退刀槽，保证相对长度 $35_{-0.1}^{\ 0}$ mm	560	—	—	手动
12	倒角 $C1$、$C2$	450	—	—	手动
13	粗、精车 M30×2–5g 外三角形螺纹	100	—	—	自动
14	切断，总长留 0.5 mm 余量	450	—	—	手动
15	掉头，垫铜皮装夹 ϕ40 mm 外圆处，用磁力百分表找正，适当加紧	—	—	—	手动
16	精车端面，保证总长 60±0.1 mm	1 120	0.1	0.1	自动
17	倒角 $C1$	450	—	—	手动

三、相关知识与技能

1. 三角螺纹

螺纹零件广泛应用在机械产品中，其功能是连接和传动。例如，车床主轴与卡盘的连接；刀架上螺钉对刀具的紧固，丝杠与螺母的传动等。螺纹的种类很多，按牙型分有三角形螺纹、梯形螺纹、矩形螺纹等（如图 2-8-2 所示）。各种螺纹又有右旋、左旋之分（如图 2-8-3 所示）。普通螺纹（也称为米制螺纹）是应用最广泛的一种三角形螺纹。在车床上可加工各种类型和直径的螺纹，多用于单件、小批量生产。

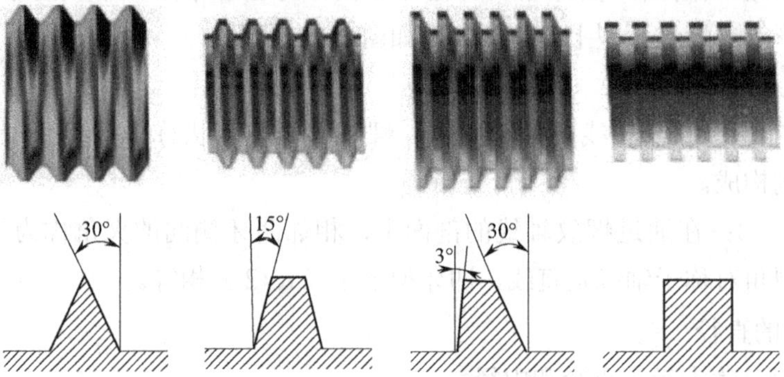

图 2-8-2　螺纹的种类

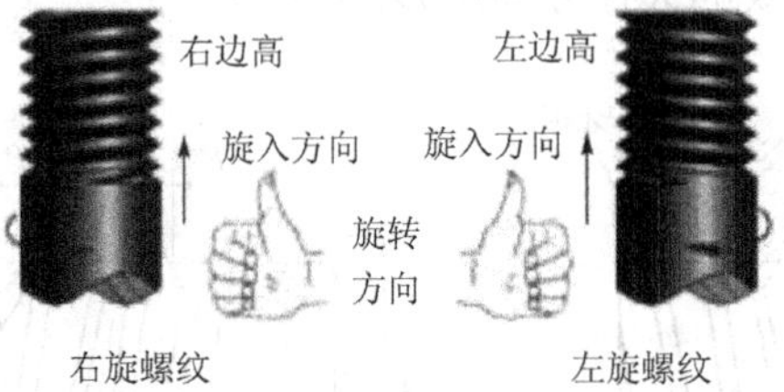

图 2-8-3　螺纹的旋向

2. 螺纹螺旋线的形成及车削原理

螺旋线可以看成是直角三角形 *ABC* 围绕圆柱体旋转一周后，斜边 *AC* 在圆柱表面上形成的曲线，如图 2-8-4（a）所示；螺纹的形成是指螺纹牙型的形成，实际加工时，是从圆柱形毛坯上切出螺纹的齿沟来获得螺纹牙型，如图 2-8-4（b）所示。

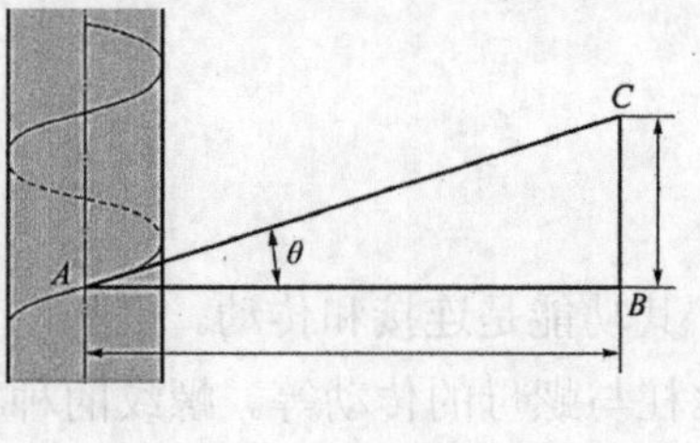

（a）

（b）

图 2-8-4　螺纹的形成

3. 普通三角形螺纹的主要参数

通过螺纹轴线剖面，获得螺纹的轮廓形状称为螺纹牙型。常见的牙型有三角形、梯形和锯齿形等。螺纹牙型上的主要参数如图 2-8-5 所示。

1）螺纹牙型及牙型角

牙型：在通过螺纹轴线的断面图上，螺纹的轮廓形状称为螺纹牙型，它由牙顶、牙底和两牙侧构成。

牙型角（α）：在通过螺纹轴线的剖面上，相邻两牙侧间的夹角称为牙型角。大多数螺纹的牙型角对称于轴线垂直线，即牙型半角（$\alpha/2$）相等。

2）螺纹的直径

螺纹直径分为大径、小径和中径。

大径：与外螺纹的牙顶或内螺纹牙底相重合的假想圆柱面的直径称为大径，大径即为公称直径，内、外螺纹的大径分别用 D、d 表示。

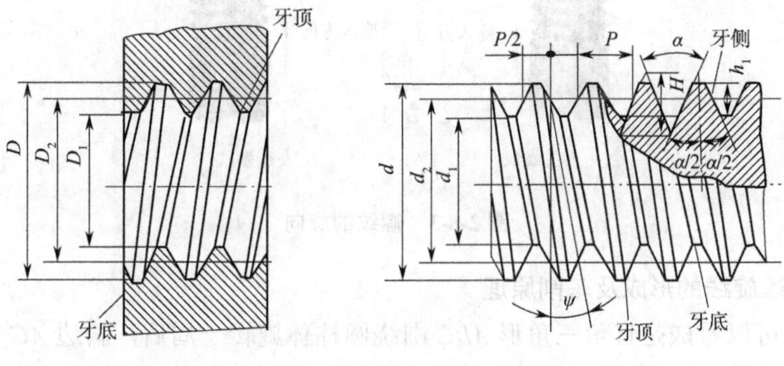

（a）　（b）

图 2-8-5　三角形螺纹主要参数

（a）内螺纹　（b）外螺纹

小径：与外螺纹牙底或内螺纹牙顶相重合的假想圆柱面的直径称为螺纹小径，内、外螺纹的小径分别用 D_1、d_1 表示。

中径：它是一个假想圆柱的直径，即在大径和小径之间，其母线通过牙型上的沟槽和凸起宽度相等的假想圆柱面的直径称为中径，内、外螺纹的中径分别用 D_2、d_2 表示。

3）螺纹的线数（n）

螺纹的线数是指形成螺纹时螺旋线的条数。螺纹有单线和多线之分：沿一条螺旋线形成的螺纹叫作单线螺纹，沿两条或两条以上在轴向等距分布的螺旋线所形成的螺纹叫作多线螺纹。

4）螺纹的螺距（P）与导程（P_h）

螺距（P）：螺距是指螺纹上相邻两牙在中径线上对应两点之间的轴向距离，用 P 表示。由于 P 在中径线上不好测出，实际工作中，测量螺纹时往往在螺纹大径的牙顶处进行。在普通螺纹中，螺纹大径相同时，按螺距的大小分粗牙螺纹和细牙螺纹，细牙普通螺纹的螺距比粗牙普通螺纹的螺距要小。

导程（P_h）：同一螺旋线上相邻牙在中径线上对应两点间的轴向距离。

5）螺纹的旋向

螺纹有右旋和左旋之分，顺时针旋转时旋入的螺纹为右旋螺纹，其螺纹特征是左低右高；逆时针旋转时旋入的螺纹为左旋螺纹，其螺纹特征是左高右低。实际应用中的螺纹绝大部分为右旋。

6）原始三角形高度（H）

牙型两侧相交而得的尖角的高度即为原始三角形高度。

7）牙型高度（h）

在螺纹牙型上，牙顶到牙底之间，垂直于螺纹轴线的距离即为牙型高度。

8）螺纹升角（ψ）

在中径圆柱上，螺旋线的切线与垂直于螺纹轴线的平面之间的夹角即为螺纹升角。

4. 普通三角形螺纹的尺寸计算

普通三角形螺纹牙型如图 2-8-6 所示，普通三角形螺纹主要参数的尺寸计算如表 2-8-3 所示。

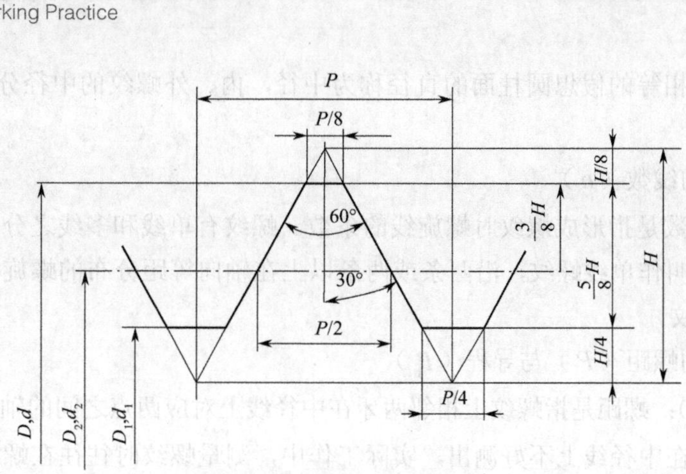

图 2-8-6　普通三角形螺纹牙型

表 2-8-3　普通三角形螺纹的尺寸计算

名称		代号	计算公式
外螺纹	牙型角	α	$\alpha = 60^\circ$
	原始三角形高度	H	H=0.866P
	牙型高度	h	$h=\frac{5}{8}H=\frac{5}{8}\times 0.866P=0.5413P$
	中径	d_2	$d_2=d-2\times\frac{3}{8}H=d-0.6495P$
	小径	d_1	$d_1=d-2h=d-1.0826P$
内螺纹	中径	D_2	$D_2=d_2$
	小径	D_1	$D_1=d_1$
	大径	D	D=d= 公称直径
螺纹升角		ψ	$\tan\psi=\frac{nP}{\pi d_2}$

5. 刀具的选择与安装

1）螺纹刀具的选择

（1）外螺纹车刀。

螺纹加工必须保证螺纹牙型和螺距的精度，并使相配合的螺纹具有相同的中径，否则加工出来的螺纹不能旋合。为了获得正确的牙型，必须正确刃磨车刀，螺纹车刀切削部分的形状必须磨成与螺纹牙型完全一致，米制螺纹车刀刀尖角为 60°，使用样板检验时，刀尖应与样板配合无缝。外螺纹车刀角度如图 2-8-7 所示。

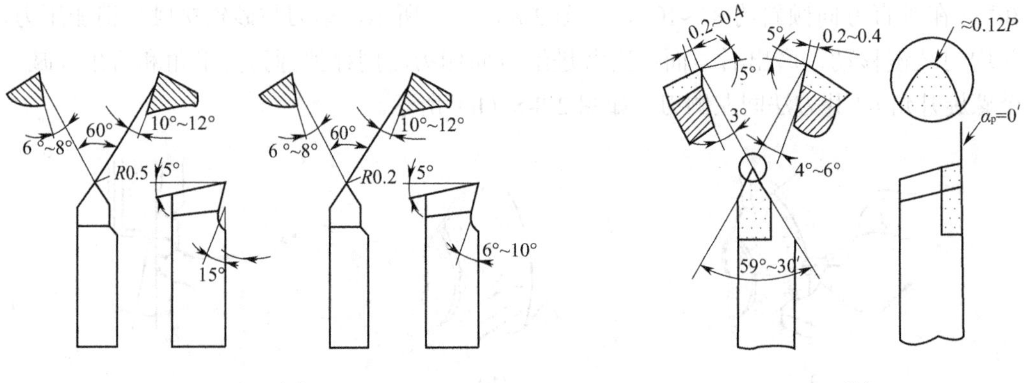

图 2-8-7　外螺纹车刀

（a）高速钢外螺纹车刀　（b）硬质合金外螺纹车刀

高速钢外螺纹车刀，粗车加工时常采用 5° ~15° 的正前角，可使切削顺利且可减小表面结构参数值。但螺纹车刀的前角会使加工出的螺纹牙型角产生误差，这种误差对要求不高的螺纹可以忽略不计，对于精度要求较高的螺纹，螺纹刀刃磨时需对牙尖角进行修正，所以精车螺纹时，应使用前角为零的螺纹车刀。硬质合金外螺纹车刀高速切削时，牙型角会扩大。因此，刀尖角需要减少 0.5°，螺纹车刀的工作后角一般为 3° ~5° 。

因受螺纹升角的影响，进刀方向一侧的刃磨后角应等于工作后角加上螺纹升角。另一侧的刃磨后角应等于工作后角减去螺纹升角。不过三角形螺纹的升角一般比较小，影响也小，在加工大螺距螺纹时才考虑。

（2）三角形螺纹车刀的刃磨。

螺纹车刀的刀尖角因受螺纹牙型角的限制，刀头体积小，刃磨较困难，其刃磨要求如下：

①当螺纹车刀背（纵向）前角 γ_o =0° 时，刀尖角等于牙型角；当螺纹车刀背前角 γ_o >0° 时，刀尖角必须修正。

②螺纹车刀两侧切削刃必须是直线。

③螺纹车刀切削刃应具有较小的表面结构参数值。

④螺纹车刀两侧后角不相等，应根据车刀进给方向在两侧后角基础上分别加减一个螺旋升角。

螺纹车刀刃磨步骤如下。

①粗磨前刀面。

②磨两侧后刀面，并初步形成两刃夹角。其中，先磨进给方向侧刃，并注意控制刀尖半角和后角（加一个螺纹升角），刃磨时双手握刀，使刀柄与砂轮外圆水平方向成

30°，在垂直方向倾斜约 8°~10°，如图 2-8-8（a）所示，车刀与砂轮接触后稍加压力，并均匀缓慢移动，磨出后刀面；磨背进给方向侧刃，同样把握刀尖半角和后角（减一个螺旋升角），其方法同左侧刃，如图 2-8-8（b）所示。

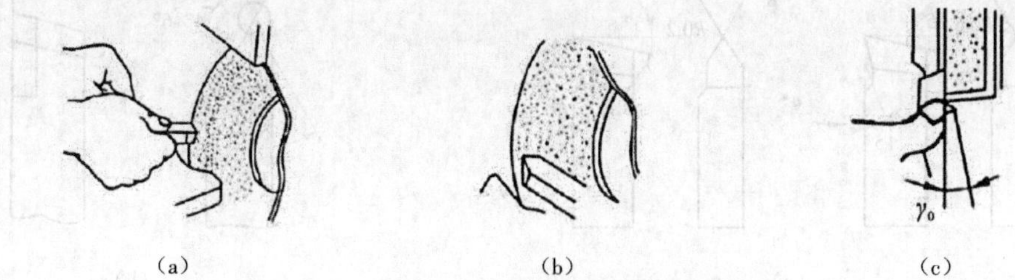

图 2-8-8　螺纹车刀的刃磨

（a）刃磨左侧刃（b）刃磨背进给方向侧刃（c）精磨前刀面

③按前角值精磨前刀面。将车刀前刀面在砂轮水平方向倾斜大约 10°~15°，在垂直方向倾斜约 10°~15°，且注意使左侧切削刃略低于右侧切削刃，前刀面与砂轮接触后稍加压力刃磨，逐渐磨至靠近刀尖处，如图 2-8-8（c）所示。

④精磨后刀面。刀尖角用螺纹车刀样板来测量，如图 2-8-9 所示。测量时，刀杆底平面应与样板平面平行，观察刀刃与样板间的透光判断刃磨的刀尖角是否正确。

⑤修磨刀尖，刀尖侧棱宽度约为 0.1P（螺距）。

⑥用油石研磨刀尖处的前后角，并注意保持刃口锋利。

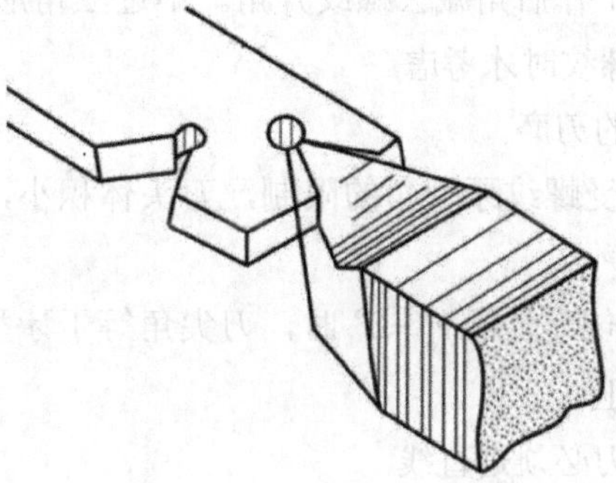

图 2-8-9　用样板检查刀尖角

2）螺纹车刀的安装

螺纹车刀的安装正确与否对加工螺纹的牙型有很大影响，若安装不正确，即使刀具刃磨再好，螺纹牙型角也难以满足加工精度要求。对于三角形螺纹，牙型半角必须对称。

安装外螺纹车刀时，使刀尖角对称线与工件轴线垂直，刀尖与工件中心等高，如图 2-8-10 所示。装夹时，用螺纹样板对刀。将样板平靠工件外圆，螺纹车刀的两侧切削刃与样板的角度槽对齐，做透光检查，若车刀歪斜，用铜棒轻敲刀柄，使车刀位置对准样板。对好后，紧固车刀，并且再复查一次，以防拧紧刀架螺钉时车刀移动。加工中因故调换或刃磨螺纹车刀后，再次装夹时要重新对刀，并使刀尖运动轨迹在原螺旋槽内。

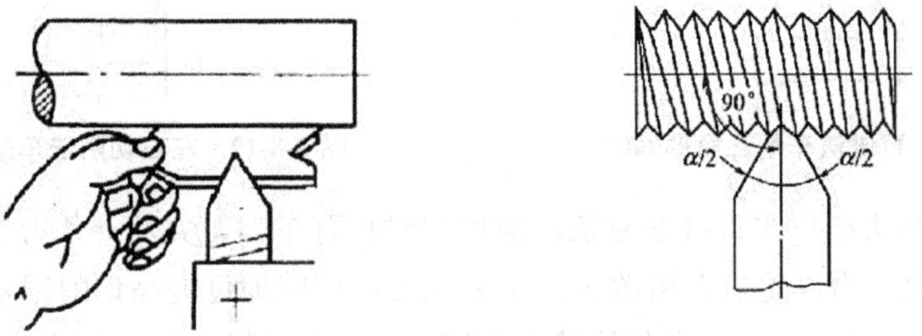

图 2-8-10 螺纹车刀的正确安装

6. 三角形螺纹的车削方法

三角形螺纹的车削方法有两种：即低速车削法与高速车削法。用高速钢车刀低速车削三角形螺纹，能获得较高的螺纹精度和较低的表面结构参数值，但这种车削方法生产效率较低，成批车削时不宜采用，适用于单件或特殊规格的螺纹加工。用硬质合金车刀高速车削螺纹，生产效率较高，螺纹表面结构参数值也较小，是目前机械制造业中被广泛采用的方法。

1）低速车削三角形螺纹

低速车削三角形螺纹一般用高速钢车刀，分粗车与精车两步。粗车时切削速度可选择 10~15 m/min，精车时切削速度可选择 5~10 m/min。

车削三角形螺纹的进给方法有三种，应根据工件的材料、螺纹外径的大小及螺距的大小来决定，下面分别介绍三种进给方法。

（1）直进法。用直进法车削，如图 2-8-11 所示。车螺纹时，螺纹车刀刀尖及左右两侧刃都直接参与切削工作。每次进给由中滑板做横向进给，随着螺纹深度的加深，背吃刀量相应减少，直至把螺纹车削好为止。这种车削方式操作较简便，车出的螺纹牙型正确，但由于车刀的两侧刃同时参加切削，排屑较困难，刀尖容易磨损，螺纹表面结构参数值较大，当背吃刀量较深时容易产生“扎刀”现象。因此，这种车削方法适用于螺距小于 2 mm 或脆性材料的螺纹车削。

（2）左右切削法。左右切削法，如图 2-8-12 所示。车螺纹时，除了用中滑板刻度控制螺纹车刀的横向进给外，同时使用小滑板刻度使车刀左右微量进给。采用左右切削法车削螺纹时，要合理分配切削余量，粗车时可顺着进给方向偏移，一般每边留精车余量 0.2~0.3 mm。精车时，为了使螺纹两侧面都比较光洁，当一侧面车好以后，再将车刀偏移到另一侧面车削。粗车时切削速度取 10~15 m/min，精车时切削速度小于

6 m/min，背吃刀量小于 0.05 mm。

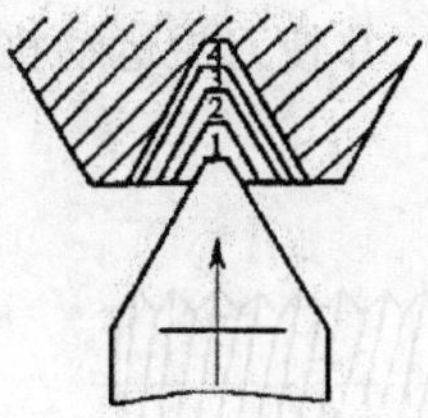

图 2-8-11　直进法车削三角形螺纹

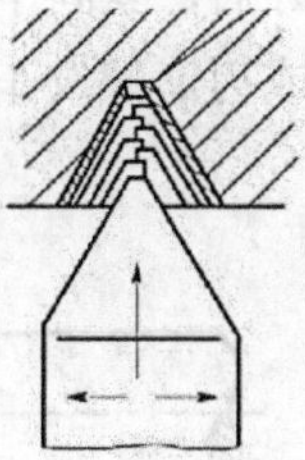

图 2-8-12　左右切削法车削三角形螺纹

这种车削法操作比直进法复杂，但切削时只有车刀刀尖及一条刃参加切削，排屑较顺利，刀尖受力、受热有所改善，不易扎刀，相应地可提高切削用量，能取得较小的表面结构参数值。由于受单侧进给力的影响，左右切削法有增大牙型误差的趋势，适用于粗、精车除矩形螺纹外的各种螺纹，可加大切削用量、提高切削效率。

（3）斜进法。斜进法与左右切削法相比，小滑板只向一个方向进给，如图 2-8-13 所示。斜进法操作比较方便，但由于背离小滑板进给方向的牙侧面表面结构参数值较大，因此只适用于粗车螺纹。在精车时，必须用左右切削法才能使螺纹的两侧面都获得较小的表面结构参数值。采用高速钢车刀低速车螺纹时要加注切削液，为防止“扎刀”现象，最好采用如图 2-8-14 所示的弹性刀柄螺纹车刀。使用这种车刀时，当切削力超过一定值时，车刀能自动让开，使切屑保持适当的厚度，粗车时可避免“扎刀”现象，精车时可降低螺纹表面结构参数值。

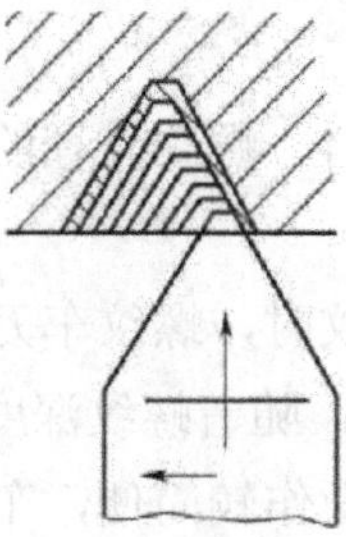

图 2-8-13　斜进法车削三角形螺纹

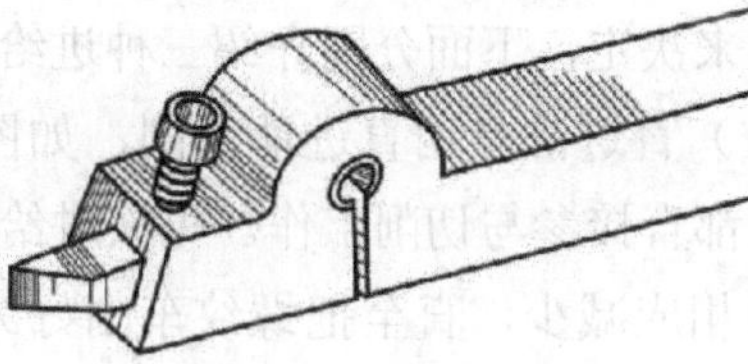

图 2-8-14　弹性刀柄螺纹车刀

2）高速车削三角形螺纹

高速车削三角形螺纹使用的车刀为硬质合金车刀，切削速度一般取 50~70 m/min，车削时只能用直进法进给，使切屑垂直于轴线方向排出。用硬质合金车刀高速车削螺纹时，背吃刀量开始可大些，以后逐渐减小，车削到最后一次时，背吃刀量不能太小（一般在 0.15~0.25 mm），否则会使螺纹两侧面表面结构参数值较大，表面成鱼鳞片状，严重时还会产生振动。

7. 三角形外螺纹的测量

测量螺纹时，应根据不同的质量要求和生产批量的大小，选择不同的测量方法。常见的测量方法有单项测量法和综合检测法两种。

1）单项测量法

（1）螺纹顶径的测量。螺纹顶径是指外螺纹的大径和内螺纹的小径，一般用游标卡尺或千分尺测量。

（2）螺距（或导程）测量。车削螺纹前，先用螺纹车刀在工件外圆上划出一条很浅的螺旋线，再用钢直尺、游标卡尺或螺纹样板对螺距（或导程）进行测量，如图 2-8-15 所示。

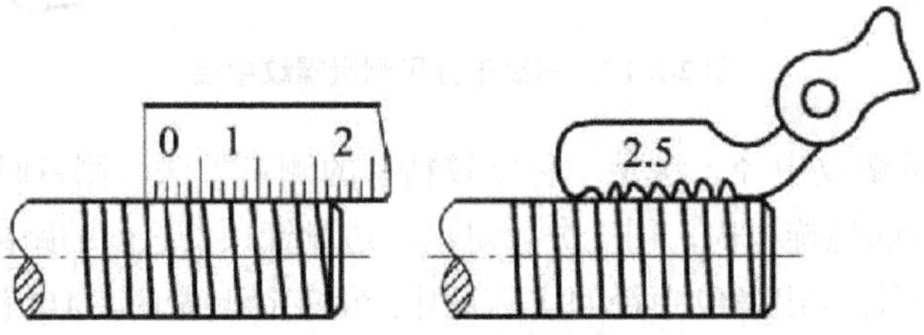

图 2-8-15　车削前的螺距测量

完成车削螺纹后螺距（或导程）的测量也可用同样的方法，如图 2-8-16 所示。用钢直尺或游标卡尺测量时，最好量 5 个或多个牙的螺距（或导程），然后取其平均值。螺纹样板又称螺距规或牙规，有米制和英制两种。测量时将螺纹样板中的钢片沿着通过工件轴线的方向嵌入螺旋槽中，如完全吻合，则说明被测螺距是正确的。

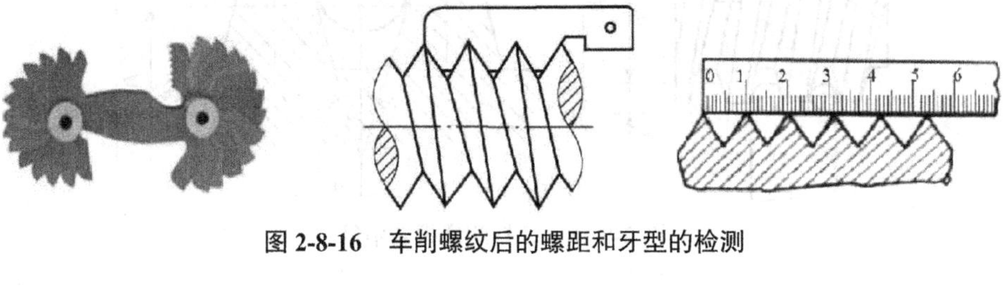

图 2-8-16　车削螺纹后的螺距和牙型的检测

（3）牙型角的测量。一般螺纹的牙型角可以用螺纹样板或牙型角样板来检验，如图 2-8-16 所示。

（4）中径的测量。

①用螺纹千分尺测量螺纹中径。三角形螺纹的中径可用螺纹千分尺来测量，其使用方法与一般千分尺相似。只是螺纹千分尺有两个可以调整的测量头（上测量头、下测量头）。测量时，两个与螺纹牙型角相同的测量头正好卡在螺纹牙侧，由图 2-8-17 可知，*ABCD* 是一个平行四边形，因此，测得的尺寸 *AD* 便是螺纹中径。图 2-8-17 所示为

使用螺纹千分尺测量螺纹中径。

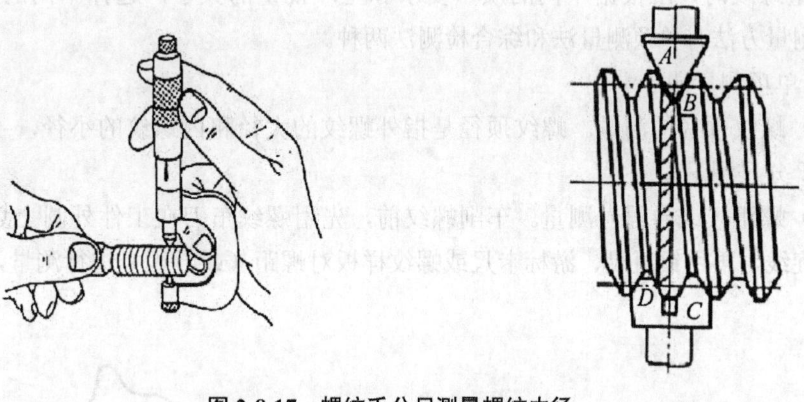

图 2-8-17　螺纹千分尺测量螺纹中径

②用三针测量螺纹中径。这是一种比较精密的测量方法。测量时将三根量针放置在螺纹两侧相对应的螺旋槽内，用千分尺量出两边量针顶点之间的距离 M，如图 2-8-18 所示。根据 M 值可计算出螺纹中径的实际尺寸。用三针测量时，M 值和量针直径 d_D 的计算公式见表 2-8-4 所示。

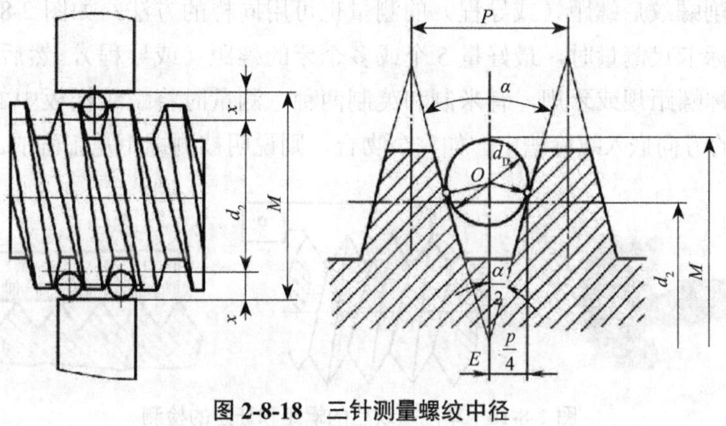

图 2-8-18　三针测量螺纹中径

测量时所用的三根直径相等的圆柱形量针，是由量具制造厂专门制造的（在没有量针的情况下，可以用直径相等的钻头柄部代替）。量针直径 d_D 不能太大或太小。最佳量针直径是指量针横截面与螺纹中径处牙侧相切时的量针直径。量针直径的最大值、最佳值和最小值可由表 2-8-4 中的公式计算出。选用量针时，应尽量接近最佳值，以便获得较高的精度。

表 2-8-4 M值和量针直径 d_D 的计算公式

螺纹	牙型角 α	M值的计算公式	量针直径 d_D		
			最大值	最佳值	最小值
普通螺纹	60°	$M = d_2 + 3d_D - 0.866P$	$1.01P$	$0.577P$	$0.505P$
英制螺纹	55°	$M = d_2 + 3.1d_D - 0.961P$	$0.894P$–0.029	$0.564P$	$0.481P$–0.016

2）综合测量

用螺纹环规可综合检测三角形外螺纹。螺纹环规分为通规 T 和止规 Z。先检查螺纹的直径、螺距、牙型和表面结构，再检查尺寸精度。当通规能通过而止规不能通过时，说明精度符合要求。用螺纹环规检查三角形外螺纹时，以拧上工件时的松紧程度来确定螺纹是否合格，如图 2-8-19 所示。螺纹精度要求不高时，也可以用标准螺母检查。

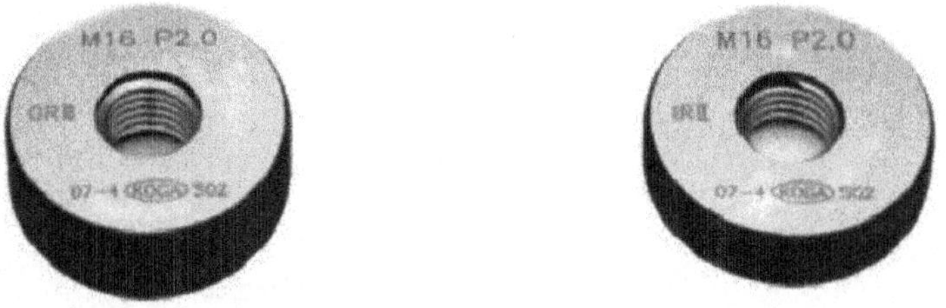

图 2-8-19 螺纹环规

8. 注意事项

（1）选择好车削用量。车螺纹时的走刀速度较快，主轴的转速不宜过高。一般粗车时选切削速度为 13~18 m/min，每次切削深度为 0.15 mm 左右，计算好进刀次数，留精车余量 0.2 mm；精车时，切削速度为 5~10 m/min，每次进刀 0.02~0.05 mm。

（2）工件和主轴的相对位置固定。从顶尖上取下工件测量时，不得松开卡箍；重新安装工件时，必须使卡箍与卡盘的相对位置不变。

（3）若切削中途换刀，须重新对刀。由于传动系统存在间隙，对刀时，应先使车刀沿切削方向走一段距离，停车后再进行对刀。此时移动小滑板使车刀切削刃与螺纹槽相吻合即可。

（4）为保证每次走刀时，刀尖都能正确落在已经车削过的螺纹槽内，当丝杠的螺距不是零件螺距的整数倍时，不能在车削过程中打开开合螺母，应采用正反车法。

（5）车削螺纹时，禁止用手触摸工件，禁止用棉纱擦拭旋转的螺纹。螺纹的测量主要是测量螺距、压力角和螺纹中径。螺距一般用钢直尺测量，压力角一般用样板测量，也可用螺距规同时测量螺距和压力角。只有中径是靠加工过程中的正确操作来保证的，因此螺纹中径常用螺纹千分卡尺测量。

在加工工件时，根据图纸要求，工件的加工余量需要经过几次走刀才能切除。为了提高生产率，保证工件尺寸精度和表面结构，可把车削加工分为粗车和精车。这样可以根据不同阶段的加工，合理选择切削参数。有时根据需要在粗车和精车之间再加半精车，其车削参数介于两者之间。

思考与练习

（1）螺纹的分类有哪些？

（2）三角螺纹的主要参数包括哪些？

（3）简述三角螺纹车刀的安装方法。

（4）简述三角螺纹车削方法有哪些。

（5）简述三角形外螺纹的测量方法有哪些。

项目 3

铣工

【项目描述】

铣削加工是使用旋转的多刃刀具切削工件，是高效率的加工方法。铣削用的机床有卧式铣床或立式铣床，也有大型的龙门铣床。铣削加工适用于加工平面、沟槽、各种成形面等。

通过铣削加工的基础训练，使学生了解铣床的加工原理和加工范围，掌握铣削加工基本技能，通过任务的实施，激发学生的学习兴趣，培养学生的工程意识，提高解决问题的能力。

【学习目标】

（1）了解铣削加工的工艺特点、加工范围。

（2）了解常用铣床的主要部件及其作用。

（3）了解铣床常用刀具和附件的结构、用途及安装调整方法。

（4）掌握铣削的操作要领和简单零件的铣削加工方法，了解用分度头进行简单分度的方法以及铣削加工所能达到的尺寸精度、表面粗糙度值范围。

（5）掌握在铣床上正确安装工件、刀具的方法，能完成铣平面、斜面、台阶面、沟槽及铣键槽等的操作。

任务一 铣工一般知识

一、铣削加工的作用及主要工作内容

机械零件一般都是由毛坯通过各种不同的加工方法达到所需的形状和尺寸，铣削加工就是其中最常用的切削加工方法之一。所谓铣削，就是在铣床上以铣刀旋转做主运动，工件做进给运动的切削加工方法。铣削加工的主要特点是：用多刀刃的铣刀来进行切削，效率较高；加工范围广，可以加工各种形状较复杂的零件；加工精度较高，其公差等级一般为IT9~IT7。

表面结构参数值 Ra 为 12.5~1.6 μm。铣削加工的基本工作内容如图 3-1-1 所示。

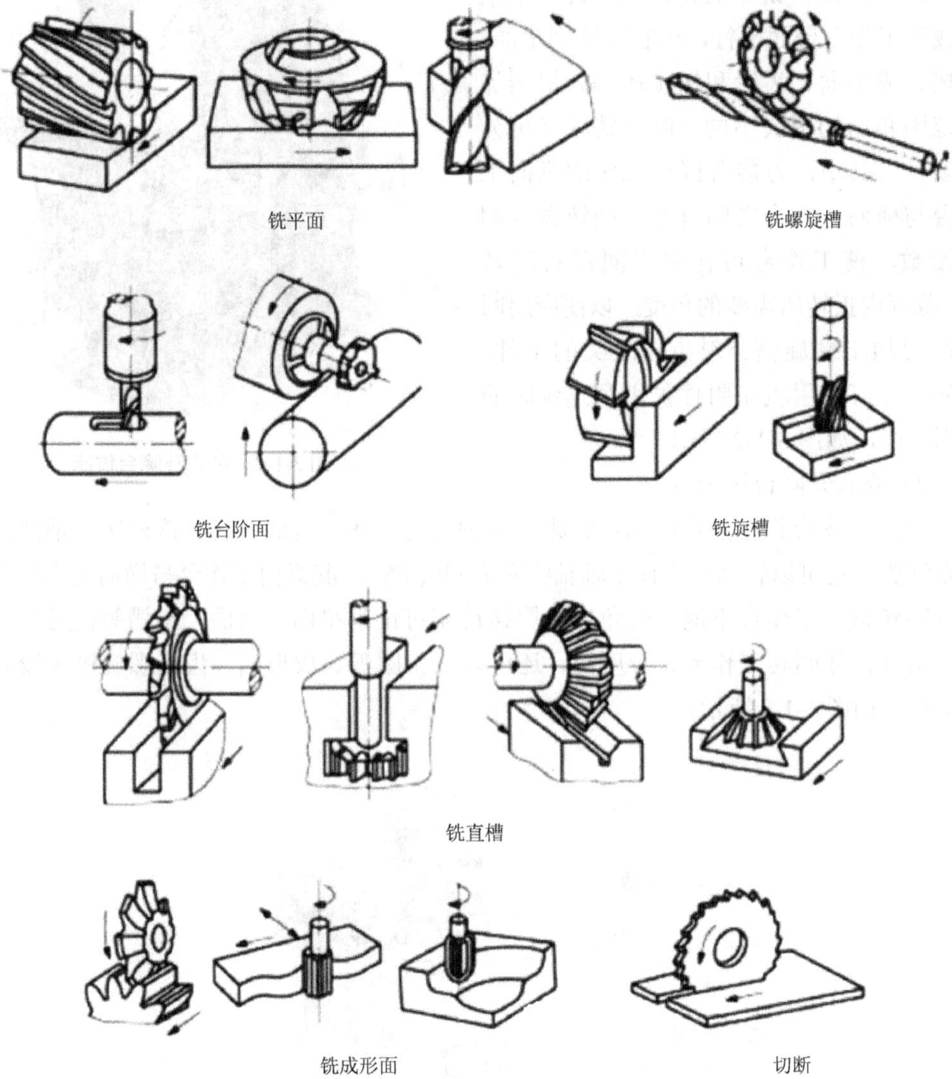

图 3-1-1 铣削的加工内容

二、铣床的基本结构

1. 铣床的种类

铣床是机械制造行业的重要设备，它的用途广泛，类型很多，目前常用的铣床有卧式升降台铣床和立式升降台铣床。

1）卧式升降台铣床

卧式升降台铣床的特征是：铣床主轴轴线与工作台台面平行，用它可铣削平面、沟槽、成形面、齿轮和螺旋槽等。根据加工范围和结构形式不同，卧式铣床又可分为多种，其中，万能升降台铣床的纵向工作台与横向工作台之间有个一回转盘并刻有度数，使工作台可在水平面左右回转45°范围内扳转所需要的角度，以便应用圆盘铣刀加工螺旋槽、斜齿轮之类的工件。另外，还可利用机床附件扩大卧式铣床的应用范围，如图3-1-2所示。

图 3-1-2　卧式升降台铣床

2）立式升降台铣床

立式升降台铣床的特征是：铣床主轴轴线与工作台台面垂直。该机床主轴部分称为立铣头，它可以转动角度使主轴轴线左右回转45°；其纵向工作台与横向工作台连接没有回转盘，工作台不能扳转角度。用该铣床可铣削平面、斜面、沟槽等；另外利用机床附件，如回转工作台、分度头，还可以加工圆弧、成形面、齿轮螺旋槽等较复杂的工件，如图3-1-3所示。

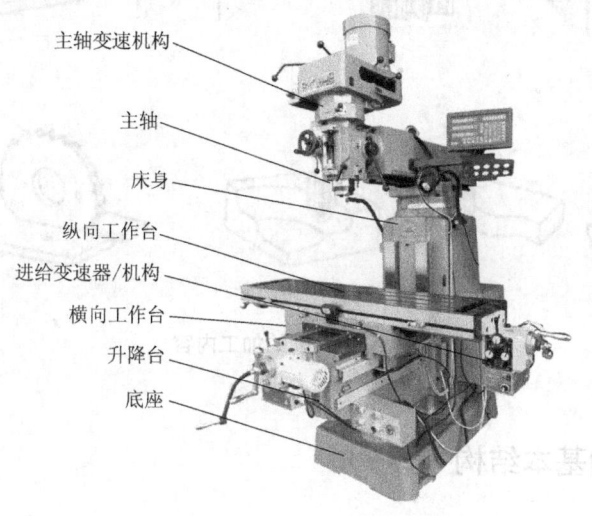

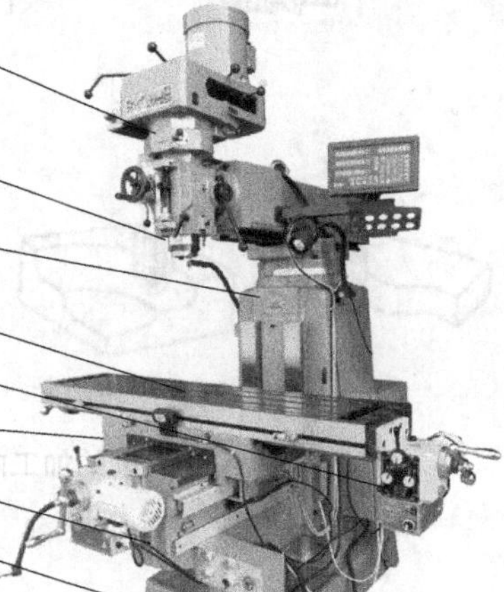

图 3-1-3　X6330 型万能立式铣床

2. 铣床的基本结构

铣床的种类虽然很多，但各类铣床的基本结构大致相同。现将图3-1-3所示的

X6330型万能立式铣床的基本部件及其功能作简略介绍。

1）主轴变速机构

主轴变速机构的作用是将主电动机的额定转速通过齿轮变速获得不同转速，并将其传递给主轴，以适应铣削的需要。

2）主轴

主轴是前端带锥孔的空心轴，铣刀刀轴就安装在锥孔中。主轴材料选用优质结构钢，并经过热处理和精密切削加工制造而成。

3）床身

床身是机床的主体，机床大部分部件都安装在床身上。床身一般用优质灰铸铁铸成，并经过精密的切削加工和时效处理。床身的前壁有燕尾形垂直导轨，升降台可沿导轨上下移动。

4）纵向工作台

纵向工作台是安装工件和带动工件做纵向移动的结构。工作台面上有三条T形槽，可用T形螺钉安装、固定工具和夹具，工作台前侧有一条长槽，用来安装固定极限自动挡铁和自动循环挡铁。纵向工作台台面、导轨部分和T形槽直槽部分的精度要求都很高。

5）进给变速机构

进给变速机构的作用是将进给电动机的额定转速，通过齿轮变速获得不同转速，并将其传递给进给机构，实现工作台以各种不同速度移动，以适应铣削的需要。

6）横向工作台

横向工作台安装在纵向工作台和升降台之间，用来带动纵向工作台做横向移动。

7）升降台

升降台安装在床身前侧垂直导轨上，中部用丝杠与底座螺母相连接。升降台主要用来支撑工作台，并带着工作台上下移动。工作台及进给部分的电动机、传动装置都安装在升降台上，因此升降台的刚性和精度要求都很高，否则在铣削过程中会产生很大振动，影响工件的加工精度。

8）底座

底座是整部机床的支承部件，具有足够的刚性和强度。底座内腔盛装切削液，供切削时冷却润滑用。

三、铣削的基本运动

铣削是以铣刀的旋转运动为主运动，以工件的直线或旋转运动或铣刀直线运动为进给运动的切削加工方法。铣削时工件与铣刀的相对运动称为铣削运动，它包括主运动和进给运动。

1. 主运动

主运动是形成机床切削速度或消耗主要动力的运动。铣刀的主运动是旋转运动。

2. 进给运动

进给运动是使工件切削层材料相继被切削，从而加工出完整表面所需要的运动。铣削运动中，工件的移动或转动、铣刀的移动等都是进给运动。另外，进给运动按运动方向可分为纵向进给、横向进给和垂直进给三种。

四、铣床操作规程

（1）操作之前必须将铣床各部位擦干净。

（2）检查铣床各部位机构和运动部件是否完好，检查各手柄和旋钮是否处于合理的位置。

（3）工作台和主轴部件不能用硬物大力敲击，工作台上不准乱放工具、毛坯等杂物。

（4）开机时，必须注意工作物与铣刀不得接触，工作台来回移动松紧应均匀；开动铣床使主轴回转，检查齿轮是否甩油；进行铣床主轴和进给系统的变速检查，使主轴和工作台进给由低速到高速运动，检查运动是否正常。

（5）不能超负荷工作。工作中，如发现铣床有异常现象和不规则响声，应立即停机并及时修理；未修复不得使用。

（6）开自动走刀时必须先检查行程限位器是否可靠；工作台自动进给时，应脱开手动进给离合器，以防手柄随轴旋转伤人；不准两个进给方向同时起动自动进给；自动进给时，不准突然变换进给速度；自动进给完毕，应先停止进给，再停止主轴（刀具）旋转。

（7）铣床运转时不得离开机器，不准做与操作内容无关的事；刀具、量具、工具放置稳妥、整齐、合理，有固定的位置，便于操作时取用，用后应放回原处。

（8）加工完毕，将铣床擦干净，用软布和毛刷清除切屑和油污，切忌用压缩空气吹，以免细小的切屑和灰尘等杂物嵌入运动部分。若无异常，对车床各部位加注润滑油。认真擦拭工具、量具和其他附件，使各物件归位，清扫工作场地，关闭电源。

（9）铣床不使用时，各手柄应置于空挡位置，各方向进给紧固手柄应松开，工作台应处于各方向进给的中间位置，导轨面应适当涂润滑油。

（10）毛坯、半成品和成品应分开放置。半成品、成品应堆放整齐、轻拿轻放，严防碰伤已加工表面。图样、工艺卡片应放置在便于阅读的位置，并注意保持其清洁和完整。工作场地周围应保持清洁、整齐，避免堆放杂物，防止绊倒。

五、铣工安全知识

（1）实习前按规定使用防护用品，所穿戴工作服的袖口要扎好，女学员戴好工作帽，不得将头发和辫子留在帽外，防止衣角、袖口、发辫卷入旋转中的机件中。操作铣床时不准戴手套。

（2）铣削过程中，不可用手触摸离旋转中的铣刀很近的工件表面，否则容易切伤手指。清理切屑时要使用铁钩子，铁屑很薄很锋利，不可用手拉拨。

（3）铣削过程中，禁止用棉丝擦拭工件或旋转中的铣刀杆等运动部件。测量工件时要停车，切忌在切削中测量工件尺寸。不准采取用手抓或用嘴吹的方法清除切屑。

（4）铣削时，操作者要戴好防护眼镜。

（5）工、量、卡具要放稳放好，防止落下伤人。工件要夹持得牢固可靠，防止切削中工件松脱而发生事故。

（6）使用砂轮时，要站在砂轮侧面位置，防止砂轮碎裂后飞出伤人。

（7）操作时不要站在切屑流出的方向。

（8）铣床上的防护罩等防护装置不可随意拆卸，防止传动带、齿轮等露在外面而发生伤害事故。

（9）在机动快速进给时，一定要把手轮离合器打开。

（10）铣床上照明限用 36 V 低压电，要按规定用电。

思考与练习

（1）铣工主要工作内容包括哪些？

（2）铣工的安全操作技术有哪些？

（3）铣工文明生产必须做到哪些方面？

任务二　铣刀与夹具

铣削所用的铣刀是多刀刃的刀具，能进行金属切削，获得具有一定表面结构的工件，对铣刀的材料及其几何形状都必须有一定的要求。

一、铣刀切削部分的材料

1. 基本要求

制造铣刀切削部分的材料应具备以下基本要求。

1）硬度

在常温下，刀具切削部分必须具有足够的硬度才能切入工件。由于在切削过程中会产生大量的热量，因而要求刀具材料在高温中仍能保持其硬度，并能继续进行切削。这种具有高温下仍能保持硬度的性质称为红硬性。

2）强度和韧性

刀具在切削过程中要承受很大的冲击力，因此要求刀具切削部分的材料具有足够的强度、韧性，能在承受冲击和振动的条件下继续进行切削，不易崩刃、碎裂。

2. 常用材料

具有上述性能的材料很多，常用的铣刀材料有高速工具钢和硬质合金两种。

（1）高速工具钢（简称高速钢、锋钢等）是以钨（W）、铬（Cr）和钒（V）为主要元素的合金工具钢，淬火硬度一般为 62~65 HRC。高速工具钢的强度较高，韧性也较好，能磨出锋利的刃口，且具有良好的工艺性能，锻造、焊接、切削加工和刃磨都较容易。

常用的高速工具钢牌号有 W18Cr4V、W6Mo5Cr4V2 等。

（2）硬质合金。硬质合金是以金属碳化物 WC（碳化钨）、TiC（碳化钛）和 Co（钴）为主的金属黏结剂经粉末冶金工艺制造而成的合金，其常温硬度高、耐磨性好、耐高温，在 800~1 000 ℃仍能保持良好的切削性能。硬质合金刀具切削速度可比高速钢刀具高 4~8 倍。但其抗弯强度低，冲击韧性差，切削刃不易刃磨得很锋利。

二、铣刀的种类及选择

铣刀是多齿刀具，因其结构复杂，一般由专业工厂生产。由于其同时参与切削的齿数多，并能用较高的切削速度切削，故生产效率较高。铣刀种类很多，划分方式多样，主要有按用途划分、按组合方式划分、按齿背形状划分等。在铣削加工中，应根据铣床的情况和加工需要合理地选择和使用铣刀。

（1）铣平面用铣刀。如图 3-2-1 所示，图（a）为圆柱铣刀，用于卧式铣床；图（b）为套式端铣刀，用于卧式铣床或立式铣床；图（c）为机夹铣刀，用于立式铣床。

(a) (b) (c)

图 3-2-1 铣平面用铣刀

(a) 圆柱铣刀 (b) 套式端铣刀 (c) 机夹铣刀

（2）铣沟槽用铣刀。如图 3-2-2 所示，键槽铣刀、立铣刀多用于立铣加工；盘形槽铣刀、三面刃铣刀和锯片铣刀多用于卧式铣床加工。

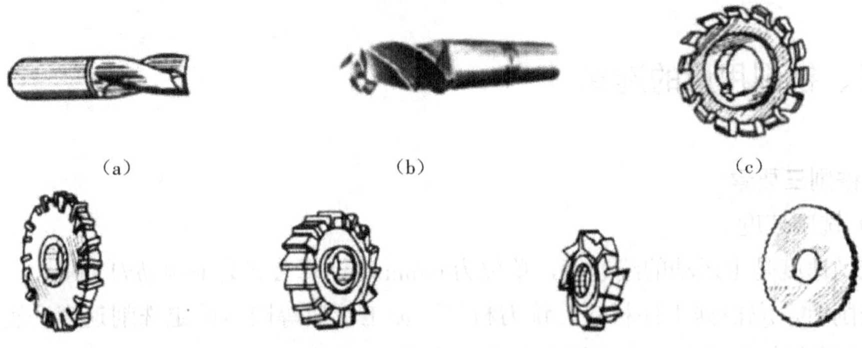

(a) (b) (c)

(d) (e) (f) (g)

图 3-2-2 铣沟槽用铣刀

(a) 键槽铣刀 (b) 立铣刀 (c) 盘形槽铣刀 (d) 镶齿三面刃铣刀

(e) 直齿三面刃铣刀 (f) 错齿三面刃铣刀 (g) 锯片铣刀

（3）铣成形面用铣刀。图 3-2-3 所示铣刀常用来铣削半圆、齿轮或其他成形面等。

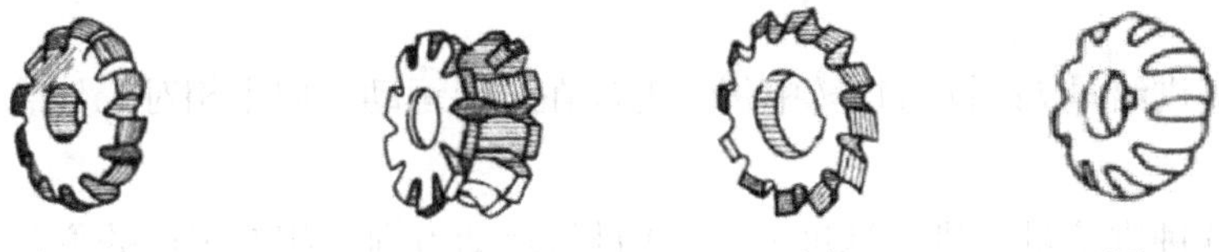

(a) (b) (c) (d)

图 3-2-3 铣成形面用铣刀

(a) 凸半圆铣刀 (b) 凹半圆铣刀 (c) 齿轮铣刀 (d) 成形铣刀

（4）铣成形沟槽用铣刀。图 3-2-4 所示铣刀常用来铣削 T 形槽、燕尾槽、半圆键键槽等。

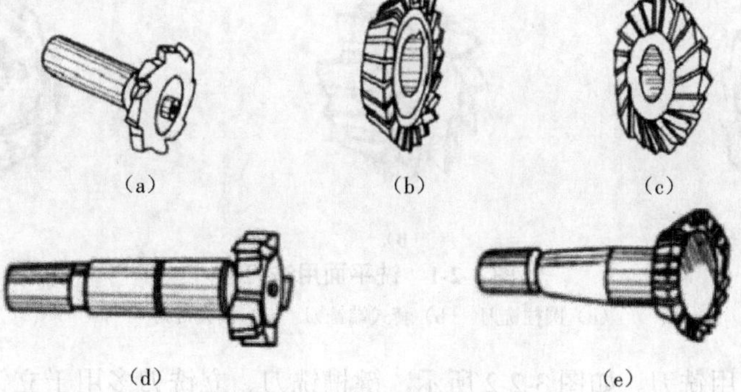

图 3-2-4 成形沟槽用铣刀

（a）T 形槽铣刀 （b）燕尾槽铣刀 （c）半圆键键槽铣刀 （d）单角铣刀 （e）双角铣刀

三、铣削用量的选择

1. 铣削三要素

1）铣削速度 v

铣削速度是主运动的线速度，单位为 m/min，计算公式为 $v=\pi dn/1\,000$。

铣削时，应根据工件材料、铣刀材料、加工性质等因素确定铣削速度，然后根据所用铣刀直径按计算公式确定铣床主轴的转速。

2）进给量 f

进给量就是铣削中，工件相对于铣刀在进给方向上所移动的距离。

进给量根据具体情况，有三种表述和度量的方法。

（1）每转进给量 f_n：铣刀每回转一周在进给运动方向上相对工件的位移量，单位为 mm/r。

（2）每齿进给量 f_z：铣刀每转中每一刀齿在进给运动方向上相对工件的位移量，单位为 mm/z。

（3）每分钟进给量（进给速度）v_f：在进给运动方向上相对工件的位移量，单位为 mm/min。

三种进给量的关系为：$v_f = f_n \cdot n = f_z \cdot z \cdot n$

铣削时，根据加工性质先确定每齿进给量 f_z，然后根据铣刀的齿数 z 和铣刀的转速 n 计算出每分钟进给量 v_f，并以此对铣床进给量进行调整。

3）背吃刀量 a_p（铣削深度）

一次铣削进给过程中待加工表面与已加工表面之间的垂直距离，即铣削深度。

2. 铣削用量的顺序及选择

应当优先采用较大的铣削深度，其次是选择较大的进给量，最后才是根据刀具寿命要求，选择适宜的铣削速度。

1）铣削深度的选择

在铣削加工中，一般根据工件的切削尺寸来选择铣刀。应尽量一次进给铣去全部的加工余量，只有当工件的加工精度较高时，才分粗铣和精铣，具体数值见表 3-2-1。

表 3-2-1 粗铣和精铣铣削深度 单位：mm

工件材料	高速钢铣刀		硬质合金铣刀	
	粗铣	精铣	粗铣	精铣
铸铁	5~7	0.5~1	10~18	1~2
软钢	<5	0.5~1	<12	1~2
中硬钢	<4	0.5~1	<7	1~2
硬钢	<3	0.5~1	<4	1~2

2）每齿进给量的选择

粗加工时，进给量应尽量取大一些；精加工时，一般选取较小的进给量。具体数值参考表 3-2-2。

表 3-2-2 每齿进给量的选择 单位：mm

刀具名称	高速钢铣刀		硬质合金铣刀	
	铸铁	铸钢	铸铁	铸钢
圆柱铣刀	0.12~0.2	0.1~0.15	0.2~0.5	0.08~0.2
立铣刀	0.08~0.15	0.03~0.06	0.2~0.5	0.08~0.2
套式面铣刀	0.15~0.2	0.06~0.1	0.2~0.5	0.08~0.2
三面刃铣刀	0.15~0.25	0.06~0.08	0.2~0.5	0.08~0.2

3）铣削速度的选择

在铣削深度和每齿进给量确定后，应在保证刀具寿命的前提下确定铣削速度。铣削速度可在推荐的范围内选取，并根据实际情况进行调整，具体数值见表 3-2-3。

表 3-2-3 铣削速度的选择 单位：m/min

工件材料	高速钢铣刀	硬质合金铣刀
20 钢	20~45	150~190
45 钢	20~35	120~150
40Cr	15~25	60~90
HT150	14~22	70~100

续表

工件材料	高速钢铣刀	硬质合金铣刀
黄铜	30~60	120~200
铝合金	112~300	400~600
不锈钢	16~25	50~100

注：①粗铣时取小值，精铣时取大值；
②工件材料强度和硬度高时取小值，反之取大值；
③刀具材料耐热性好取大值，耐热性差取小值。

四、夹具

在铣床上安装工件，主要夹具有平口虎钳、三爪定心卡盘，通常根据工件的形状和大小，用压板、螺栓、垫铁和挡铁直接安装在工作台上，当生产批量较大时，也可采用专用夹具或组合夹具安装工件。

安装平口虎钳时，应擦净钳座底面和铣床工作台面。一般情况下，平口虎钳在工作台面上的位置，应处在工作台长度方向的中心偏左、宽度方向的中心，以方便操作。

五、工件的装夹

用平口虎钳装夹工件铣削垂直面时，用机用平口虎钳装夹工件只需把工件基准面与固定钳口紧密贴合即可。为了使工件基准面与固定钳口紧密贴合，往往在活动钳口与工件之间放置一根圆棒。对于工件上高低不平或与基准面不平行的毛坯面，若不放圆棒，夹紧后工件的基准面与固定钳口不一定会很好地贴合，如图 3-2-5 所示。

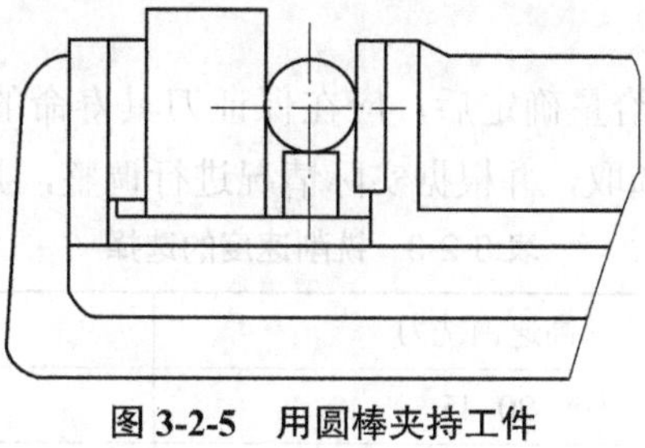

图 3-2-5　用圆棒夹持工件

工件的基准面靠向钳体导轨面时，在工件与导轨之间要垫以平行垫铁，为了使工件基准面与导轨面平行，稍紧后可用铝或铜锤轻击工件上面，并用手试着移动垫铁，当其不松动时，工件与垫铁贴合良好，然后夹紧，如图 3-2-6 所示。

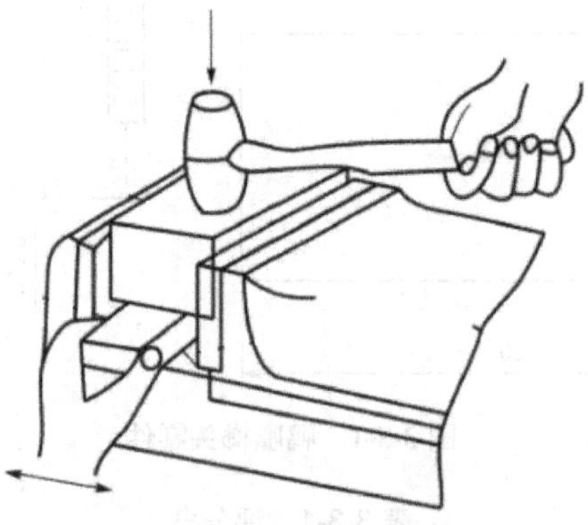

图 3-2-6 用平行垫铁装夹工件

思考与练习

(1) 铣刀材料类型有哪些?

(2) 铣刀的种类及选择方法是什么?

(3) 简述铣削用量的选择方法。

(4) 简述夹具的使用及工件的装夹方法。

任务三 铣平面

一、图样与技术要求

如图 3-3-1 所示,工件毛坯尺寸为 120 mm×50 mm×25 mm,材料为 45 钢,仅加工工件的上表面。评分表见表 3-3-1。

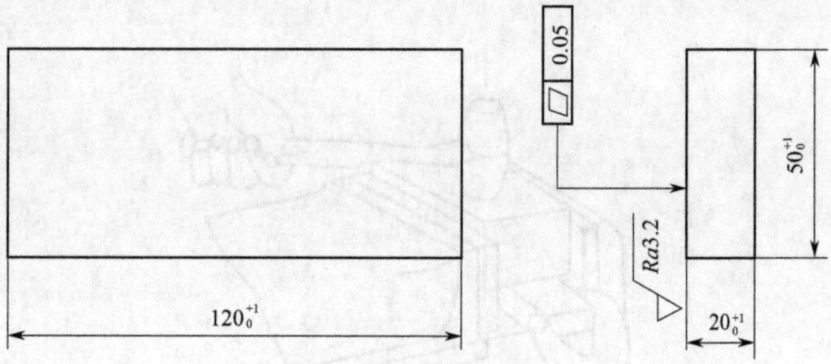

图 3-3-1　鸭嘴榔头零件

表 3-3-1　评分表

技术要求	配分	实测结果	得分
平面度 0.05 mm	30		
表面结构参数 *Ra*3.2 μm	30		
长度 20_{0}^{+1} mm	20		
规范操作	10		
安全文明生产	10		

二、加工工艺分析

（1）读零件图（如图 3-3-1 所示），检查毛坯尺寸。

（2）安装平口虎钳。

（3）选择并安装端铣刀。

（4）选择并调整切削用量（取主轴转速 n=118 r/min，进给速度 v_f=47.5 mm/min 或 f=60 mm/min，背吃刀量度 a_p=2 mm）。

（5）安装并校正工件（应垫铜片）。

（6）对刀，调整铣削宽度（即切削层深度），自动进给铣削工件。

（7）铣削完毕后，停车，降落工作台并退出工件。

（8）测量并卸下工件。

（9）由教师对组内成员的加工工件和表现进行评价，结合评分表 3-3-1 给出分数。

三、相关知识与技能

1. 圆周铣和端铣

在工件上用铣刀加工平面的方法称为铣平面。铣平面是铣工最重要的工作之一，也是进一步掌握铣削其他复杂零件表面的基本技能。铣平面可以在卧式铣床上铣削，也可以在立式铣床上铣削。平面质量的好坏，取决于平面的平面度和表面结构。平面的铣削方法主要有圆周铣和端铣两种，如图 3-3-2 所示。

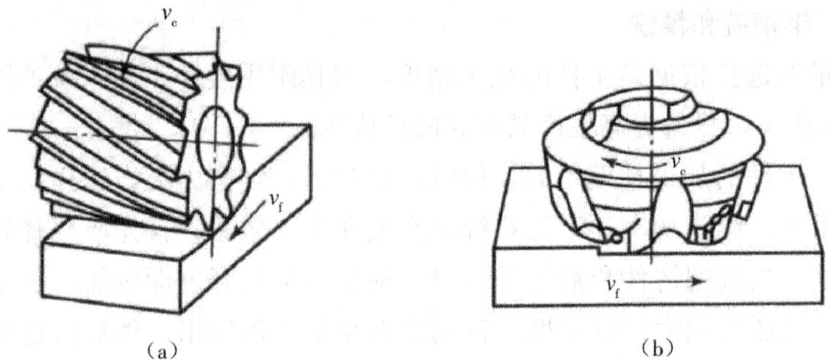

图 3-3-2 圆周铣和端铣

(a) 圆周铣 (b) 端铣

2. 顺铣和逆铣

根据铣刀旋转方向和工件进给方向之间的关系，铣削过程可分为顺铣和逆铣。当铣刀旋转方向和工件进给方向相同时称为顺铣；反之，称为逆铣，如图 3-3-3 所示。

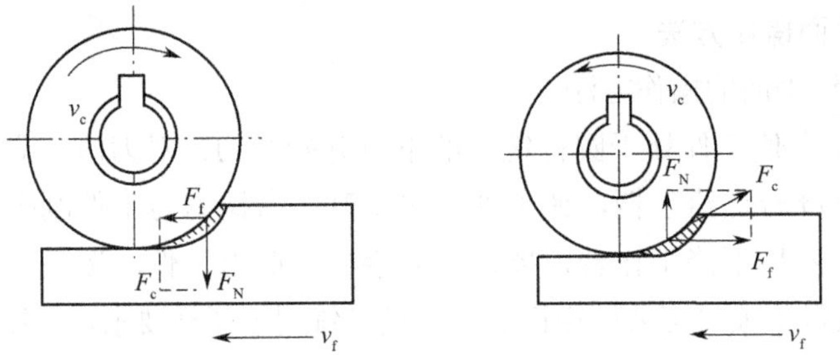

图 3-3-3 顺铣和逆铣

(a) 顺铣 (b) 逆铣

顺铣的优点：顺铣时，切削厚度由最大开始，刀具磨损小，耐用度高。垂直切削力向下，夹紧可靠，因此铣削出的工件表面比较光滑。

顺铣的缺点：顺铣时，铣削力在进给方向的分力与工件的进给方向相同，由于工作台丝杠螺母存在间隙，当进给力逐渐增大时，铣削力会拉动工作台而产生蹿动，造成进给不均匀，严重时会使铣刀崩刃。

逆铣的优点：逆铣时，由于进给力作用，使丝杠与螺母传动时没有窜动现象，铣削过程较平稳。

缺点：逆铣时，切削厚度由零逐渐增大，由于刃口圆角半径的影响，开始切削时前角为负值，刀齿在工作表面上挤压、滑行，造成工件表面加工硬化严重，并加剧了刀齿的磨损。另外由于垂直切削方向向上，与工件的夹紧力和工件重力相反，这个作用力有把工件从工作台抬起的趋势，加剧了振动，影响工件的夹紧和表面结构。

3. 粗铣、半精铣和精铣

铣削用量的选择依据是工件的加工精度、刀具耐用度和工艺系统的刚度。在保证产品质量的前提下，尽量提高生产效率和降低成本。

粗铣时，工件的加工精度不高，但铣削余量大，切削力大。选择铣削用量应主要考虑铣刀耐用度、铣床功率、工艺系统的刚度和生产效率。首先应选择较大的铣削深度和铣削宽度。当铣削铸件和锻件毛坯时，应使刀尖避开表面硬层。加工铣削宽度较小的工件时，可适当加大铣削深度。铣削宽度尽量一次铣出，然后再选用较大的每齿进给量和较低的铣削速度。

当工件表面结构要求为 *Ra*3.2~6.3 μm 时，应采用半精铣。半精铣的铣削深度为 0.2~2 mm。精铣时，为了获得较高的尺寸精度和较小的表面结构参数值，铣削深度应取小些，铣削速度可适当提高，每齿进给量宜取小值。

一般情况下，选择铣削用量的顺序是：先选大的铣削深度，再选每齿进给量，最后选择铣削速度。铣削宽度尽量等于工件加工面的宽度。

4. 铣削平面操作方法

1）圆周铣平面的操作方法

移动工作台使工件处于圆柱铣刀的下方开始对刀，对刀时，启动机床，铣刀旋转后，再摇动升降台进给手柄，使工件缓慢上升，当铣刀与工件刚刚接触时记下升降刻度盘刻度值。然后下降工作台，摇动纵向手柄，退出工件，按毛坯件实际尺寸调整铣削层深度。余量不大时可采用逆铣法一次进给铣削至图样要求，否则应分粗铣和精铣。

检测工件表面结构：铣削完毕后卸下工件，用游标卡尺或千分尺测量工件各部分尺寸，并用铣削表面结构样板对比检测工件的表面结构。

2）端铣平面的操作

对刀操作方法与圆柱铣刀对刀操作基本相同，所不同的是端铣刀对刀时使用端面切削刃切痕，而圆柱铣刀对刀时用圆周切削刃切痕。

对刀后，应采用不对称逆铣法，如图 3-3-4 所示。

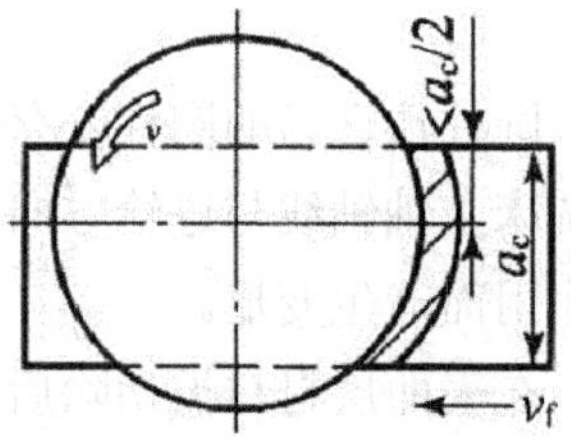

图 3-3-4 不对称逆铣

端铣平面检测工件表面质量的方法与圆周铣平面相同。

检测工件平面度：铣好平面后，一般都用刀口形直尺来检验其平面度。对平面度要求较高的平面，则可用标准平板来检验，检验时在标准平板的平面上涂红丹粉或龙胆紫溶液，再将工件的平面放在标准平板上进行对研，对研几次后把工件取下，观察平面的着色情况，若着色均匀而细密，则表示平面的平面度较好，如图 3-3-5 所示。

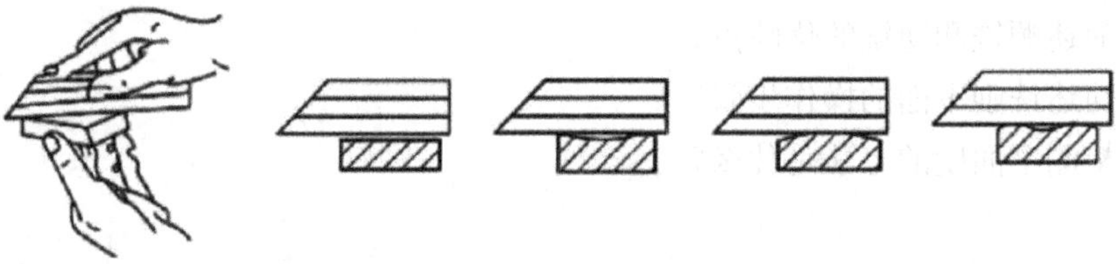

图 3-3-5 用刀口形直尺来检测工作平面度

5. 平面铣削的质量分析

平面的铣削质量不仅与铣削时所用的铣床、夹具和铣刀的质量有关，还与铣削用量和切削液的选用等诸多因素有关。

1）影响工件表面结构的因素

（1）铣刀磨损，刃口变钝。

（2）铣削时进给量太大。

（3）铣削时切削层深度太大。

（4）铣刀的几何参数选择不当。

（5）铣削时切削液选择不当。

（6）铣削时有振动。

（7）铣削时有积屑瘤产生，或因进给传动系统间隙过大，而造成工作台产生窜动现象。

（8）铣削时有扎刀现象。

（9）铣削过程中因进给停顿，铣削力突然减小，而使铣刀突然下沉进而在工件加工面上切出凹坑（称为“深啃”）。

2）影响工件平面度的因素

（1）用圆周铣铣削平面时，圆柱形铣刀的圆柱度存在误差。

（2）用端铣铣削平面时，铣床主轴轴线与进给方向不垂直。

（3）工件受夹紧铣削力的作用而产生变形。

（4）工件自身存在内应力，在表面层材料被切除后产生变形。

（5）铣床工作台进给运动的直线性误差。

（6）铣床主轴轴承的轴向和径向间隙过大。

（7）铣削中，由铣削热引起工件的热变形。

（8）铣削时，由于圆柱形铣刀的宽度或端铣刀的直径小于被加工面的宽度而使接刀产生接刀痕。

思考与练习

（1）平面的铣削方法有哪些？

（2）简述顺铣和逆铣的优缺点？

（3）简述铣削平面的操作步骤？

（4）检测平面度的方法是什么？

任务四　铣垂直面和平行面

一、图样与技术要求

如图 3-4-1 所示，工件毛坯尺寸为 130 mm×70 mm×60 mm，材料为 45 钢。评分表见表 3-4-1。

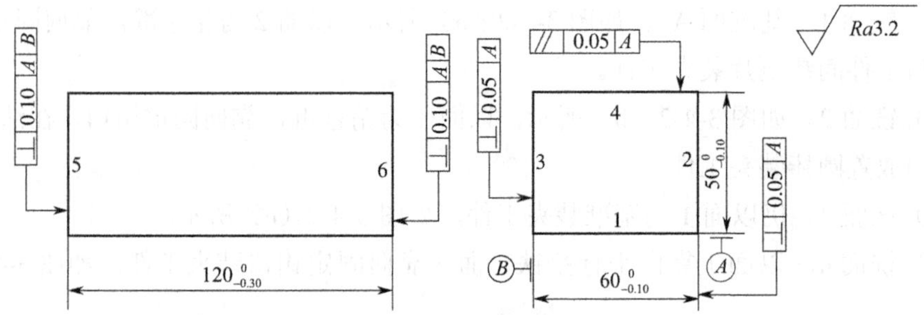

图 3-4-1　长方体零件一

表 3-4-1　评分表

技术要求	配分	实测结果	得分
垂直度公差 0.1 mm	12		
垂直度公差 0.05 mm	10		
平行度公差 0.05 mm	10		
长度 $120^{\ 0}_{-0.3}$ mm	10		
宽度 $60^{\ 0}_{-0.1}$ mm	10		
高度 $50^{\ 0}_{-0.1}$ mm	10		
表面结构参数 $Ra3.2$ μm	18		
规范操作	10		
安全文明生产	10		

二、加工工艺分析

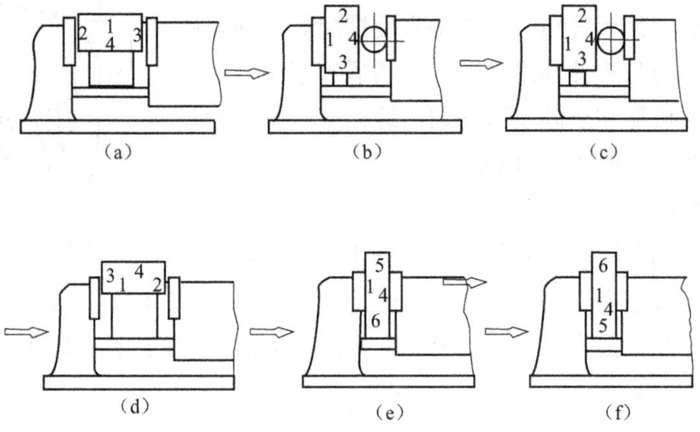

图 3-4-2　长方体工件的铣削顺序

（1）铣面 1（基准面 A）。如图 3-4-2（a）所示，以面 2 为粗基准，靠向固定钳口，两钳口与工件间垫铜片装夹工件。

（2）铣面 2。如图 3-4-2（b）所示，以面 1 为精基准，靠向固定钳口，在活动钳口与工件间放置圆棒装夹工件。

（3）铣面 3。仍以面 1 为基准装夹工件，如图 3-4-2（c）所示。

（4）铣面 4。以面 1 靠向平行垫铁、面 3 靠向固定钳口装夹工件，如图 3-4-2（d）所示。

（5）铣面 5。调整平口虎钳。面 1 靠向固定钳口，用 90° 角尺找正面 2 与平口虎钳钳体导轨面垂直。装夹工件，如图 3-4-2（e）所示。

（6）铣面 6。面 1 靠向固定钳口，面 5 靠向平口虎钳钳体导轨面装夹工件，如图 3-4-2（f）所示。

三、相关知识与技能

1. 垂直面的铣削

两相邻表面相交成 90° 的平面称为垂直面。在卧式铣床上用圆柱铣刀和在立式铣床上用端铣刀铣出的平面，都与工作台台面平行。因此，只需将工件基准面装夹成与工作台台面垂直即可铣出垂直面。至于加工方法，除了工件的装夹有要求外，其他均与铣平面基本相同。和基准面相互垂直的平面称为垂直面。如图 3-4-3 所示。

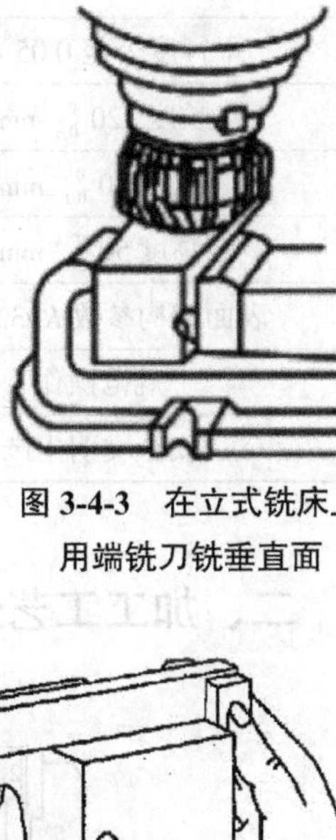

图 3-4-3　在立式铣床上用端铣刀铣垂直面

2. 垂直度检验方法

检验工件垂直度时常使用宽座直角尺、90° 圆柱角尺或万能角度尺等。使用直角尺检验垂直度情况时，使宽座直角尺底座的一边与工件基准面贴合，然后观察角尺另一边与检测面间的接触和缝隙是否均匀，这样来判断被测量垂直面的正确性，如图 3-4-4 所示。

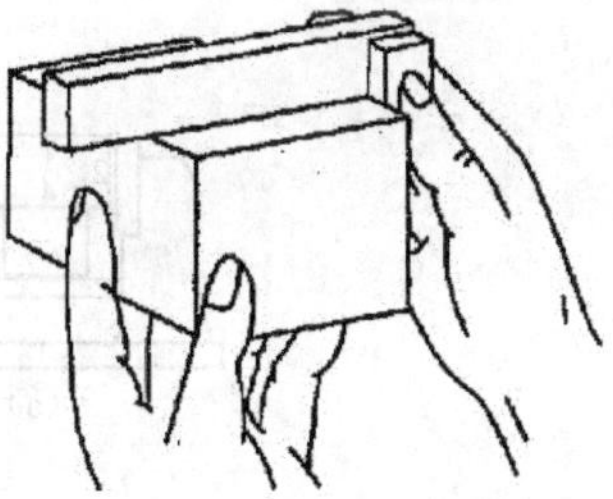

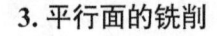

图 3-4-4　直角尺检验垂直度

3. 平行面的铣削

平行面是指与基准面平行的平面。铣平行面时，除了平行度、平面度要求外，还有两平行面之间的尺寸精度要求。在立式铣床上用端铣刀铣平行面，当工件有台阶时，可直接用压板将工件装夹在立式铣床的工作台台面上，使基准面与工作台台面贴合，如图 3-4-5 所示。

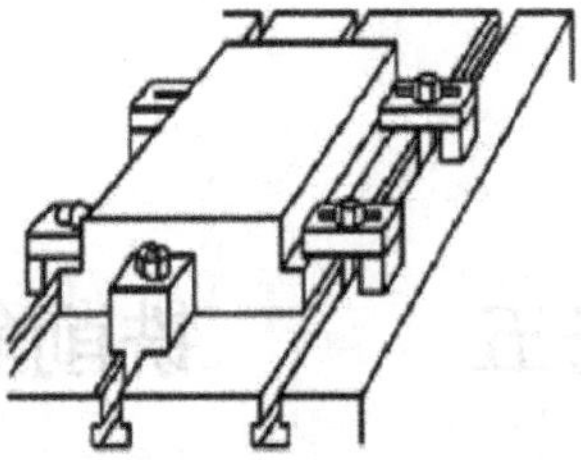

图 3-4-5　立式铣床上用端铣刀铣平行面

4. 平行度和尺寸精度的检测

平行度的检测如图 3-4-6 所示，将基准面贴合在平板上，用百分表来测量。加工好的工件，在对其尺寸精度和平行度同时进行检测时，可用千分尺或游标卡尺测量工件的四角及中部，检查所有的尺寸是否都在图样所规定的尺寸范围，并观察各部分尺寸的差值，这个差值就是平行度误差。

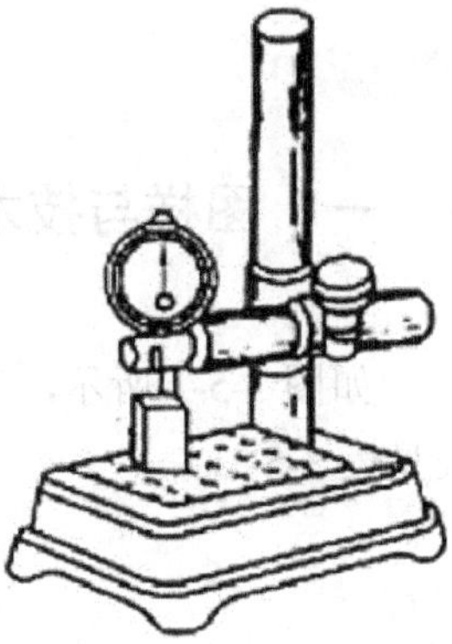

图 3-4-6　平行度的检测

5. 垂直面和平行面的质量分析

1）影响垂直度的主要因素

铣削时，影响垂直度的主要因素有下列几个方面。

（1）平口虎钳固定钳口面与工作台面不垂直，铣出的平面与基准面不垂直。

（2）基准面没有与固定钳口贴合。

（3）基准面的平面度误差大，影响工件装夹时的位置精度。

（4）夹紧力太大，使固定钳口向外倾斜。

2）影响平行面之间尺寸精度的因素

铣削时，影响平行面之间尺寸精度的因素有下列几个方面。

（1）调整铣削层深度时看错刻度盘，手柄摇过头，没有消除丝杠螺母副的间隙而直接退回，造成尺寸铣错。

（2）读错图样上标注的尺寸，测量错误。

（3）工件或平行垫铁的平面未擦净，由于有杂物而使尺寸发生变化。

（4）精铣对刀时切痕太深，调整铣削层深度时没有去掉切痕，使尺寸铣小。

思考与练习

（1）简述铣削垂直面的方法。

（2）简述铣削平行面的方法。

（3）简述平行度和垂直度的检测方法。

任务五　铣削斜面

一、图样与技术要求

如图 3-5-1 所示，工件毛坯尺寸为 110 mm×60 mm×30 mm，材料为 45 钢。评分表见表 3-5-1。

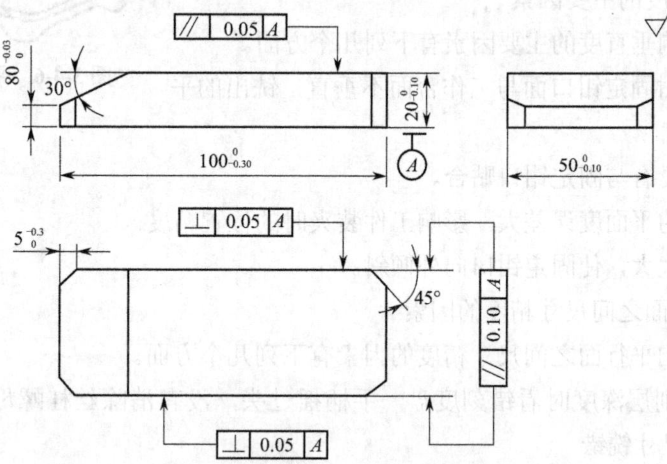

图 3-5-1　长方体零件二

表 3-5-1　评分表

技术要求	配分	实测结果	得分
平行度公差 0.05 mm	10		
平行度公差 0.1 mm	5		
垂直度公差 0.05 mm	5		
高度 $20^{0}_{-0.1}$ mm	5		
宽度 $50^{0}_{-0.1}$ mm	5		

续表

技术要求	配分	实测结果	得分
长度 $100_{-0.3}^{0}$ mm	5		
$8_{0}^{+0.03}$ mm	5		
$5_{0}^{+0.3}$ mm	5		
30° 角	10		
45° 角	10		
表面结构参数 $Ra3.2$ μm	20		
规范操作	10		
安全文明生产	5		

二、加工工艺分析

（1）铣削工件平面至尺寸。

（2）铣削 30° 斜面。

①找正平口虎钳，固定钳口。

②选择并安装铣刀（选择直径 80 mm 的镶齿端铣刀）。

③装夹并找正工件（工件基准面与工作台台面平行）。

④铣削用量（取 n=150 r/min，v_f=60 mm/min，a_p 分次达到要求）。

⑤调转立铣头角度为 α=30°。

⑥对刀，铣削工件（对刀，调整背吃刀量 a_p 后紧固纵向进给，用横向进给分数次走刀铣出 30° 斜面）。

（3）铣削 45° 斜面。

①换装直径 20~25 mm 的立铣刀。

②调转立铣头角度为 α=45°。

③将工件基准面（底面）靠向平口虎钳固定钳口，装夹工件。

④对刀，调整背吃刀量 a_p（即切削深度），铣出 45° 斜面。

⑤分次装夹工件，铣出其余三个 45° 斜面。

（4）由教师对组内成员加工的工件和表现进行评价，结合评分表 3-5-1 给出分数。

三、相关知识与技能

1. 铣斜面的方法

铣斜面与铣平面的原理一致，只是工件的切削位置或对工件的安装位置有相应的改变，以使斜面能达到准确的斜度。斜面的铣削方法主要有以下几种。

1）把铣刀转成所需的角度铣斜面

通常在装有立铣头的卧式铣床上或在立式铣床上，根据工件的斜度要求，将立铣头转动一定的角度，使工作台横向进给来加工斜面，如图 3-5-2 所示。

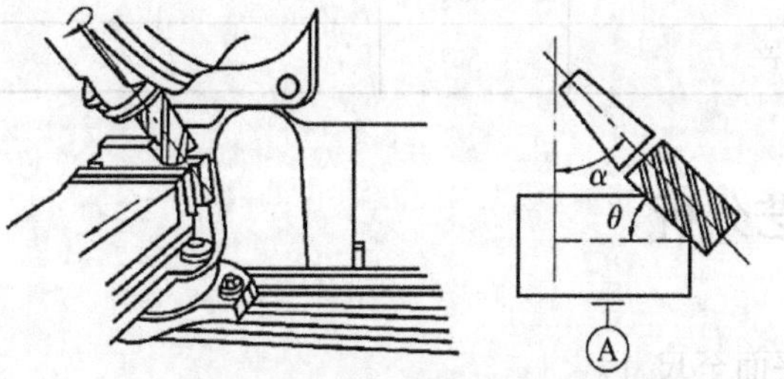

图 3-5-2　工件的基准面与工作台台面平行用圆周刃铣削斜面

2）使用角度铣刀铣斜面

直接用带角度的铣刀铣削斜面，其铣刀角度要和工件的斜面角度相等。但由于角度铣刀的刀刃宽度限制，该方法仅适用于尺寸较小的工件，如图 3-5-3 所示。

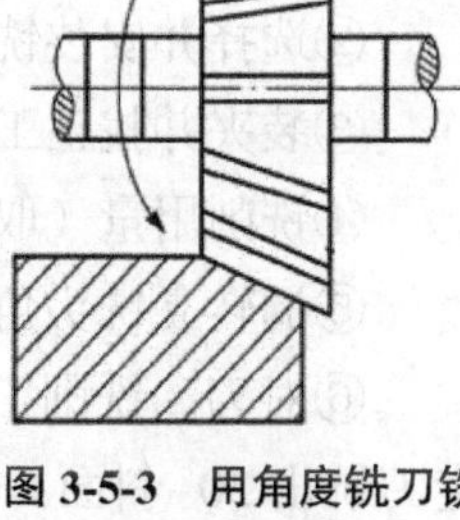

图 3-5-3　用角度铣刀铣削斜面

3）把工件转成所需的角度铣斜面

改变工件的安装位置，使斜面转到水平位置，采用辅助线、分度头等夹紧工件后，用铣削平面的方法来铣斜面，如图 3-5-4 所示。装夹工件的方法如图所示。

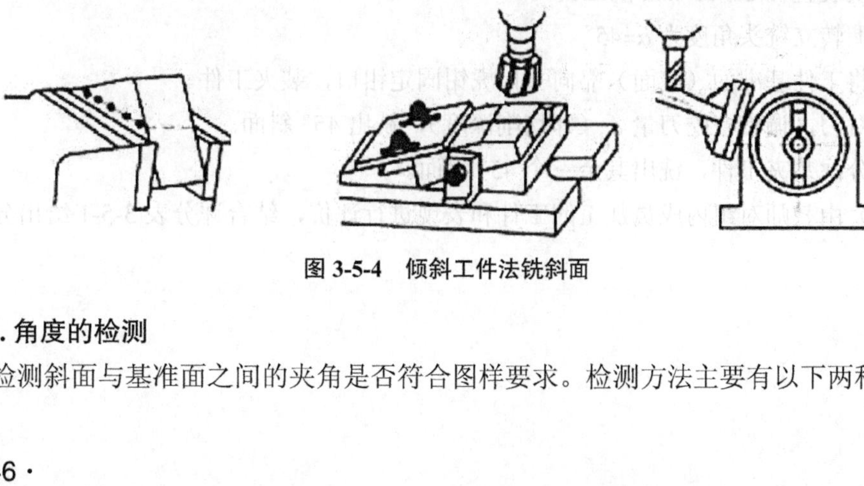

图 3-5-4　倾斜工件法铣斜面

2. 角度的检测

检测斜面与基准面之间的夹角是否符合图样要求。检测方法主要有以下两种。

（1）用万能角度尺检测。当工件精度要求不高时，可用万能角度尺来直接量得斜面与基准面之间的夹角。

（2）用正弦规检测。当工件精度要求较高时，可用正弦规配合百分表和量块来检测斜面与基准层面之间的夹角。

3. 斜面的质量分析

1）影响斜面尺寸精度的因素

（1）看错刻度或摇错手柄转数，以及没有消除丝杠螺母副的间隙。

（2）测量不准，将尺寸铣错。

（3）铣削过程中，工件有松动。

2）影响斜面角度的因素

（1）立铣头转动角度不准确。

（2）按划线装夹工件时划线不准确，或铣削时工件产生位移。

（3）采用圆周铣时，铣刀圆柱度误差大（如有锥度）。

（4）用角度铣刀铣削时，铣刀角度不准。

（5）工件装夹时，平口虎钳钳口、钳体导轨面及工件表面未擦净。

3）影响斜面表面质量的因素

（1）进给量过大。

（2）铣刀不锋利。

（3）机床、夹具刚性差，铣削中有振动。

（4）铣削过程中，工作台进给或主轴回转时突然停止，“啃伤”工件表面。

（5）铣削钢件时未使用切削液，或切削液选用不当。

思考与练习

（1）铣斜面的方法有哪些？

（2）角度检测的量具有哪些？

（3）铣斜面影响斜面角度的因素有哪些？

任务六　铣削台阶

一、图样与技术要求

如图 3-6-1 所示，工件毛坯尺寸为 90 mm×50 mm×40 mm，材料为 45 钢。评分表见表 3-6-1。

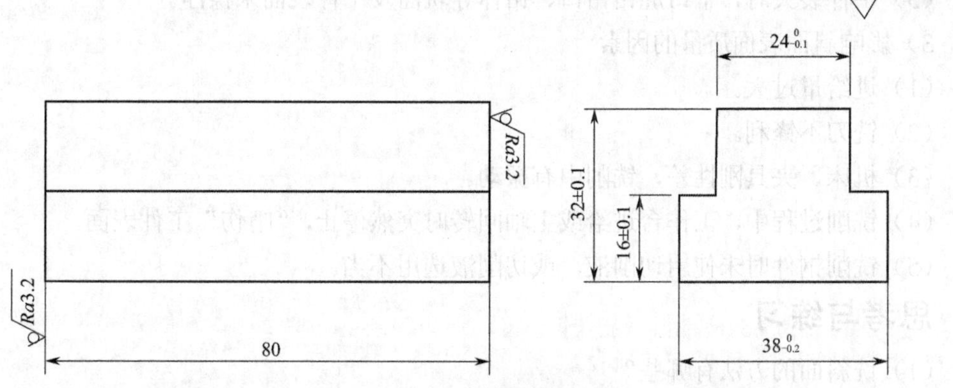

图 3-6-1　台阶零件

表 3-6-1　评分表

技术要求	配分	实测结果	得分
32±0.1 mm	20		
16±0.1 mm	20		
$24^{\ 0}_{-0.1}$ mm	15		
$38^{\ 0}_{-0.2}$ mm	15		
表面结构参数 *Ra*3.2 μm	20		
规范操作	5		
安全文明生产	5		

二、加工工艺分析

（1）安装平口虎钳，校正固定钳口与工作台纵向进给方向平行。

（2）选择并安装端铣刀（套式端铣刀尺寸为 63 mm×40 mm）。

（3）调整切削用量（取 n=118 r/min，v_f=60 mm/min）。

（4）铣削四面至尺寸。

（5）铣削两台阶面至尺寸。

（6）测量尺寸，卸下工件。

（7）由教师对组内成员加工的工件和表现进行评价，结合评分表 3-6-1 给出分数。

三、相关知识与技能

1. 台阶的形式及工艺要求

常见台阶零件的形式有单台阶和双台阶（T 形键）以及其他形式的台阶。台阶的工艺要求包括以下几个。

（1）尺寸精度：台阶上与其他零件相配合的表面，其尺寸精度一般要求较高。

（2）形状位置精度：如各表面的平面度、台阶侧面与基准面的平行度，以及双台阶对中分线的对称度。

（3）表面结构：台阶两侧配合面一般表面结构要求都较低。

2. 铣削台阶的方法

1）划线对刀

在装夹前，用高度尺划出台阶尺寸线，然后安装在平口虎钳内，开动机床，目测铣刀的侧齿刃落在边缘线上，缓慢上升垂向工作台，使铣刀在工件表面切出刀痕，观察切痕是否处于边缘线之间，若有偏差，则重新调整工作台。

2）用端铣刀铣削台阶

宽度较宽且深度较浅的台阶，常使用端铣刀在立式铣床上铣削。端铣刀刀杆刚度大，铣削时切削厚度变化小，切削平稳，加工表面质量好，生产效率较高。铣削时，所选用的端铣刀的直径应大于台阶宽度，一般可按 D=（1.4~1.6）×B 选取，如图 3-6-2 所示。

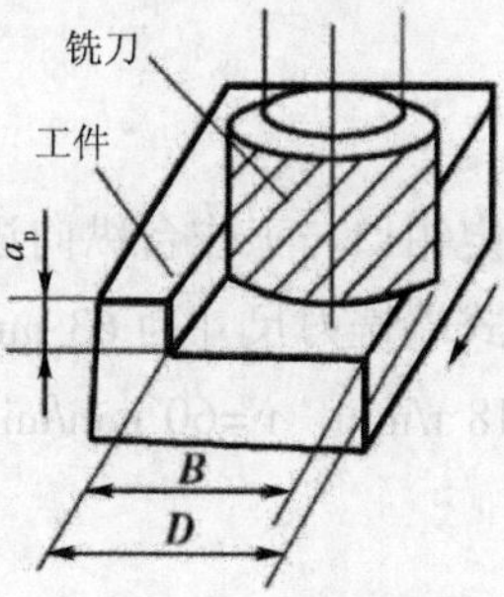

图 3-6-2　用端铣刀铣削台阶

3）深度较深的台阶用立铣刀铣削

用立铣刀圆周刃铣削台阶时，先调整到要求的台阶深度，台阶的宽度可分数次铣成。由于立铣刀的强度较弱，允许的切削用量应小些，如图 3-6-3 所示。

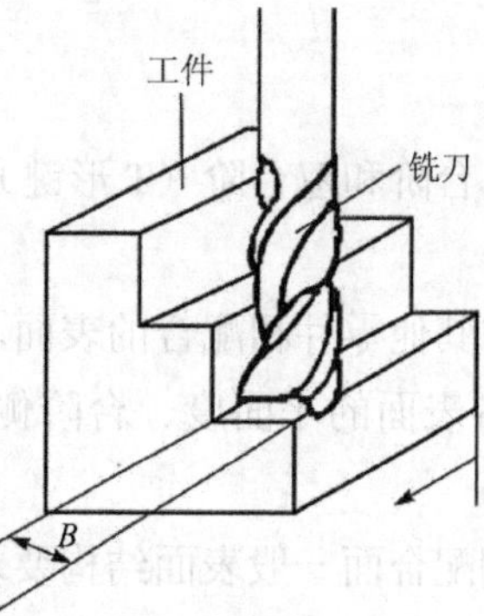

图 3-6-3　用立铣刀铣削台阶

3. 台阶的测量

台阶的宽度和深度一般可用游标卡尺、深度游标卡尺来测量。两边对称的台阶，当台阶深度较深时，可用千分尺测量，如图 3-6-4（a）所示；台阶深度较浅时，可用极限量规测量，如图 3-6-4（b）所示。

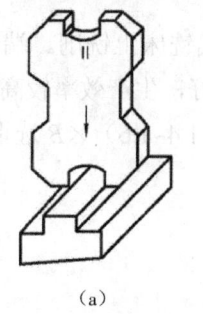

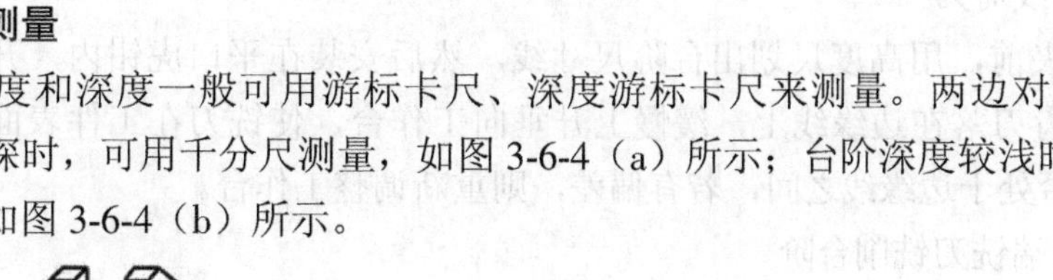

（a）　　（b）

图 3-6-4　台阶的测量

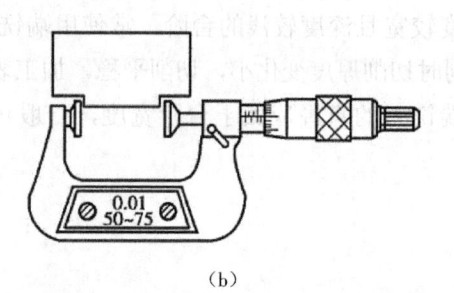

（a）用公法线千分尺测量台阶　（b）用极限量规测量台阶

4. 台阶的质量分析

1）影响台阶尺寸的因素

（1）工作台移动调整尺寸时摇得不准。

（2）台阶尺寸测量不准。

（3）铣削中，铣刀受力不均匀出现“让刀”现象。

（4）铣刀摆差太大。

（5）工作台零位不准，用三面刃铣刀铣台阶时会使台阶上部尺寸变小。

2）影响台阶形状、位置精度的因素

（1）平口钳固定钳口未校正，或用压板装夹时工件未校正，铣出的台阶产生歪斜。

（2）工作台零位不准，用三面刃铣刀铣削时，不仅会将台阶铣成上窄下宽，而且会将侧面铣成凹面。

（3）立铣头零位不准，纵向进给用立铣刀铣削时，台阶底面会产生凹面。

3）影响台阶表面结构的因素

（1）铣刀变钝。

（2）铣刀摆差太大。

（3）铣削用量选择不当，尤其是进给量过大。

（4）铣削钢件时没有使用切削液或切削液使用不当。

（5）铣削时振动太大，未使用的进给机构没有紧固，工作台产生窜动现象。

思考与练习

（1）用端铣刀铣削台阶时刀具直径确定方法是什么？

（2）台阶的宽度和深度一般可用哪些量具测量？

（3）如何铣削较深的台阶？

任务七 铣削槽

一、图样与技术要求

如图 3-7-1 所示，工件毛坯尺寸为 110 mm×60 mm×30 mm，材料为 45 钢。评分表见表 3-7-1。

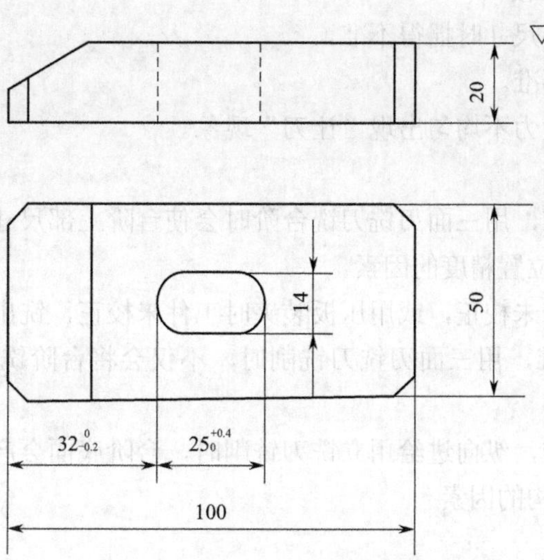

图 3-7-1　封闭槽零件

表 3-7-1　评分表

技术要求	配分	实测结果	得分
$32_{-0.2}^{0}$ mm	20		
$25_{0}^{+0.4}$ mm	20		
14 mm	20		
表面结构参数 $Ra3.2$ μm	20		
规范操作	10		
安全文明生产	10		

二、加工工艺分析

（1）固定平口虎钳，校正钳口与铣床主轴轴线垂直。

（2）安装万能立铣头，使立铣头主轴轴线与工作台台面垂直。

（3）在工件上划出槽的尺寸、位置线和钻孔位置线。

（4）安装并校正工件。

（5）安装 $\phi13$ mm 的钻头，钻落刀孔。

（6）选择并安装铣刀（选用 $\phi14$ mm 的立铣刀）。

（7）对刀，锁紧横向进给。

（8）分数次铣出封闭槽。

（9）测量尺寸，卸下工件。

（10）由教师对组内成员加工的工件和表现进行评价，结合评分表 3-7-1 给出分数。

三、相关知识与技能

在铣床上可加工的沟槽种类有很多，常见的有直角沟槽、V 形槽、燕尾槽、T 形槽、圆弧槽和各种键槽等。直角沟槽比较常见，其又可以分为敞开式、半封闭式和封闭式三种，如图 3-7-2 所示。

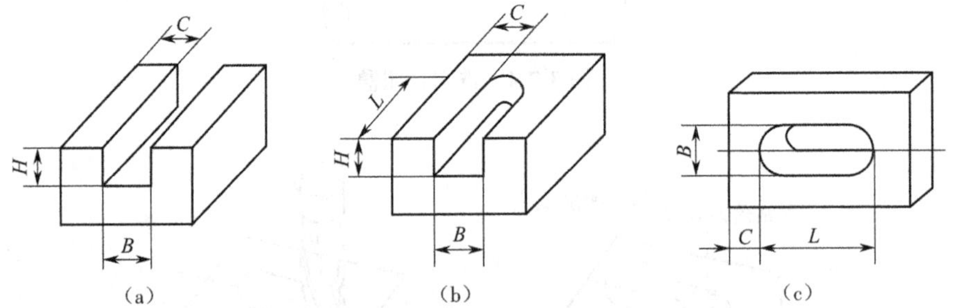

图 3-7-2　直角沟槽的种类

（a）敞开式　（b）半封闭式　（c）封闭式

1. 直角沟槽的铣削

1）铣削半封闭槽

一般用立铣刀铣削半封闭槽时，选择的立铣刀直径应不大于槽的宽度。由于立铣刀刚度较差，铣削时易产生偏让引起受力过大使铣刀折断，故在加工较深的沟槽时，应分几次铣削，以达到要求的深度。槽深铣好后，再扩铣沟槽两侧，扩铣时应避免顺铣，以免损坏铣刀，损伤工件，如图 3-7-3 所示。

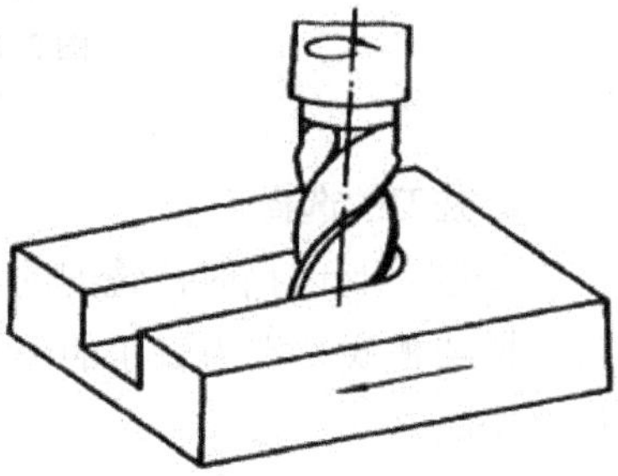

图 3-7-3　铣削半封闭槽

2）铣削封闭槽

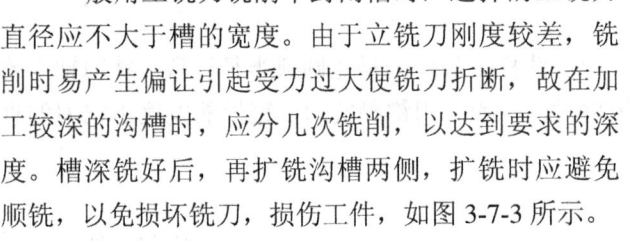

铣削封闭键槽时可选用键槽铣刀，也可用立铣刀。用键槽铣刀铣削时，应先将工件垂直进给移向铣刀，采用一定的吃刀量，使工件纵向进给切至键槽的全长，再垂直进给吃刀，最后反向纵向进给，经多次反复直到完成键槽的加工，如图 3-7-4 所示。用立铣刀铣削封闭槽时，由于立铣刀的端面中心附近没有切削面，不能垂直进给切削工件，因此要预钻落刀孔，如图 3-7-5 所示。落刀孔的深度应略大于沟槽的深度，其直径

应小于所铣槽宽度的 0.5~1 mm。铣削时，应分几次进给，每次进给都由落刀孔一端铣向另一端，槽深达到要求后，再扩铣两侧。铣削时，不使用的进给机构应紧固（如使用纵向铣削时，应锁紧横向进给机构；反之，则锁紧纵向进给机构），扩铣两侧时应避免顺铣。

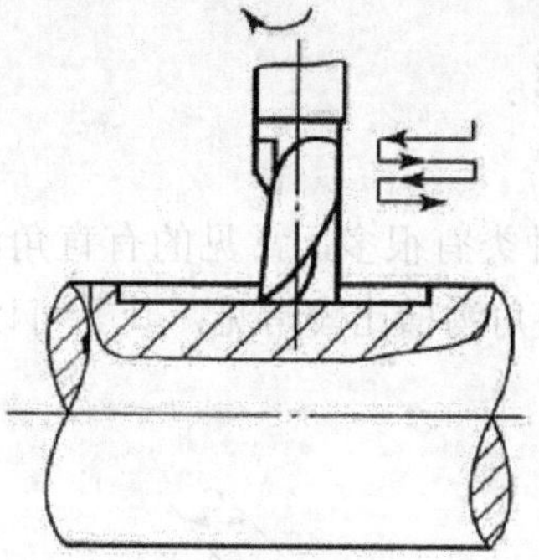

图 3-7-4　铣削封闭槽

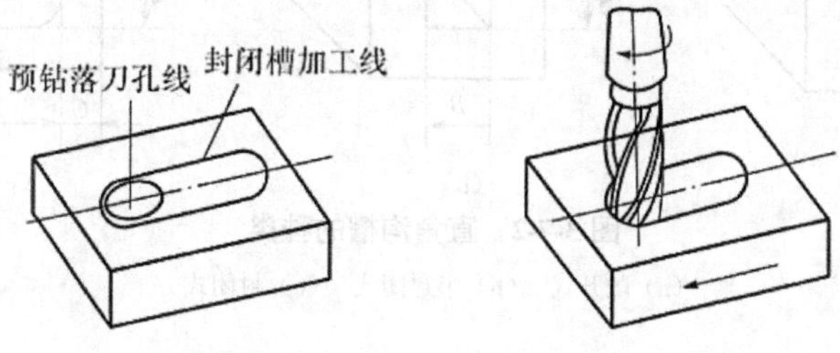

图 3-7-5　用立铣刀铣封闭槽的准备工作

（a）划出槽及落刀孔加工线　（b）钻出落刀孔

2. 铣刀的选择

铣削半封闭槽时，铣刀的半径应与图样上规定的槽底圆弧半径一致。铣削轴上的封闭槽或槽底一端为直角的半封闭槽时，一般采用键槽铣刀，键槽的宽度由铣刀的直径来保证。

3. 工件的装夹与校正

铣键槽时，工件的装夹方法很多，一般用平口钳或 V 形架装夹，应使工件的轴线与工作台的进给方向一致并与工作台台面平行。

1）用平口虎钳装夹工件

用平口虎钳装夹工件，装夹简单，适用于单件生产。为保证铣出的轴槽两侧面和底面都平行于工件的轴线，必须使工件的轴线既平行于工作台的纵向进给方向，又平行于工作台台面。用平口虎钳装夹工件时，应使用百分表找正固定钳口与工作台进给方向平行，装夹工件时还应找正工件上母线与工作台台面平行，如图 3-7-6 所示。

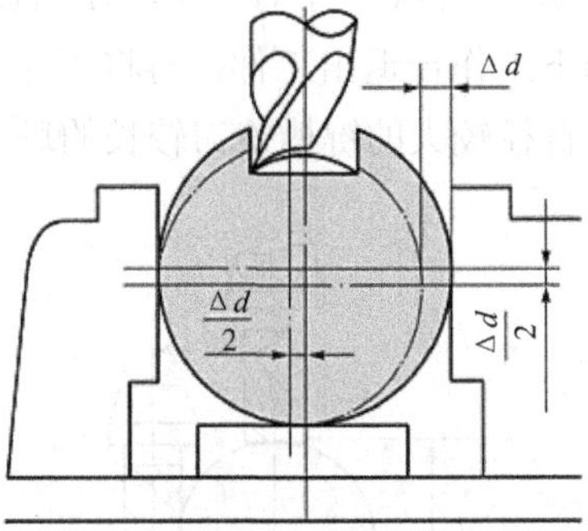

图 3-7-6 用平口虎钳装夹轴类零件

2）用 V 形架和压板装夹工件

用 V 形架和压板装夹工件是铣床上常用的方法之一。其特点是工件中心必定在 V 形架的角平分线上，对中性好，当工件直径变动时，不影响键槽的对称度。铣削时虽然铣削深度有改变，但变化量一般不会超过槽深的尺寸公差，如图 3-7-7 所示。

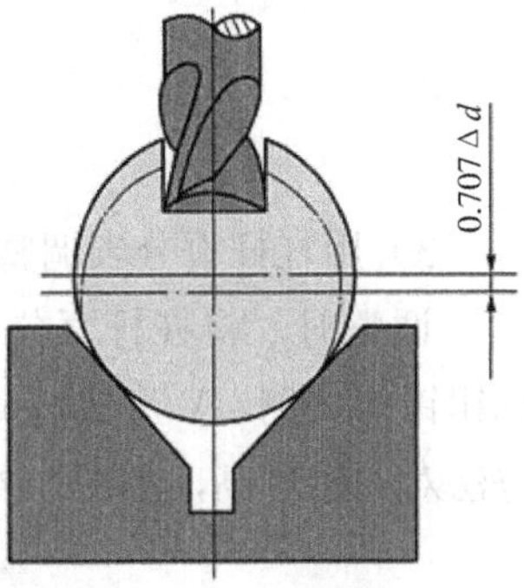

图 3-7-7 用 V 形架装夹轴类工件

4. 刀具的调整

为保证轴上键槽的对称度，必须调整铣刀的位置，使键槽铣刀的轴线通过工件的轴线。常用的调整方法有以下几种。

1）按切痕调整工件对称中心

键槽铣刀切痕调整方法与盘形槽铣刀切痕调整方法相同，只是键槽铣刀铣出的切痕是一个边长等于铣刀直径的四方形小平面。

对中时，使铣刀在旋转时落在小平面的中间位置即可，这种方法对中精度不高，但使用简便，是最常用的一种方法，如图 3-7-8 所示。

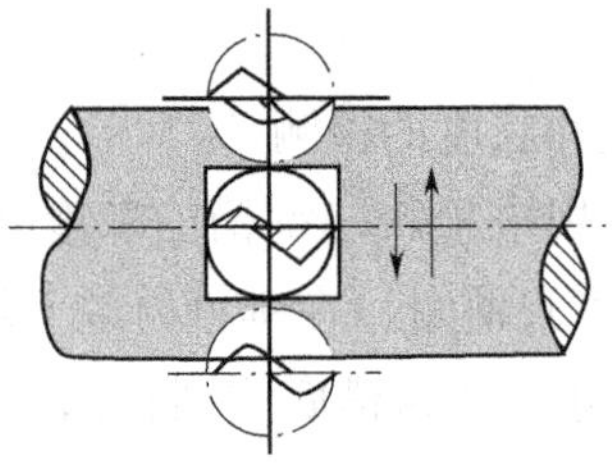

图 3-7-8 键槽铣刀切痕对中心

2）按侧面调整工件对称中心

调整时，先在工件侧面贴一薄纸，开动机床，使回转的铣刀逐渐靠向工件，当铣刀的刀刃接触到薄纸后，降下工作台退出工件，再将工作台横向移动一个距离 A，这种方法对中精度较高，适用于直径较大的键槽铣刀较长的场合。如图 3-7-9 所示。

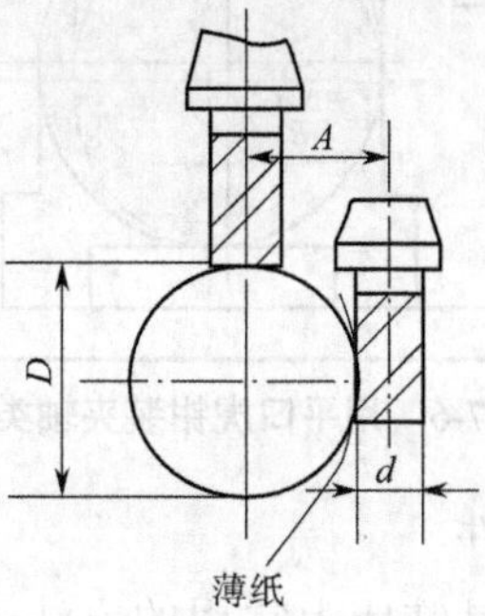

图 3-7-9　键槽铣刀切痕对中心

3）用杠杆百分表调整铣刀位置中心

调整时，将杠杆百分表固定在立铣头主轴上，用手转动主轴，观察百分表在平口虎钳钳口两侧、V 形架两侧和角尺两侧的读数，横向移动工作台使两侧读数相同。这种方法对中精度高，如图 3-7-10 所示。

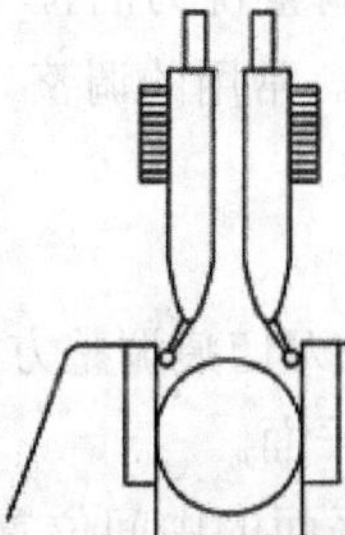

图 3-7-10　用杠杆百分表对中心

5. 轴上键槽的检测

1）轴上键槽的宽度、深度和长度的检测

轴上键槽的宽度通常用塞规来检查，键槽的深度和长度一般用游标卡尺来检测。对封闭式键槽的槽深可用游标卡尺测量，在键槽内放一块比键槽深度略大的矩形量块，量得的尺寸减去矩形量块的尺寸，即为槽底到圆柱面的尺寸。宽度大于千分尺测量杆直径的键槽，可用千分尺直接测量，如图 3-7-11 所示。

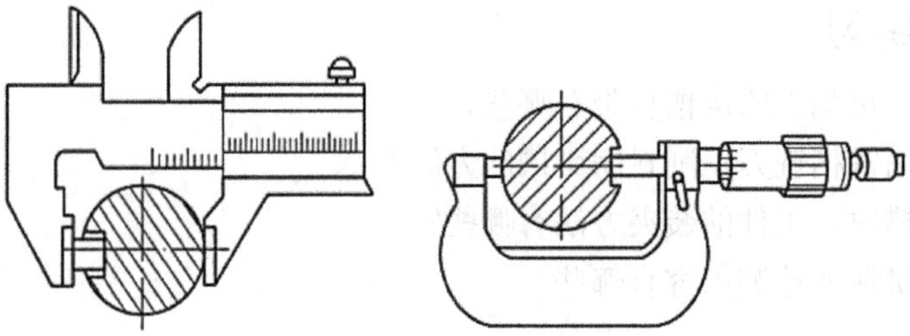

图 3-7-11 轴上键槽深度的测量

2）轴上键槽对称度的检测

当键槽窄而浅的时候，可在槽内紧密插入键，增大测量面。对宽而深的键槽，可直接测量槽的侧面。检测时，把工件安放在 V 形架或两顶尖之间，并一起放在平板上，使其能做定轴旋转。先使键槽处在一侧，用百分表将键的上平面校到与平板平行，并记下百分表读数，然后将工件转过 180°，用同样方法检测，得到另一个读数。两个读数的差值即为对称度误差，如图 3-7-12 所示。

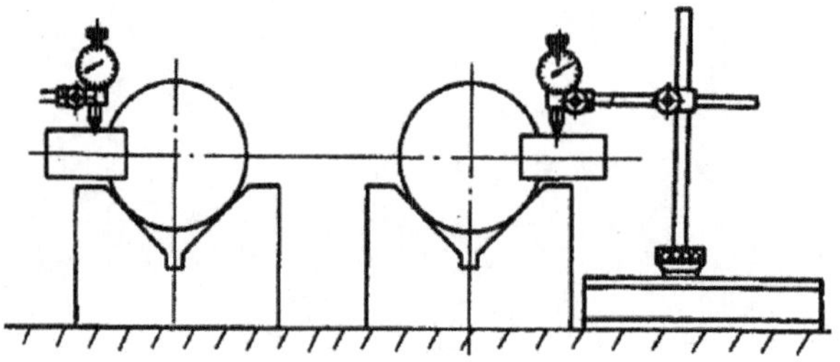

图 3-7-12 键槽对称度的检测

6. 轴上键槽的质量分析

1）影响尺寸精度的因素

（1）没有通过试铣来检查铣刀尺寸，就直接铣削工件，造成尺寸误差。

（2）用键槽铣刀铣键槽，铣刀径向圆跳动过大。

（3）铣削时，吃刀深度过大，进给量过大，产生“让刀”现象，将槽铣宽。

2）影响对称度的因素

（1）铣刀对称中心调整不准，铣削中，铣刀让刀量太大。

（2）成批生产时，工件外圆尺寸误差太大。

（3）轴槽两侧扩铣余量不一致。

3）影响键槽两侧面与轴线平行度的因素

（1）工件外圆直径圆柱度超差。

（2）用平口虎钳或 V 形架装夹工件时，平口虎钳或 V 形架没有找正好。

思考与练习

（1）铣床上可加工的沟槽种类有哪些？

（2）简述用键槽铣刀铣削封闭键槽的方法？

（3）铣键槽时，工件的装夹方法有哪些？

（4）轴上键槽的检测内容有哪些？

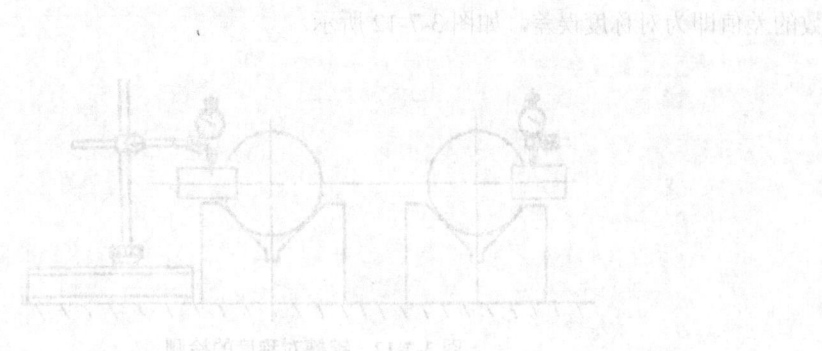

NUMERICAL CONTROL TECHNOLOGY SERIES OF LUBAN WORKSHOP

Metalworking Practice

LIU RUI SU CHEN

天津大学出版社
TIANJIN UNIVERSITY PRESS

Preface

The book of Metalworking Practice was prepared to coordinate with the theoretic and practical training and teaching of "Luban Workshop" in Madagascar, carry out exchanges and cooperation, improve the international influence of China' s vocational education, innovate the international cooperation mode of vocational schools, and export excellent vocational education resources of China on the basis of the implementation of the cooperation on vocational education between Tianjin and the countries along the Belt and Road.

This book is written in both Chinese and English for Luban Workshop project in Madagascar. With numerical control machining equipment of Luban Workshop as the carrier, it reflects the position needs on the basis of actual situation of enterprises and in compliance with the National Occupational Standards and the concept of "taking occupational standards as basis, enterprises needs as orientation, and occupational ability as the core", highlighting the new knowledge, new technology, new process, and new method, and centering on the training of occupational ability. There are three items and 22 tasks in total, trying to explain the profound things in a simple way and facilitate teaching.

The content of this book goes from simple to deep, combines theory with practice, and arranges knowledge points and skill points together in task-driven form with appropriate cases. Based on the principles of scientificity, practicability and universality, this book is in line with the status quo of the mechanical curriculum system of vocational education. Metalworking practice is a highly practical and basic course of technology for majors in Mechanics, which lays the foundation for students to establish the concept of machinery manufacturing and production process, and acquire basic machinery manufacturing skills.

This practice tutorial presents students initial access to the manufacturing process of machinery, and enables them to acquire basic knowledge of common materials and metal machining process in mechanical manufacturing, get familiar with common machining methods of mechanical parts and the main equipment and tools used, preliminarily master the basic operation skills of common machine tools and possess certain operation skills, so as to lay the foundation for theory study of related courses and future production technology work. This book can be used as a teaching book for majors in mechanical processing technology and electromechanical technology application, and electromechanical related majors in vocational colleges, as well as a post training textbook for related industries and the self-study material for enterprise employees.

This book was edited by Liu Rui, Su Chen of Tianjin Technician Institute of Mechanical & Electrical Technology, and Du Yang, Zhang Zhiying and Zhang Jinfeng of Tianjin Technician Institute of Mechanical & Electrical Technology also participated in the preparation. We would like to express our gratitude for the great help ! Limited by the experience and level of the editor, defects are unavoidable in this book, please do not hesitate to raise your valuable opinions.

Editors

Contents

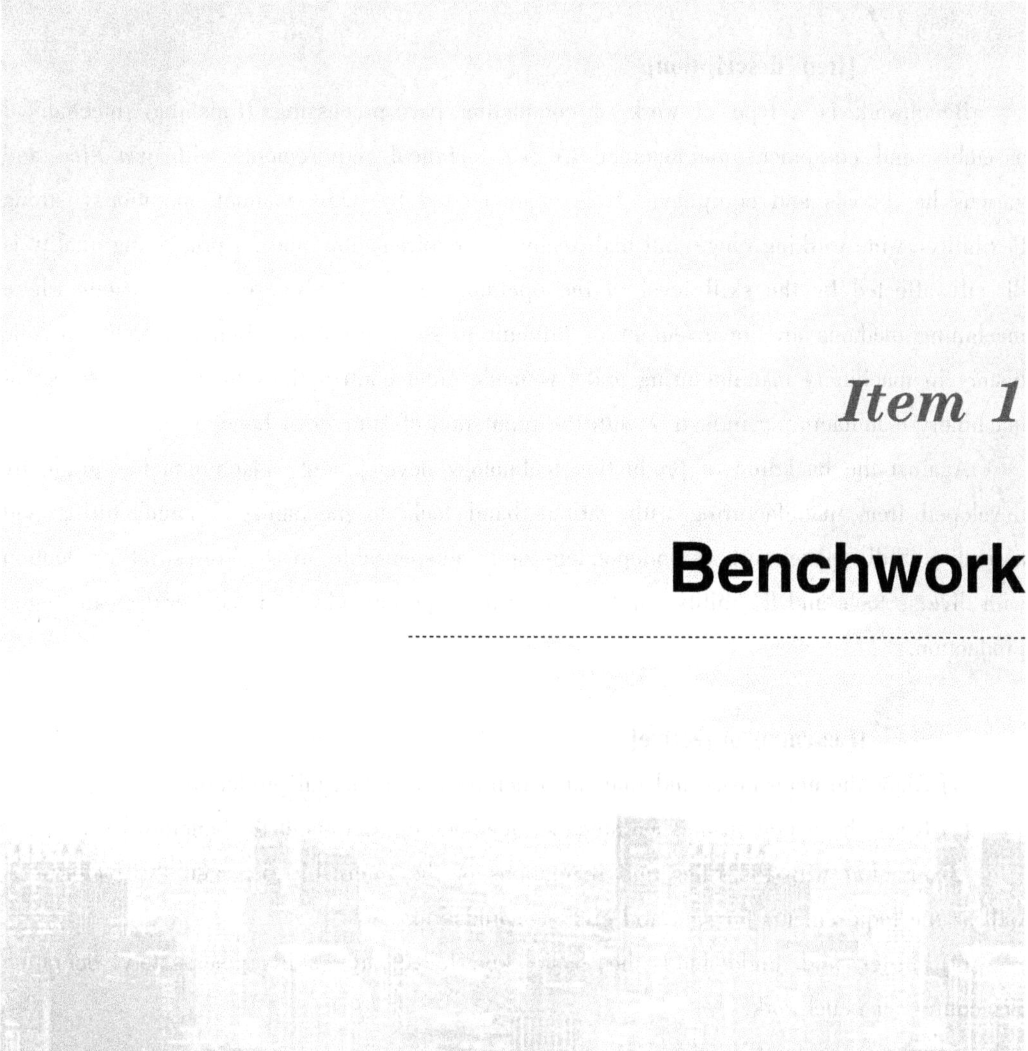

Item 1

Benchwork

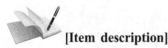

[Item description]

Benchwork is a type of work of conducting part processing，finishing，mechanical assembly and equipment maintenance as per technical requirements with jaw vice and various hand tools and equipment. It is characterized by many manual operations，strong flexibility，wide working range and high technical requirements，and its processing quality is directly affected by the skill level of the operators. It is mainly used in occasions where machining methods are inconvenient or difficult to solve problems. Benchwork has a long history in machinery manufacturing and has made great contributions to the development of machinery manufacturing industry，with the reputation of "universal benchwork" .

Against the backdrop of production technology development，machining has gradually developed from manufacturing with various hand tools to mechanized manufacturing and assembly. Still，benchwork is independent and indispensable in the industrial production with diverse tools and flexibility，and occupies an important place in the world of automatic production.

[Learning objective]

(1) Know the main tasks and roles of benchwork in industrial production;

(2) Know the operation and maintenance methods of equipment for benchwork;

(3) Familiar with the rules and regulations of the internship site and the factory，as well as the requirements for safe and civilized production;

(4) Master and understand the basic knowledge and skills and safety operation procedures for benchwork.

Task I General knowledge of benchwork

I. Function and typical job contents of benchwork

Benchwork is a type of work for processing，finishing，assembly of workpiece with related tools. The manufacturing process of any mechanical product usually includes the phase of blank manufacturing, part processing and manufacturing, component assembly, complete machine assembly, and commissioning and trial operation. Among them, a lot of work must be done by methods of benchwork. Benchwork is an indispensable method used in mechanical manufacturing, which is of a wide work scope, mainly including the following aspects:

1. Parts processing

Processing tasks that are not suitable or cannot be solved by mechanical methods can be completed by methods of benchwork, such as scribing, finishing (such as scraping, grinding, template filing and mould manufacturing, etc.), inspection and repair during part processing, etc.

2. Making and repair of tools

Benchwork can be used to make and repair various tools, fixtures, measuring tools, molds and special equipment.

3. Mechanical assembly

Benchwork can be used to assemble the parts as per the technical requirements of assembly, and make them qualified functional parts or mechanical equipment after adjustment, inspection and commissioning.

4. Equipment maintenance

When the mechanical equipment is of failure or damage during operation or its function is affected due to loss of accuracy after long-term use, maintenance can be carried out by methods of benchwork.

Although many heavy tasks have been finished by methods of machining with the increasing development of the machinery industry, processing of parts with high accuracy and complex shape and the installation, commissioning and maintenance of equipment are difficult to complete mechanically. Then benchwork of virtuosity is required to complete those tasks. Therefore, benchwork is an indispensable type of work in the machinery manufacturing industry.

II. Operation skills necessary for benchwork

1. Scribing

As the first process of parts processing, scribing is closely related to the processing allowance of parts. When scribing, the benchworker shall be familiar with the drawing first, then reasonably use the scribing tool, and draw the processing limit of the part and the centerline of various holes on the workpiece to be processed as per the scribing steps, which can be used as the basis for part clamping and processing.

2. Sawing

Sawing is used to divide materials or saw grooves that meet technical requirements on workpiece. When sawing, the saw blade and sawing method must be selected correctly according to the material properties and shape of the workpiece, so that the sawing operation can be carried out smoothly to meet the specified technical requirements.

3. Filing

Filing is to use files of various shapes to cut and shape the workpiece, so that the workpiece can achieve higher precision and more accurate shape. Filing is one of the main operation methods of benchwork. It can be used to process the outer plane, curved surface, inner and outer corners, grooves and various shapes of surfaces of the workpiece.

4. Chiseling

Chiseling is the most basic operation of the benchwork, which is to use simple tools such as chisel and hammer to cut and cut off the workpiece. Chiseling is mainly used in occasions where the processing requirements of parts are not high or the task cannot be completed by machining. At the same time, chiseling also requires the operator to have skilled hammering skills.

5. Hole processing (including drilling, counterboring, countersinking and reaming)

Drilling, counterboring, countersinking and reaming are four methods of rough processing, semi-finishing and finishing by methods of benchwork, which are selected according to the processing requirements and conditions of holes. Among them, the drilling, counterboring, and countersinking of benchwork shall be carried out on the drilling machine, and reaming can be completed by hand or by machine. To master the operation skills of drilling, counterboring, countersinking and reaming, cutting performance of the tools used, as well as the structural performance of drilling machines and some fixtures shall be known to choose the cutting amount reasonably. Mastering the specific methods of grinding and manual operation of drill bit is the key to ensure the quality of hole processing.

6. Thread processing (including tapping and thread die cutting)

Tapping is a method of processing internal thread on the inner cylindrical surface of workpiece with tap, and thread die cutting is a method of processing external thread on the outer cylindrical surface of workpiece with round die. Threads processed by method of benchwork are usually triangular threads with smaller diameters or threads that are not suitable for processing on machine tools.

7. Scraping and grinding

Scraping is a method used by benchworkers for workpiece finishing. By scraping, not only can higher geometric accuracy, dimensional accuracy, contact accuracy and transmission accuracy be obtained, but also the workpiece surface structure can be compacted by the extrusion of the scraper during scraping, thus improving the mechanical properties, wear resistance and corrosion resistance of the material.

Grinding is the most precise processing method. It produces micro-cutting with the abrasive by sliding and rolling between the grinding tool and the workpiece, which makes the workpiece achieve high dimensional accuracy and low surface roughness.

8. Correction and bending

The correction and bending are to correct the deformed or defective raw materials and parts by using the plastic deformation of the metal materials and using appropriate methods, to eliminate defects such as deformation, or to bend the raw materials into the shape required by the drawings with special tools and calculate the length of the materials before bending.

9. Assembly and repair

Assembly is to combine parts into components or complete machines through appropriate connections as required by the drawings. Repair is to adjust the machine or parts that have been used for a long time or whose accuracy and performance have decreased or even damaged due to improper operation, so as to restore them to the original required accuracy and performance.

10. Measurement

In order to ensure the processing accuracy and that the requirements of parts will be met in the production process, first of all, necessary measurement and inspection shall be carried out on the products. In the process of parts processing and assembly, benchworkers often use flat plates, vernier calipers, micrometer, dial indicator and level gauge to measure and inspect parts. These are all measurement skills that benchworkers must master.

In addition, the benchworker must also know and master the general knowledge of heat treatment of metal materials, and master the manufacturing and heat treatment methods of some tools of benchwork, such as hammer, chisel, sample punch, scriber, scribing compass, and scraper.

III. Tools and measuring tools commonly used in the basic operation of benchwork

Common tools include scriber, tosecan, scribing compass, sample punch, dividing head and flat plate for scribing, hammer and chisel for chiseling, various files for filing, saw bow and saw blade for sawing, various drills, countersinks and reamers for hole processing, various taps, dies and tap holders for tapping and thread die cutting, scrapers for scraping the flat and curved surfaces, and various wrenches and screwdrivers, etc.

Commonly used measuring tools include steel ruler, knife straightedge, vernier caliper, micrometer, 90° square, universal angle meter, feeler gauge, dial indicator, etc.

IV. Safe and civilized production system

1. Internship code

(1) When entering the internship factory, work clothes, work shoes, and work caps shall be worn, and notes and various internship supplies shall be taken.

(2) Before class, each class shall be in the forefront of the internship workshop on a class-by-class basis to check the actual number of people, and queue up to enter the internship workshop 10 minutes in advance. No slippers, gloves or scarves are allowed to enter the workshop. The work clothes shall be kept clean and tidy.

(3) When entering the internship workshop, no loud noise, fighting, random running and stranding are allowed. No snacks or littering. When walking in the internship teaching place, keep a single row and walk right.

(4) After entering the internship factory, please follow the arrangement of the intern instructor and workshop management staff, accept the instruction of the intern instructor, listen to the class carefully, take notes in class, study hard, and take every skill practice seriously.

(5) Implement the production process standard of internship and carry out the production according to the plan.

(6) During the internship period, please stick to your own station, and do not leave your own post and visit other' s at work hours. You are not allowed to leave the internship workshop without the approval of the teacher on duty.

(7) Strictly abide by the work schedule, and do not be late, leave early, or absent from school. If something really happens, please ask for leave from the intern instructor on duty first.

(8) Strictly abide by the safety operation rules of each type of work, and practice according to the related operation process standard.

(9) Respect teachers. If any communication needed, be polite and talk with teachers modestly. Mutual understanding is needed to enhance friendship. Do not speak at will in class. If there are any questions, raise your hand first for indication and speak after the teacher agrees.

(10) Internship workpiece, tools, instruments and meters, books and other internship supplies shall be placed in an orderly manner to keep the bench, equipment and

surrounding ground clean and tidy, free of debris, stains and dust, and hygiene responsibility area shall be cleaned up on time.

(11) Without the teacher' s approval, articles irrelevant to the internship shall not be brought into the internship workshop, and public property shall not be taken out of the internship workshop.

(12) The equipment shall not be used until the equipment performance, correct operation specifications and operating procedures are mastered and approved by the intern instructor on duty. Maintenance shall be carried out immediately after the equipment is used.

(13) In case of problems or failures of instruments, meters, tools, measuring tools, machine tools and equipment during the internship, the intern instructor on duty shall be immediately reported for situation explanation and they shall not be handled without permission.

(14) Abide by the rules and regulations of the school and internship factory.

(15) Violation of the above provisions shall be punished as per relevant management regulations. If any losses are caused, the responsible person shall make compensation.

2. Benchworker safety operation rules

(1) Wear work clothes, caps and other protective articles before work.

(2) Before work, check whether the tools used are full and complete. File, scraper, hammer shall be fitted with firm wood handle. No quenching cracks, curling and flying burr are allowed at the hammering places of tools such as punches and chisels.

(3) When using a hammer, the direction of swinging hammer shall be selected to prevent the hammer head from falling off or iron filings from flying out and hurting people. When chiseling the workpiece, no one is allowed to stand on its opposite side. The protective net shall be provided at the fixed operation place, and no gloves shall be worn when holding the hammer.

(4) When using hand drill, hand grinding wheel and all portable electric tools, foot shall be put on the insulating plate and insulating gloves and protective glasses shall be worn.

(5) Before using hand grinding wheel and flexible shaft grinding wheel, you must check whether the grinding wheel is in good condition, carefully check whether there is any leakage phenomenon, and avoid leakage operation. It is not allowed to use the grinding wheel until it is in normal operation.

(6) When drilling on drilling machine, it is forbidden to wear gloves, and it is not allowed to clean cuttings by hand wiping or blowing by mouth.

(7) Before disassembling the equipment and adjusting the operating parts of the equipment, the power supply must be cut off. If the safety device on the equipment is not

repaired, it is strictly prohibited to carry out commissioning; after maintenance or commissioning of the equipment, careful inspection must be carried out, and tools or workpiece are not allowed to be left in the machine tool to prevent accidents.

(8) Electrical appliances are not allowed to be disassembled and assembled without the permission of the electrician.

(9) Reasonable use of work card measuring tools, and no mixing is allowed.

(10) When using grinding wheels, drilling machines, welding machines and lifting equipment, you must be familiar with their operation rules and strictly abide by them.

[Thinking and practice]

(1) What are the typical job contents of benchwork ?

(2) What are the main operating skills of benchwork ?

(3) What are the precautions during benchwork operation ?

Task II Practice environment for benchwork

I. Layout and requirements for site of benchwork

Benchwork site refers to the fixed work place of benchworkers. Reasonable benchwork site shall be provided as required by safe and civilized production to facilitate work, ensure product quality and safe production.

1. Reasonable layout of main equipment

The bench shall be placed in a place with suitable light and of easy working. When the benches are used face to face, a protective net shall be installed between the two benches. Grinder and drilling machine shall be set at the edge of the site. Especially the grinder must be installed in a safe and reliable position.

2. Place blanks and workpiece correctly

Blanks and workpieces shall be placed neatly separately and put on the workpiece shelf as far as possible to avoid bumping and damage.

3. Place tools and measuring tools reasonably

Commonly used tools and measuring tools shall be placed near the working position, and shall not be piled up at will to avoid damage. When working on the bench, for the convenience of access, the tools and measuring tools taken by the left hand shall be placed on the left side, and that by the right hand shall be placed on the right side. They shall be arranged in order respectively. Tools, measuring tools and workpieces shall not be mixed and their edges shall not be extended beyond bench edge. Tools and measuring tools shall be cleaned, maintained and repaired in time and properly placed after use.

4. The workplace shall be kept clean

After the practice, the equipment shall be cleaned and lubricated as required, and the workplace shall be cleaned.

II. Common equipment and tools for benchwork operation

Most of the equipment commonly used in bench processing is relatively simple, mainly including bench, bench vice, grinder, bench drill, vertical drilling machine and radial drilling machine.

1. Bench

It is mainly used for installing vice and storing common hand tools, measuring tools and fixtures. There are two types of benches: multi-person single row and multi-person double row.

Because the operator operates face to face, a protective plate or a protective net must be provided in the middle of the bench of double row. The bench is mostly made of cast iron and solid wood. Its surface is generally rectangular or hexagonal with its length and width determined by the work site and the work needs. The height is generally 800-900 mm. After the bench vice is installed, the appropriate height of bit of the vice can be obtained (generally equal to the elbow of the person). As shown in Fig. 1-2-1.

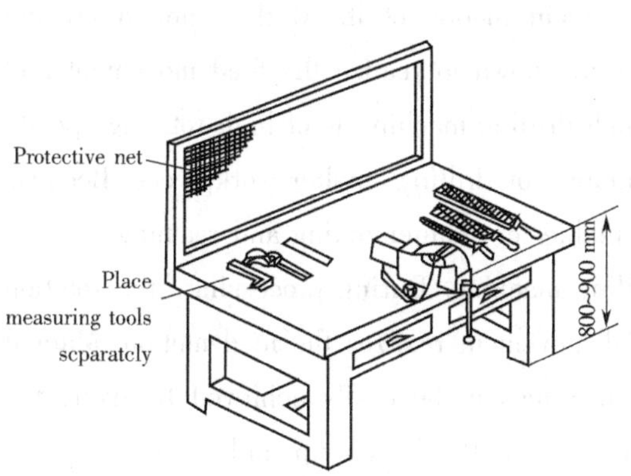

Fig. 1-2-1 Bench

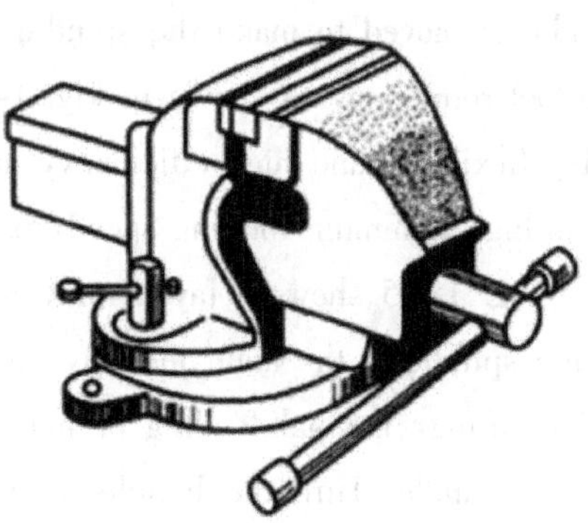

Fig. 1-2-2 Rotary bench vice

2. Bench vice

It is a general fixture for clamping workpieces, and its specifications are expressed by the width of the bit of the vice. The commonly used specifications are 100 mm, 125 mm, 150 mm, etc. The common structure type of bench vice is rotary, as shown in Fig. 1-2-2. The body of the rotary vice can be rotated, which can make the workpiece rotate to a suitable working position.

3. Grinder

It is used to grind tools including drill bits, chisels, and scrapers, and is composed of motors, grinding wheels, brackets and machine bodies, as shown in Fig. 1-2-3. Attention must be paid to safety when using the grinder, and the Safety Operating Rules for Use of Grinder shall be strictly observed.

Fig.1-2-3 Grinder

4. Drilling machine

Drilling machine is a piece of common hole processing equipment for benchworkers and can be divided into bench drilling machine, vertical drilling machine and radial drilling machine according to different structures.

Bench drilling machine is a small drilling machine used for processing holes, which is generally installed on the bench. It uses tools like drill bits for cutting. During the work, the workpiece is fixed and the tool rotating is the main motion of it. At the same time, the handle is moved to make the spindle move up and down to realize the feed movement and the tool retraction, as shown in Fig. 1-2-4. Bench drilling machine is of high rotating speed, using flexibility and high efficiency, and is suitable for drilling smaller workpieces. Because of its high minimum rotating speed, it is not suitable for countersinking and reaming.

Fig. 1-2-5 shows a layout of vertical drilling machine. During processing, the rotation of the spindle is the main motion, and its axial movement realizes the feed motion. Manual quick lifting, manual feeding or maneuvering feeding can be easily achieved by using the control handle. Turn the handle of the bench or move the bench up and down along the column guide rail, so as to adapt to the processing of workpieces with different heights. Vertical drilling machine is suitable for drilling, countersinking, reaming and tapping of single piece or small batch of medium-sized workpieces.

Fig. 1-2-4 Bench Drilling Machine

Fig. 1-2-5 Vertical Drilling Machine

Fig. 1-2-6 Radial Drilling Machine

Features of radial drilling machine: radial drilling machine is suitable for all kinds of holes processing of medium parts, large parts and porous parts produced in single piece, small batch and medium batch. The working range on the radial drilling machine is wide. When working, the workpiece can be pressed on the bench or placed directly on the base of it. Aligning the center of the hole on the workpiece by moving the spindle is more convenient than the vertical drilling machine. The radial drilling machine has a wide range of spindle speed and feed, which can achieve high production efficiency and processing accuracy. As shown in Fig. 1-2-6.

III. Common measuring tools for benchwork

1. Steel ruler

Steel ruler is the simplest measuring tool in common use, as shown in Fig. 1-2-7. It is mainly used to measure the length, width, height and depth of the workpiece, as well as the guide tool for drawing straight lines sometimes. Its common specifications are 150 mm, 300 mm, 500 mm and 1 000 mm.

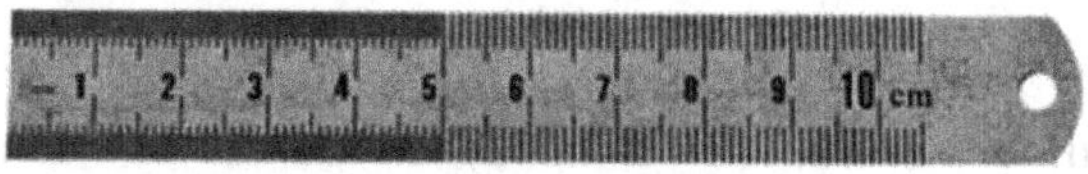

Fig. 1-2-7 Steel Ruler

2. Vernier caliper

Vernier caliper is the measuring tool of medium accuracy, which can be used to measure length, thickness, outer diameter, inner diameter, hole depth and center distance, etc. The accuracy types of vernier caliper: 0.1 mm(1/10), 0.05mm(1/20) and 0.02 mm(1/50), as shown in Fig. 1-2-8. Vernier calipers with measuring accuracy of 0.02 mm are most commonly used.

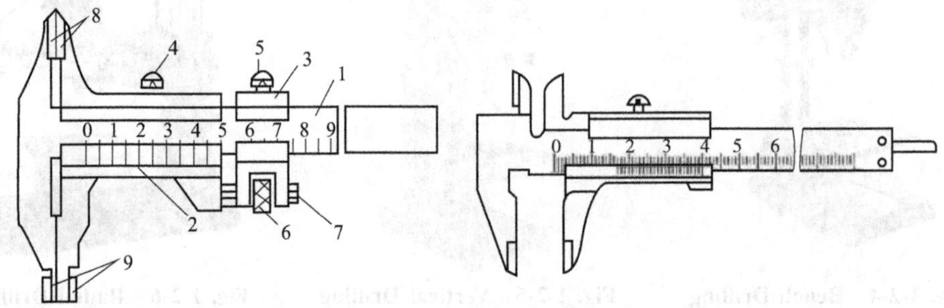

Fig. 1-2-8 Vernier Caliper

(a) Vernier caliper with fine adjustment (b) Vernier caliper with depth rod
1—Main scale; 2— Vernier scale (metric); 3—Vernier scale (imperial); 4—Locking screw; 5—Screw;
6—Fine adjustment nut; 7—Screw rod; 8—Lower jaw for outside measurement; 9—Upper jaw for inside measurement

During measurement, the two measuring jaws of the vernier caliper shall be opened to a level slightly larger than the measured dimension. First, the measuring surface of the fixed jaw shall be placed against the workpiece, and then the vernier shall be gently pushed with the thumb to keep the sliding jaw gradually close to the workpiece, and then the start reading, as shown in Fig. 1-2-9.

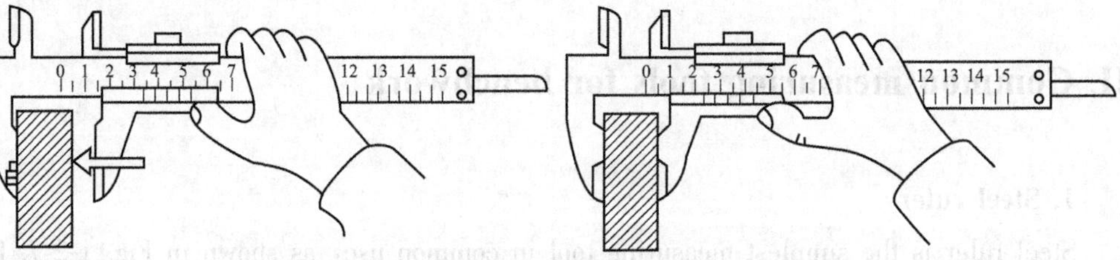

Fig. 1-2-9 Measuring Method of Vernier Caliper

The correct reading method of vernier caliper is as follows:

(1) First, read the integer value (mm) of the left of the 0 marker of the main scale.

(2) Then read the main scale where it lines up with the vernier scale’ s zero. The product of the number on vernier scale markers and the accuracy is the decimal part less than 1 mm.

(3) The measured actual dimension is finally determined by adding both the value from the main scale and the value from the vernier scale together. Here, take the reading method of the 0.02 mm vernier caliper as an example. First read integer value of the left of the 0 marker of the main scale, and multiply the value at the coincidence line between the vernier scale and the main scale by the accuracy value of 0.02. The sum of the two is the measured dimension. As shown in Fig. 1-2-10, the integer value of the left of the 0 marker of the main scale is 60, the value on the marker where the vernier scale coincides with the main scale is 4, and the reading is 60 mm + 24 mm×0.02 = 60.48 mm. When reading, the sight

line shall be perpendicular to the vernier scale marking line to avoid reading error caused by oblique view.

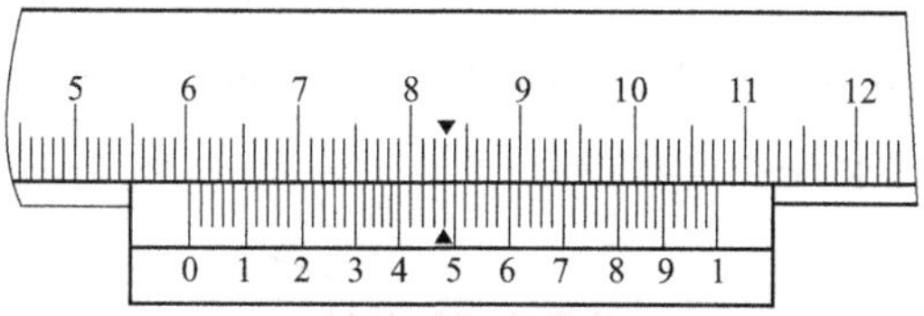

Fig. 1-2-10 Reading Method of Vernier Caliper

3. Micrometer

Micrometer is one of the most commonly used precision measuring tools in measurement. There are many kinds of micrometer, which can be divided into outside micrometer, inside micrometer, internal measuring micrometer, depth micrometer, screw thread micrometer, common normal micrometer, wire micrometer, thickness micrometer, etc. according to their different uses. According to different measuring objects, there are three kinds of micrometer commonly used: outside micrometer, inside micrometer and depth micrometer.

When the micrometer' s measuring range is within 500 mm, every 25 mm is a length stop, such as 0-25 mm, 25-50 mm, 50-75 mm, 75-100 mm, etc. When the measuring range is 500-1 000 mm, every 100 mm is a length stop, such as 500-600 mm, 600-700 mm, etc. Micrometer is divided into 0, 1 and 2 grades according to manufacturing accuracy, and its measuring accuracy is generally 0.01 mm

The outline and structure of the outside micrometer are shown in Fig.1-2-11. It consists of frame, anvil, spindle, lock, threaded sleeve, thimble, barrel, nut, joint, rotating vernier scale, spring, ratchet pawl, ratchet wheel and other parts.

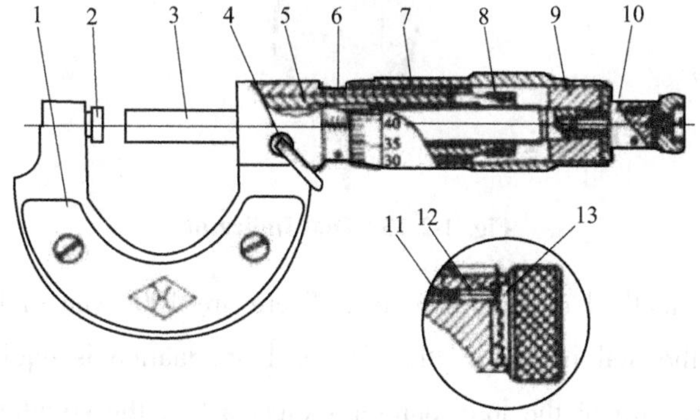

Fig. 1-2-11 Outline and Structure of Outside Micrometer

1—Frame; 2—Anvil; 3—Spindle; 4—Lock; 5—Threaded sleeve; 6—Thimble; 7—Barrel; 8—Nut; 9—Joint; 10—Rotating vernier scale; 11—Spring; 12—Ratchet pawl; 13—Ratchet wheel

The correct reading method of outside micrometer is as follows:

(1) Read the integer value (mm) and half millimeters (0.5 mm) on the graduations on

the thimble;

(2) Then, determine the position where the barrel graduation line is aligned to the reference line of the thimble, and read the decimal part less than half a millimeter;

(3) Finally, add the two readings together to obtain the measured dimensions of the workpiece, as shown in Fig. 1-2-12.

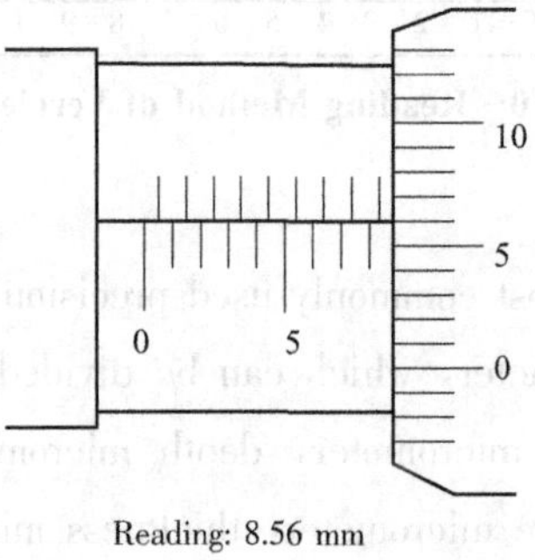

Fig. 1-2-12 Reading Method of Micrometer

4. Dial indicator

The dial indicator is an indicating measuring instrument, which is mainly used to measure the dimension, shape and position error of the workpiece, and can also be used to check the geometric accuracy of machine tools or adjust the clamping position deviation of the workpiece. The measuring range of dial indicator is generally 0-3 mm, 0-5 mm and 0-10 mm. According to the different manufacturing accuracy, the dial indicator can be divided into grade 0, grade 1 and grade 2. The outline of the dial indicator is shown in Fig. 1-2-13.

Fig. 1-2-13 Dial Indicator

Correct reading method of dial indicator: There are 100 even divisions printed on the graduation plate of the dial indicator, that is, each graduation is equivalent to 0.01mm of the spindle, so every turn of the long pointer indicates that the spindle moves by 0.01mm, and when it moves by 1mm, the long pointer rotates a round, so the measuring accuracy of the dial indicator is 0.01 mm.

When measuring, it must be fixed on a reliable clamping frame (such as a universal meter frame or a magnetic meter base). The clamping frame shall be placed stably, and the spindle must be perpendicular to the surface of the workpiece to be measured, so that the axis of the spindle is consistent with the direction of the measured dimension. When the gauge head touches the surface of the workpiece to be measured, the spindle is pushed into the tube, and the distance of the spindle is equal to the reading of the short pointer (the measured integer part) plus the reading of the long pointer (the measured decimal part).

5. Universal angle meter

Universal angle meter is a measuring tool for measuring the internal and external angles of workpieces. Its measuring accuracy is divided into 2′ and 5′, and the measuring range is 0°-320°. Its outline and structure are shown in Fig. 1-2-14.

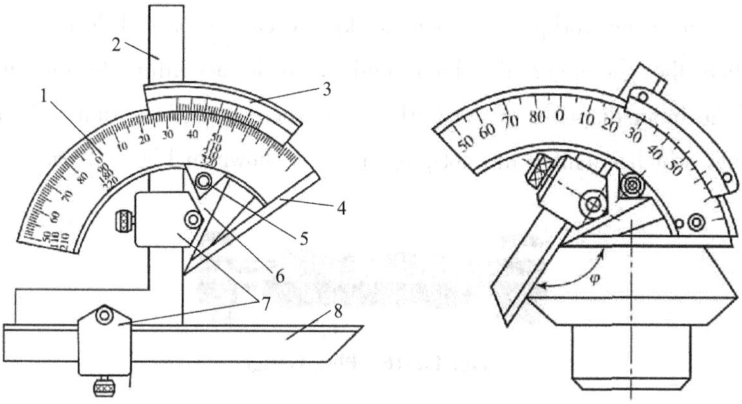

Fig. 1-2-14 Universal Angle Meter

1—Main scale; 2—Square angle scale; 3—Vernier scale; 4—Base scale; 5—Brake; 6—Sector plate; 7—Block; 8—Straight scale

When measuring, the sector gear can be rotated through pinion on the sector plate to change the angle of the angle meter and drive the main scale to rotate along the vernier scale. The square angle scale and straight scale can be used together or independently.

6. Feeler gauge

The feeler gauge is a sheet-shaped gauge used to check the clearance between two joint surfaces. It is a measuring tool composed of a group of metal sheets with different thicknesses, and its length is 50 mm, 100 mm and 200 mm, as shown in Fig. 1-2-15.

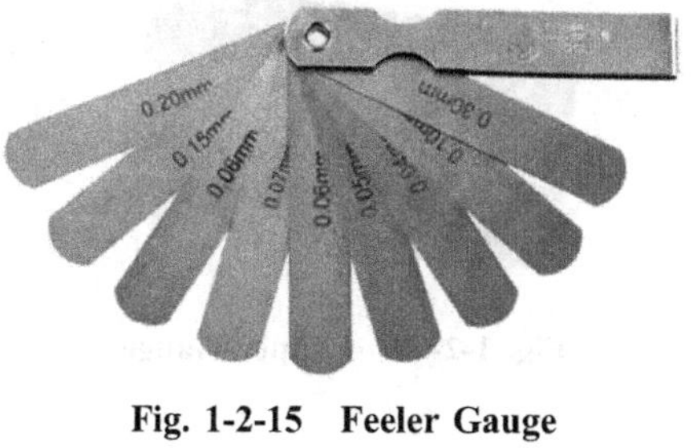

Fig. 1-2-15 Feeler Gauge

When using the feeler gauge, the number of feeler gauges shall be selected according to the dimension of the clearance, and one or several pieces of them can be overlapped and inserted into the clearance. If the feeler gauge has a slight sense of blockage when moving after inserting into the clearance, the thickness of the feeler gauge at this time is the dimension of the clearance to be measured.

The feeler gauge with small thickness is very thin and easy to bend and break. Do not apply too much force when inserting. Wipe it clean and put it in the clamp in time.

7. Plug gauge and caliper gauge

Plug gauge is a measuring tool used to check the inner diameter of workpiece. It has two measuring surfaces. The small end is made according to the minimum limit dimension of the inner diameter of the workpiece, where workpiece can be passed during bore measuring, which is called the go-gauge; the large end is made according to the maximum limit dimension of the inner diameter of the workpiece, where workpiece cannot be passed during bore measuring, which is called the not-go-gauge, as shown in Fig. 1-2-16.

Fig. 1-2-16 Plug Gauge

Caliper gauge is used to check the outer circle dimensions of shaft workpiece. It has two measuring surfaces, as shown in Fig. 1-2-17. Among them, the large end is made according to the maximum limit dimension of the shaft, and it shall pass through the shaft journal when measuring, which is called the go–gauge; The small end is made according to the minimum limit dimension of the shaft. When measuring, it does not pass through the journal, which is called the not-go-gauge. When checking the shaft workpiece with the caliper gauge, if the go–gauge can pass and the not-go-gauge cannot pass, it means that the dimension of the workpiece is within the allowable tolerance range and is qualified. Otherwise, it is unqualified.

Fig. 1-2-17 Caliper Gauge

8. Knife straight edge

Knife straight edge is a measuring tool to check the straightness and flatness of workpieces. It has the advantages of simple structure, light weight, no rust, convenient operation and high measuring efficiency. Its commonly used specifications are 125mm, 200mm, 300mm and 400mm, etc. As shown in Fig. 1-2-18.

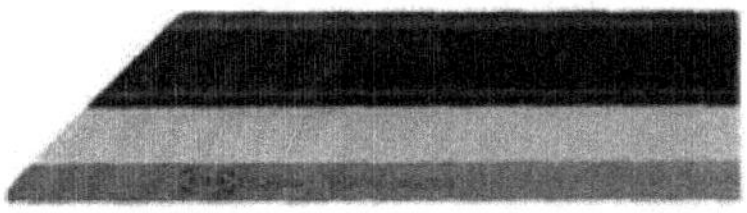

Fig. 1-2-18 Knife Straight Edge

9. Right angle ruler

Right angle ruler is a professional measuring tool, referred to as square and guiding rule for short. Square is used for vertical measurement of the workpiece and the verticality of the relative position of the workpiece, and sometimes it is also used for scribing. As shown in Fig. 1-2-19.

Fig. 1-2-19 Square

[Thinking and practice]

(1) What is the equipment commonly used for benchwork processing ?

(2) What are the measuring tools commonly used for benchwork ?

(3) Briefly describe the correct reading method of outside micrometer.

(4) What are the usage methods of plug gauge and caliper gauge ?

Task III Scribing

Scribing is one of the important processes in machining, which is widely used in single piece and small batch production. It refers to the operation process of drawing processing boundary lines or points and lines as reference on blank or semi-finished workpieces with scribing tools based on the drawings and technical requirements.

It is divided into two types: scribing on the same plane and on the different plane. Scribing on the same plane is the operation of drawing processing boundary line on a plane of the workpiece based on the drawings and technical requirements. If it is necessary to draw lines on several planes of the workpiece at different angles (generally perpendicular to each other) to clearly indicate the processing boundary line, it is scribing on the different planes. Clear lines and accurate positioning are required when scribing.

The main functions of scribing are:

(1) To determine the processing allowance of the workpiece, and find and deal with unqualified blank or semi-finished workpiece in time;

(2) To facilitate the clamping of complex workpieces on the machine tool, alignment and positioning can be carried out according to the scribed lines;

(3) When the blank error is small, it can be remedied by borrowing materials to improve the utilization rate of the blank;

(4) Blanking shall be carried out on the plates according to the lines, which can be discharged properly and the materials can be used reasonably.

The so-called material borrowing means that the allowance of each machined surface can be borrowed from each other and reasonably distributed through trial scribing and adjustment, so as to ensure that each machining surface has enough machining allowance to eliminate errors and defects after machining.

I. Scribing tool

1. Scribing platform

The scribing platform, also known as scribing plate, is made of cast iron blank, which is mainly used as the reference plane for scribing. As shown in Fig. 1-3-1.

Fig. 1-3-1 Scribing Platform

2. Height scribing ruler

Height scribing ruler, also known as height vernier caliper, is a relatively precise measuring tool and scribing tool, which can be used for scribing and measuring height. As shown in Fig. 1-3-2.

Fig. 1-3-2 Height Vernier Caliper

3. Scribing compass

Scribing compass is a tool for drawing circles and arcs, equally dividing line segments, equally dividing angles and measuring dimensions. Its using method is similar to that of compasses in drawing tools. As shown in Fig. 1-3-3.

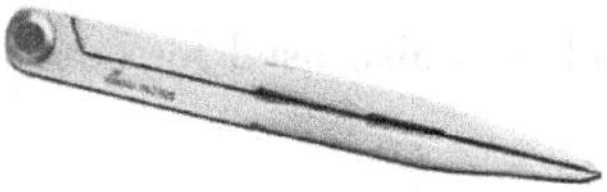

Fig. 1-3-3 Scribing Compass

4. Scriber

It is a tool for connecting lines directly on the workpiece, which is generally made of steel wire or high-speed steel, as shown in Fig. 1-3-4.

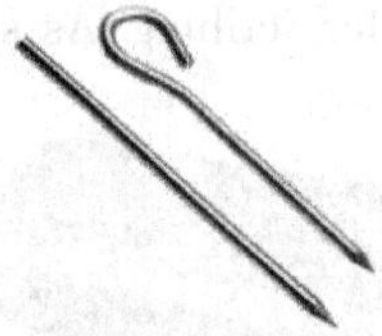

Fig. 1-3-4 Scriber

The pinpoint of the scriber shall be close to the edge of the guide tool. The upper part shall be inclined 15°-20° outward and 45°-75° towards the scribing direction, and shall be drawn at one time, as shown in Fig. 1-3-5. It cannot be repeated. In order to make the lines clear and accurate, the pinpoint shall be kept sharp, and it can be ground with an oilstone after being blunt.

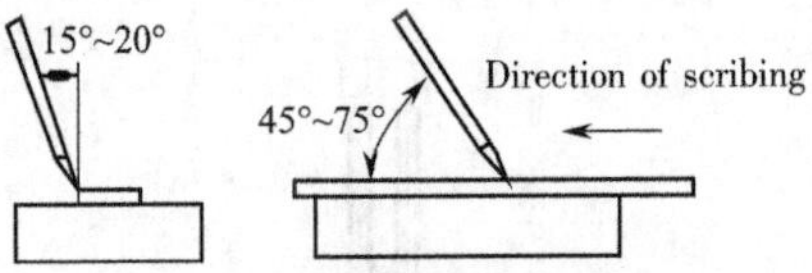

Fig. 1-3-5 Use of Scribing

5. Sample Punch

It is used to punch center points of holes on the machining lines marked on the workpiece as reinforced marks, as shown in Fig. 1-3-6.

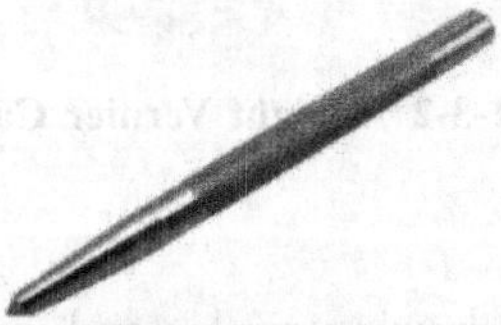

Fig. 1-3-6 Sample Punch

After the workpiece is scribed, a hammer is often used to hammer the sample punch for center points of holes, so as to prevent the scribed lines on the workpiece from being worn away during processing, and it is also used for center positioning when drawing an arc or drilling a hole.

6. Scribing block

It is an adjustable scribing tool with a scriber, which is mainly used for scribing on different planes and correcting the position of the workpiece. It consists of a base, a pole, a scriber and a locking device. Both ends of the scriber are usually divided into straight end and elbow end: the straight end is used for scribing, and the elbow end is used for aligning the position of the workpiece, as shown in Fig. 1-3-7.

Fig. 1-3-7 Scribing Block

7. V-block and cubical box

V-blocks and cubical boxes are tools for supporting, clamping and supporting thin workpieces, as shown in Fig. 1-3-8 and Fig. 1-3-9.

Fig. 1-3-8 V-block

Fig. 1-3-9 Cubical box

8. Jack

Jack is a supporting tool for scribing or detecting workpiece, which is generally used to support workpiece with irregular shapes, protruding parts or heavier parts for workpiece check, alignment and scribing. It is usually used in a group of three. As shown in Fig. 1-3-10.

Fig. 1-3-10 Jack

II. Preparation before plane scribing and the selection of scribing reference

Before scribing, you shall first understand the drawings and process requirements, make clear the scribing tasks, check whether the workpiece are qualified, then clean and color the scribing parts, determine the scribing reference and select the scribing tools.

1. Preparation before scribing

Include cleaning and coloring workpiece or blanks. Commonly used paints include lime water (blank parts) and blue oil (semi-finished parts) to better see the scribed lines.

2. Scribing alignment and material borrowing

Before scribing the blank part, it is generally necessary to place and align it first. The so-called alignment is to use scribing tools (such as scribing block, square, odd leg caliper, etc.) to adjust the supporting tools to make the relevant blank surface on the workpiece in a proper position. For blank workpiece, alignment work shall be done before scribing.

When the dimension, shape or position errors and defects of blanks are difficult to remedy by alignment and scribing, it is necessary to use the method of borrowing materials to solve them. Material borrowing means that the allowance of each surface to be machined can be borrowed from each other and reasonably distributed through trial scribing and adjustment, so as to ensure that each machining surface has enough machining allowance to eliminate errors and defects after machining.

3. Selection of scribing reference

It is the point, line and plane used to determine the position relationship between other points, lines and planes on the workpiece when scribing. It is the reference used in the scribing process. Generally, the design reference will be selected as the reference for scribing.

It is generally selected according to three types:

(1) Based on two mutually perpendicular planes (or straight lines), as shown in Fig. 1-3-11(a):

(2) Based on two mutually perpendicular center lines (or center planes), as shown in Fig. 1-3-11(b):

(3) Based on a plane and a centerline (or plane), as shown in Fig. 1-3-11(c).

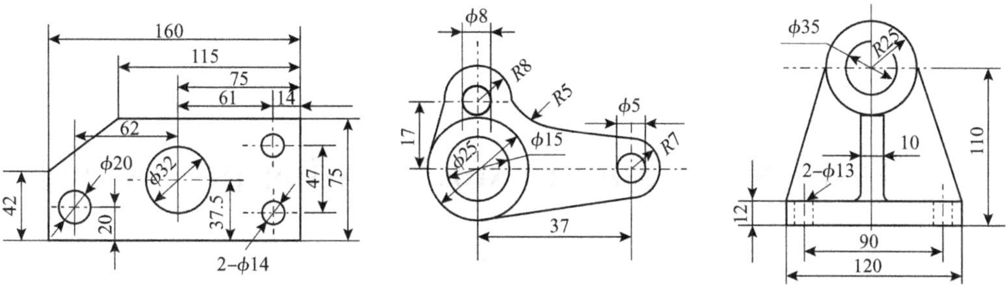

(a) (b) (c)

Fig. 1-3-11 Selection of Scribing Reference

(a) Based on two mutually perpendicular planes (b) Based on two centerlines (c) Based on a plane and a centerline

III. Steps for scribing

(1) Clearly see and understand the drawings, understand the scribing positions, and make clear the scribing requirements and process.

(2) Select a good scribing reference.

(3) Paint the parts on the workpiece that require scribing.

(4) According to the drawing and process requirements, first draw the reference line and position line, then draw the processing line, that is, draw the horizontal line first, then draw the vertical line and diagonal line, and finally draw the circle, arc and curve, and the lines drawn shall be clear and uniform.

(5) Three-dimensional workpiece shall be turned over, placed and scribed in sequence according to the above methods.

(6) After scribing, the workpiece shall be checked item by item from the reference according to the scribing sequence according to the drawings and process requirements, and the mistakes or omissions shall be corrected in time to ensure the accuracy of scribing.

(7) After checking, punch center points of holes on the processing boundary lines. The center points of holes must be corrected, the blank surface shall be properly deeper, the processed surface or sheet parts shall be lighter with thinner density. Finishing surface and soft materials may not be punched.

[Thinking and practice]

(1) Briefly describe the role of scribing.

(2) What are the common tools for scribing ?

(3) What are the two forms of scribing ?

(4) What is the function of borrowing materials ?

(5) Briefly describe the basic steps of scribing.

Task IV Square stock sawing processing

I. Drawings and technical requirements

As shown in Fig. 1-4-1, The blank dimension of the workpiece is 150 mm×25 mm×25 mm and the material is 45 steel. Scoring table see Table 1-4-1.

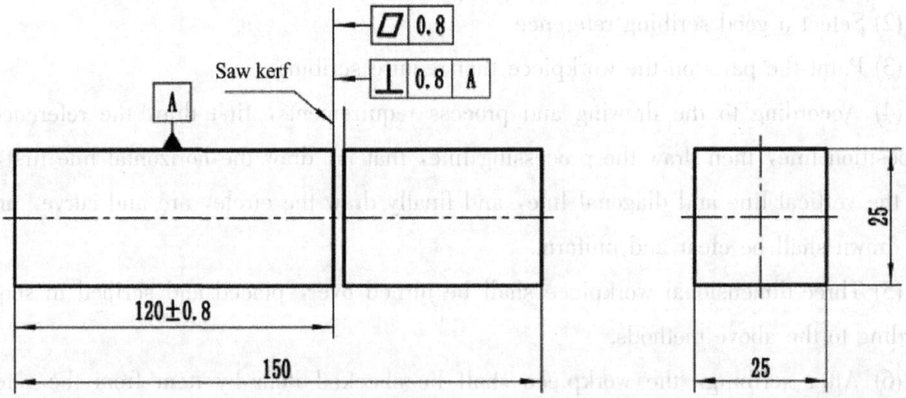

Fig. 1-4-1 Square Stock Sawing

Table 1-4-1 Scoring Table

Technical requirements	Assigned score	Measured results	Score
The flatness of the sawed surface is 0.8 mm	20		
The verticality of the sawed surface is 0.8 mm	30		
Dimension: (120±0.8)mm	20		
Correct posture	15		
Safe and civilized production	15		

II. Analysis of processing

(1) Check the incoming size.

(2) Draw a 120mm processing line according to the drawing requirements.

(3) Saw the bar by sawing method to meet the requirements of dimensional accuracy, flatness and verticality, and ensure that the saw marks are neat and the saw kerf is straight.

III. Knowledge and skills

Sawing is a processing method of cutting off or sawing grooves on materials or workpiece by hand saw or mechanical saw (sawing machine).

1. Sawing tool-hand saw

Hand saw consists of saw bow and saw blade.

(1) The saw bow is a tool for clamping and tensioning saw blades, which has two types: fixed and adjustable. The structure is shown in Fig. 1-4-2.

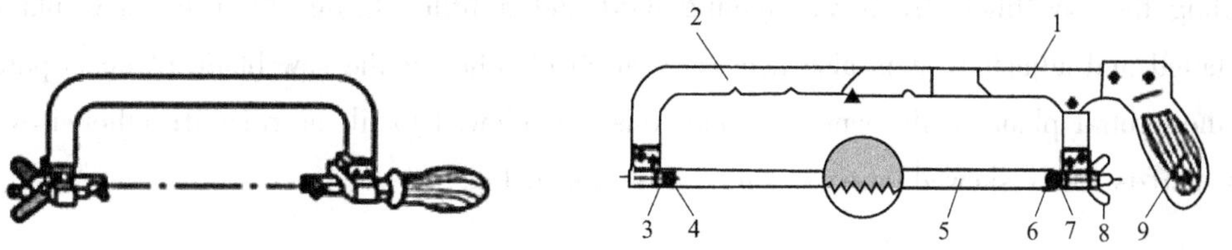

(a) (b)

Fig. 1-4-2 Structure of Hand Saw

(a) Fixed (b) Adjustable

1—Fixed part; 2—Movable part; 3—Fixed clamp; 4, 6—Pin; 5—Saw blade; 7—Movable clamp; 8—Butterfly nut; 9—Handle

(2) The saw blade is a steel strip with a toothed edge, which is the main tool for sawing, as shown in Fig. 1-4-3. Saw blades are generally made of carburizing steel by cold rolling, or carbon tool steel or alloy tool steel by heat treatment and hardening. The length of saw blade is expressed by the center distance of mounting holes at both ends. Commonly used saw blades are about 300 mm long, 12 mm wide and 0.8 mm thick.

The thickness of saw blade teeth is expressed by the number of saw blade teeth per 25mm length of saw blade. Generally, it can be divided into three types: coarse, medium and fine. The coarse tooth saw blade is suitable for sawing soft materials or larger sections, and the fine tooth saw blade is suitable for sawing hard materials or workpiece with smaller sections, which can improve cutting efficiency.

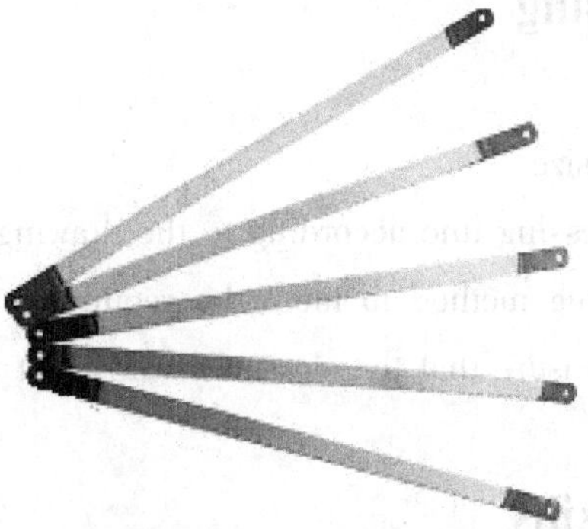

Fig. 1-4-3 Saw Blade

2. Sawing method

1) Installation of saw blade

The hand saw has a sawing effect only when it is pushed forward. Therefore, the saw blade shall be installed in such a way that the saw tooth direction of the saw blade is forward and the saw tooth faces downward. The tightness of the saw blade shall be adjusted properly and the position shall be correct. The degree of tightness can be achieved by pulling the saw blade by hand, feeling hard and a little elastic. After the saw blade is installed and adjusted, it is also necessary to check whether the saw blade plane is parallel to the central plane of the saw bow and it is not allowed to tilt or twist it; otherwise, the saw kerf is easily skewed when sawing, as shown in Fig. 1-4-4.

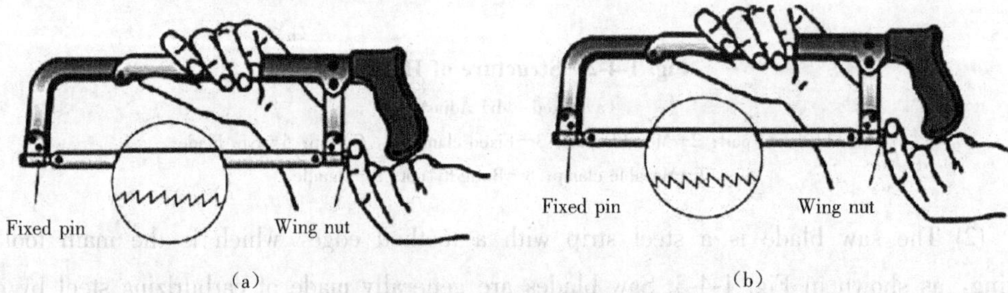

Fig. 1-4-4 Installation of Saw Blade
(a) Correct installation (b) Incorrect installation

2) Clamping of the workpiece

The workpiece shall be clamped on the left side of the vice for operation. The saw kerf of the workpiece is about 15-20 mm away from the vice jaw, and basically parallel to the side of the jaw. The workpiece shall be clamped firmly to prevent the saw blade from breaking due to loosening of the workpiece during sawing, as shown in Fig. 1-4-5.

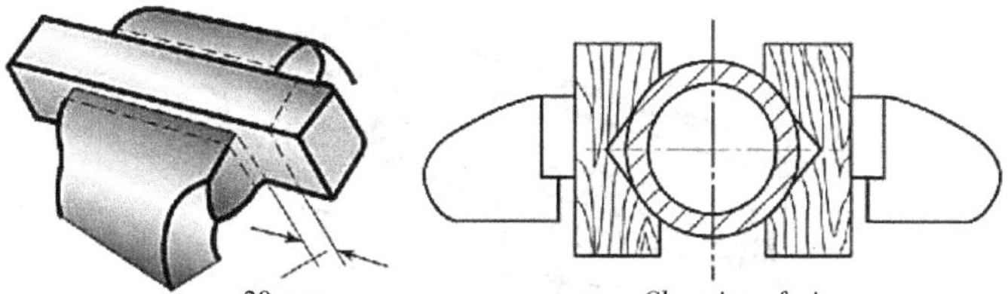

Fig. 1-4-5 Workpiece Clamping

3) Sawing operation posture

In order to give full play to greater strength, the operator must maintain the correct standing position, with his left foot half step ahead, his legs standing naturally, his body's center of gravity slightly inclined to his back foot, and his eyes falling on the cutting part of the workpiece, as shown in Fig. 1-4-6.

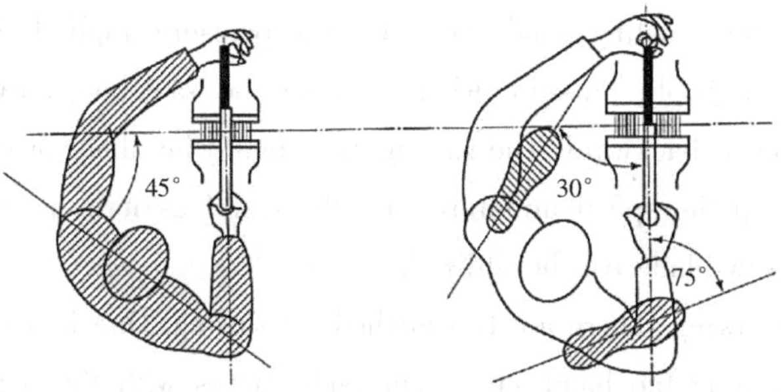

Fig. 1-4-6 Standing Posture during Sawing

4) Sawing method

(1) Start of sawing. There are two methods for start of sawing: starting sawing at the near edge and starting sawing at the far edge, as shown in Fig. 1-4-7. Usually, the start of sawing is from the far edge. Because the saw teeth are not easy to get stuck in this method, when starting sawing, the left thumb shall lean against the saw blade, so that the saw blade can be sawed at the required position correctly, and the sawing travel shall be short, with small pressure and slow speed, as shown in Fig. 1-4-8. The angle of sawing shall be about 15°, whether it is done from the far or near edge. The saw teeth will be stuck by the edge of the workpiece and cause cracking with too large sawing angle and cutting resistance, especially for starting sawing at the near edge. It is not easy to cut into the material, and easy to bring sawing defects and scratch the workpiece with too small sawing angle.

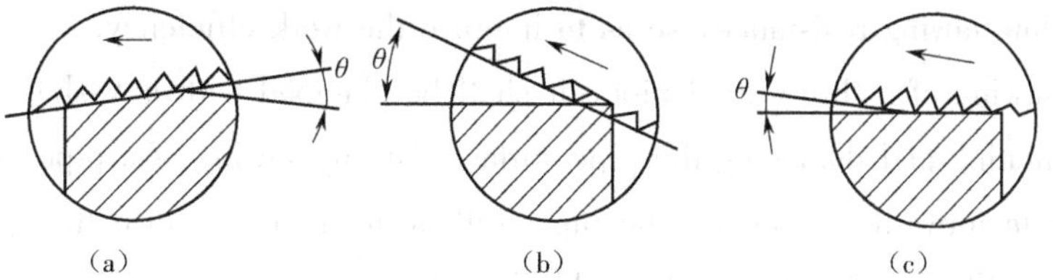

Fig. 1-4-7 Sawing Angle

(a) Start sawing at the far edge (b) A large angle at the start of sawing (c) Start sawing at the near edge

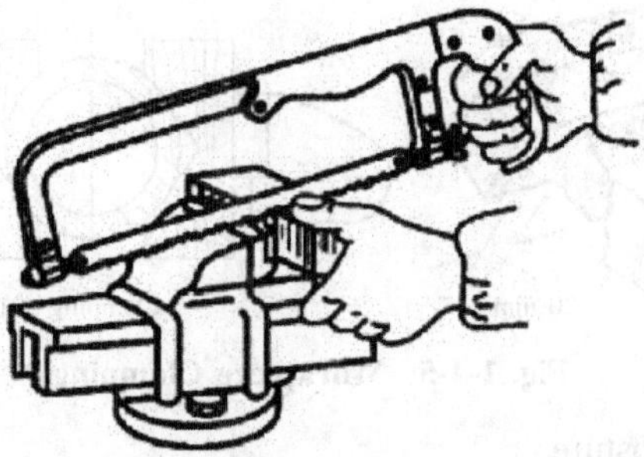

Fig. 1-4-8 Start of Sawing (thumb against the workpiece for guidance)

(2) Normal sawing. During normal sawing, the hand saw holding method is holding the handle with a full grasp by the right hand and holding the front end of the saw bow by the left hand. The right hand controls the push force and the press force of sawing, while the left hand serves as the assisting hand. Note that the pressure applied should not cause the saw blade to deviate. Apply pressure when pushing the saw bow forward, and apply no pressure when pulling it backward. The sawing rate should be about 30-40 times/min. While pushing forward, keep the speed uniform, with the travel as long as possible, so that the entire length of the saw blade can be utilized.

(3) Methods of sawing. There are two methods of sawing. One is a straight reciprocating operation: When pushing the hand saw, your body moves with the hand saw at the same time; when pulling back the saw, your body returns due to the counter-acting force of sawing, and use your both arms to control the saw blade to move in a straight manner. The straight reciprocating operation applies to thin materials and straight saw cuts. When straight saw kerf sections are required, such a method should be adopted. Another is an oscillating operation: When pushing the hand saw, your body inclines slightly forward, and while your both hands press against the hand saw, your right hand presses downward and your left hand tilts up; when returning, lift your right hand up, and your left naturally return as well. In this method, you operate naturally and are not likely to get fatigued, so it is mostly used in sawing.

5) Sawing of different materials

(1) If the sawing surface of the bar is required to be flat, it shall be continuously sawed from the start of sawing to the end. If the requirements for the sawing surface are not high, it is acceptable to skip the deeply sawn kerf of the bar and continue sawing at a position with low sawing resistance, so as to improve the work efficiency.

(2) For sawing of a thin-walled pipe, it shall be clamped with a V-shaped wooden pad to prevent flattening and damaging the pipe surface. During sawing of a pipe, it is necessary to turn an angle forward just before the pipe wall is about to be sawn through; otherwise, the sawteeth are likely to be damaged quickly, as shown in Fig. 1-4-9.

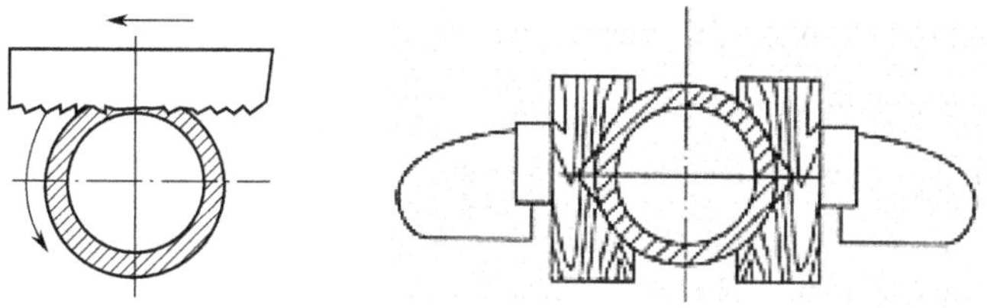

(a) (b)

Fig. 1-4-9 Sawing of Pipe

(a) Clamping of pipe (b) Sawing after turning an angle

(3) During sawing of thin material, a wooden plate is often clamped on the bench vice as a pad, and then the wooden plate is sawed together with the material, or the material is sawed transversely in an oblique pushing manner. Carry out sawing from the wide side of material whenever possible, so that the number of sawteeth sawing at the same time is large, and the sawteeth are not easily hooked and do not easily fall, as shown in Fig. 1-4-10.

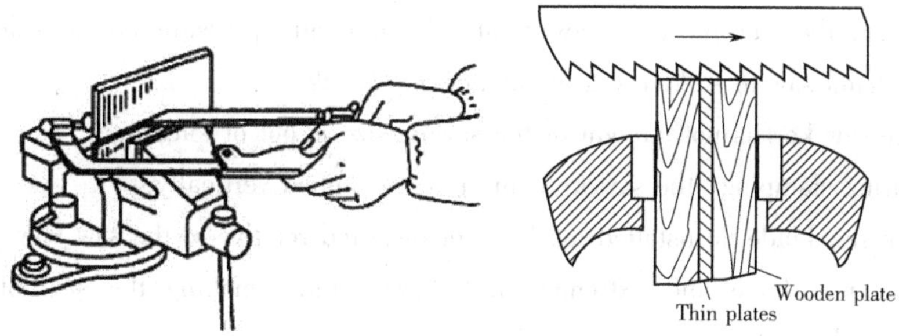

(a) (b)

Fig. 1-4-10 Sawing of Thin Plate

(a) Sawing material together with wooden plate (b) Sawing transversely in an oblique pushing manner

(4) When the saw kerf depth exceeds the saw bow height, the saw blade shall be reinstalled after being turned by 90°, so that the saw bow is turned to one side of the workpiece. When the saw bow' s height is still not enough after it is placed horizontally, sawing can be carried out with the saw blade installed with sawteeth facing toward the inside of the saw, as shown in Fig. 1-4-11.

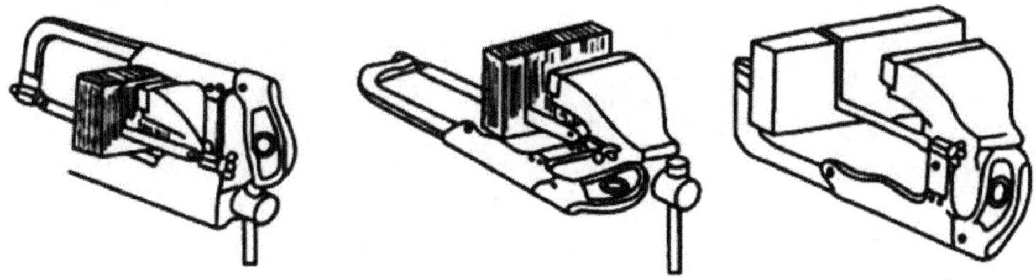

(a) (b) (c)

Fig. 1-4-11 Deep-kerf Sawing

(a) Normal sawing (b) Sawing after turning the saw blade 90° (c) Sawing after turning the saw blade 180°

6) Common problems and solutions during sawing

(1) Causes of sawteeth breaking.

① A large angle or excessive force at the start of sawing;

② Suddenly increased pressure during sawing, causing sawteeth to be hooked by workpiece edges and break;

③ Fine tooth saw blade not being used for sawing thin plates and thin-wall pipes.

(2) Causes of saw blade sticking or breaking.

① Saw blade being installed too tight or too loose;

② Workpiece being clamped not firmly or clamped not in a proper position, causing the workpiece to loosen or shake;

③ Forced correction of a deviated saw kerf;

④ Extremely fast sawing rate and excessive pressure, causing the saw blade to stick;

⑤ The new saw blade is easily stuck when it is still sawing in the old saw kerf;

⑥ When the workpiece is sawed off, the rate and pressure are not slowed down, causing the hand saw to suddenly lose balance and break.

(3) The saw kerf is not straight or the sawing size is out of tolerance.

① During clamping, the saw kerf line is not clamped vertically;

② The saw blade is installed too loose or distorted relative to the saw bow plane;

③ Incorrect force and extremely fast sawing rate, making the saw blade deviate leftward and rightward;

④ Inaccurate dimensional control or deviation of saw kerf at the start of sawing;

⑤ The operator does not observe whether the saw blade coincides with the line during sawing.

[Thinking and practice]

(1) Why is it necessary to adjust the saw blade tension properly ?

(2) Which aspects are of concern during sawing of pipes and thin plates ?

(3) Briefly describe the causes of saw blade breaking and the saw kerf not being straight.

Task V Plane chiseling

I. Drawings and technical requirements

As shown in Fig. 1-5-1, The blank size of the workpiece is 100 mm×80 mm×45 mm and the material is 45 steel. Scoring table see Table 1-5-1.

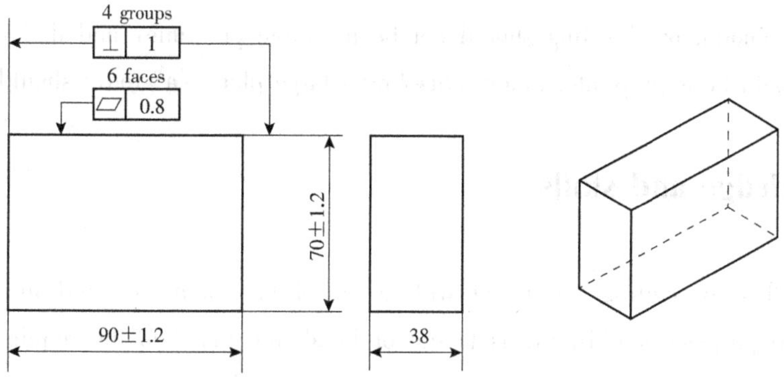

Fig. 1-5-1 Plane Chiseling

Table 1-5-1 Scoring Table

Technical requirements	Assigned score	Measured results	Score
90±1.2 mm	10		
38 mm	10		
70±1.2 mm	10		
Flatness: 0.8 mm	18		
Verticality: 1 mm	24		
Correct posture	16		
Safe and civilized production	12		

II. Analysis of processing

(1) Check the blank size.

(2) Choose a surface as the first machining surface (i.e. reference) according to the blank material. After rough chiseling, fine chiseling shall be carried out to ensure that the chiseling pattern is neat, and the chiseling surface shall be checked with the steel straight edge until the flatness requirement of 0.8 mm is met, which can be used as the reference surface for machining of the hexahedron.

(3) Scribe and chisel as required by drawings. Carry out chiseling according to the dimensions in drawings until the technical requirements are met.

(4) Observe whether the workpiece is clamped tight during chiseling; the height of the workpiece protruding above the vice jaws should be generally 10-15 mm. Meanwhile, wooden cushion blocks are placed below, and soft jaws are added for the bench vice to protect the workpiece.

(5) The amount of chiseling should not be too large per time, and the relief angle of the chisel should be appropriate. Before chiseling a large plane, a groove should be made.

III. Knowledge and skills

The method of cutting the metal with a chisel (stricken by hand force) is called chiseling. The purpose of chiseling is to cut or break the metal to the required shape and size. Chiseling is mainly used in those occasions where machining is not convenient, such as removing flanges and burrs, dividing materials, and making oil grooves.

1. Chiseling tools

The tools commonly used for chiseling are chisels and hammers

1) Chisel

The chisel consists of the chisel head, body, and cutting part, forged with carbon tool steel, heat-treated to reach the hardness of HRC 52-62, and can be used after edge grinding.

(1) Type and purpose of chisels: According to different purposes of chisels, they are generally divided into flat chisels, cross-cut chisels, and oil groove chisels.

① The flat chisel has a long cutting edge with a small arc and a flat cutting surface, as shown in Fig. 1-5-2(a). It is commonly used in plane chiseling, cutting, removing flanges and burrs, and chamfering.

② Cross-cut chisel. The cutting edge of the cross-cut chisel is short, and the two cutting surfaces gradually narrow from the cutting edge to the entire body, and the cutting edge is "cross-shaped" with the entire body width, as shown in Fig. 1-5-2(b). It is commonly used in grooves to divide curved surfaces and plates.

③ Oil groove chisel. The cutting edge of the oil groove chisel is short and the two cutting surfaces are arc-shaped, as shown in Fig. 1-5-2(c). It is mainly used for chiseling oil grooves.

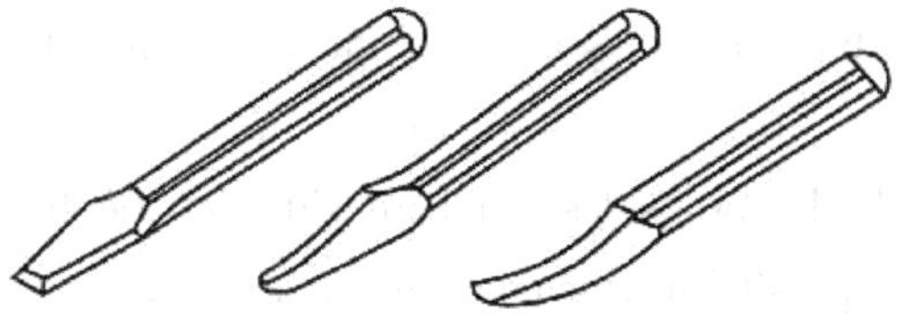

Fig. 1-5-2 Common Chisels

(a) Flat chisel (b) Cross-cut chisel (c) Oil groove chisel

(2) Geometrical angles of chisel. The chisel is generally made of carbon tool steel T7A. The cutting part is ground into a wedge shape and heat-treated to reach the hardness of HRC 52-62. The chisel's geometric angles are shown in Fig. 1-5-3.

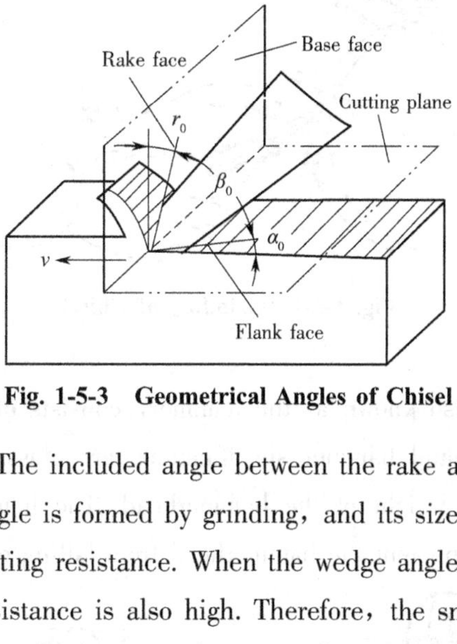

Fig. 1-5-3 Geometrical Angles of Chisel

① Wedge angle (β_0). The included angle between the rake and flank faces is called the wedge angle. The wedge angle is formed by grinding, and its size depends on the strength of the cutting part and the cutting resistance. When the wedge angle is large, the edge strength is high, but the cutting resistance is also high. Therefore, the smaller wedge angle shall be selected as far as possible on the premise of meeting the strength.

② Relief angle (α_0). The included angle between the flank face and the cutting plane is called the relief angle. The relief angle depends on the direction of the chisel being held, and it is used to reduce the friction between the flank face and the cutting plane. When the relief angle is large, the cutting depth is large, causing difficulty in cutting, while an extremely small relief angle is likely to cause the chisel to slip across the workpiece surface. Generally, a relief angle of 5°-8° is appropriate for chiseling.

③ Anterior angle (γ_o). The included angle between the front tool face and the base face is called the anterior angle. The anterior angle has an impact on both cutting force and cuttings deformation. When the anterior angle is large, both cutting force and cuttings deformation are small. Due to γ_o =90° −(β_o+α_o), the anterior angle can be determined after the wedge angle and the relief angle are determined.

(3) Grinding process of chisel

① Grind to smooth the two bevels and pay attention to maintaining the symmetry of the two bevels.

② Grind to smooth the two sides (on both sides of the bevel) and pay attention to maintaining the two sides parallel or symmetric to each other.

③ Grind the edge of the head and pay attention to maintaining the symmetry of the two planes.

During grinding, the chisel shall be dipped in water frequently to cool down to prevent the annealing hardness of the cutting part from reducing, and the height of the chisel shall be slightly higher than the grinding wheel centerline, as shown in Fig. 1-5-4.

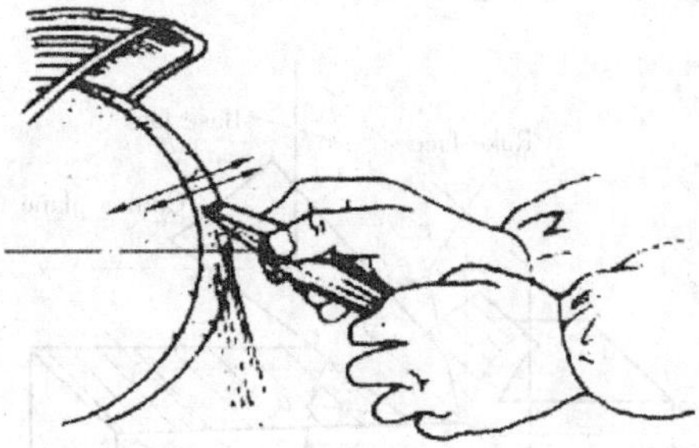

Fig. 1-5-4 Grinding of Chisel

2) Hammer

The hand hammer, also known as the hammer, consists of the hammerhead, wooden handle, and wedge. The hand hammer specifications are classified into 0.25 kg, 0.5 kg, and 1 kg according to the weight of the hammerhead. The hammer hole and the wooden handle are wedged tight to prevent the hammerhead from falling off, as shown in Fig. 1-5-5.

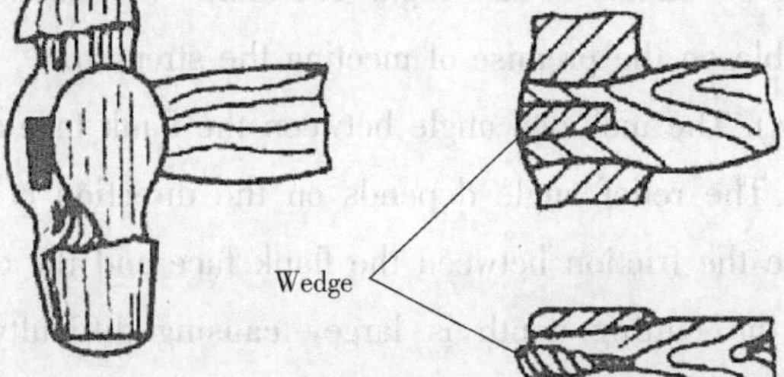

Fig. 1-5-5 Hammer

2. Chiseling method

1) Holding methods of chisel

The chisel is mainly held by the middle finger and ring finger of the left hand, with the little finger folding naturally, the index finger and the thumb are in natural contact, and the chisel head protrudes about 20 mm. Hold the chisel easily, and do not hold it too tight to avoid excessive vibration on the center of the palm during striking, or injuring the hand after the hammer misses the expected point. During chiseling, keep horizontal the forearm of the corresponding hand holding the chisel, without the elbow hanging down or being raised up, as shown in Fig. 1-5-6.

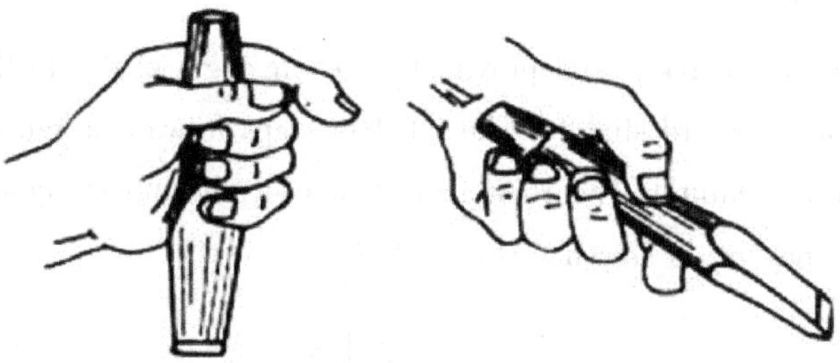

Fig. 1-5-6 Holding Methods of Chisel

(1) Forward grasping method. With the center of the palm downward, straighten the wrist, hold the chisel with the middle finger and the ring finger, with the little finger folded naturally, make the index finger and the thumb naturally straight and in loose contact, and the chisel head protrudes about 20 mm.

(2) Reverse holding method. With the center of the palm upward, the fingers naturally pinch the chisel, and the palm is suspended.

2) Holding methods of hammer

The hammer is generally held in a full grip with five fingers on the right hand, the thumb is gently pressed on the index finger, the part of the hand between the thumb and the index finger is aligned with the hammerhead direction and shall not be skewed to one side, and the tail of the wooden handle is exposed by 15-30 mm. There are two methods to hold the hammer by hand during striking: The tight holding method is to hold the hammer with five fingers from lifting the hammer to striking and keep the holding state unchanged; the loose holding method is to lift the hammer with the little finger, the ring finger and the middle finger loosely holding it, and strike the hammer with those fingers holding it tight, as shown in Fig. 1-5-7.

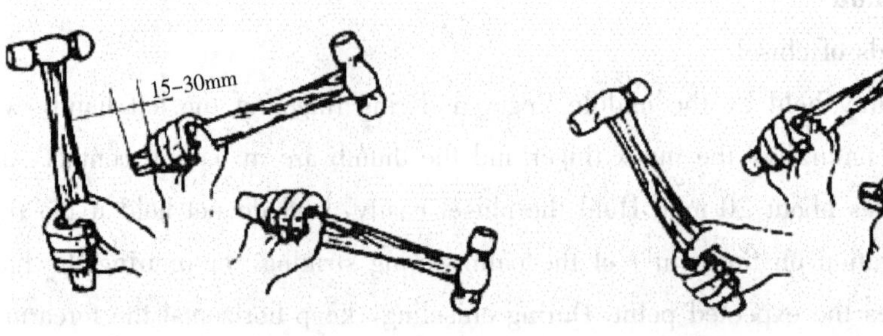

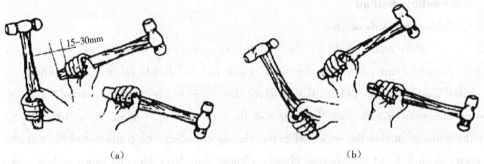

Fig. 1-5-7 Holding Methods of Hammer

(a) Tight holding method (b) Loose holding method

3) Chiseling posture

During chiseling, the body is approximately at an angle of 45° to the centerline of the bench vice, and leans forward slightly. The left foot steps forward a half step, with the knee slightly bent to keep it natural. The right foot stands steady and straight, without excessive force. The chiseling posture is shown in Fig. 1-5-8.

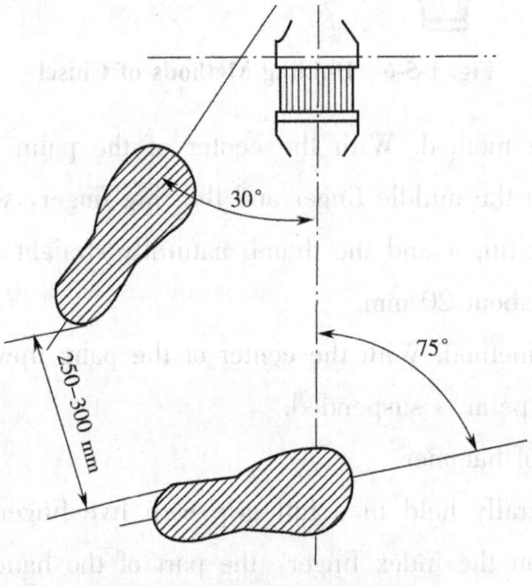

Fig. 1-5-8 Chiseling Posture

4) Methods of hammer swing

There are three methods of hammer swing, i.e. hammer swing by wrist, elbow and arm, as shown in Fig. 1-5-9. Different methods of hammer swing shall be selected according to the chiseling allowance. When the chiseling allowance is small and at the start and end of chiseling, the chiseling force is small, hammer swing by wrist, i.e. holding hammer tight with five fingers, should be selected; when the chiseling allowance is large, hammer swing amplitude is large, and the hammering force is large, hammer swing by elbow, i.e. swinging the hammer with both the wrist and elbow, should be selected; hammer swing by arm is swinging with the wrist, elbow and whole arm, with the maximum hammering force, and is generally used for occasions requiring a lot of hammering force.

(a) (b) (c)

Fig. 1-5-9 Methods of Hammer Swing

(a) Hammer swing by wrist (b) Hammer swing by elbow (c) Hammer swing by arm

5) Chiseling methods

(1) Start chiseling from the edge sharp corner of the workpiece, as shown in Fig. 1-5-10(a), or make the chisel basically perpendicular to the initial chiseling end face of the workpiece, as shown in Fig. 1-5-10(b); then tap the chisel with a hammer to accurately and smoothly start chiseling.

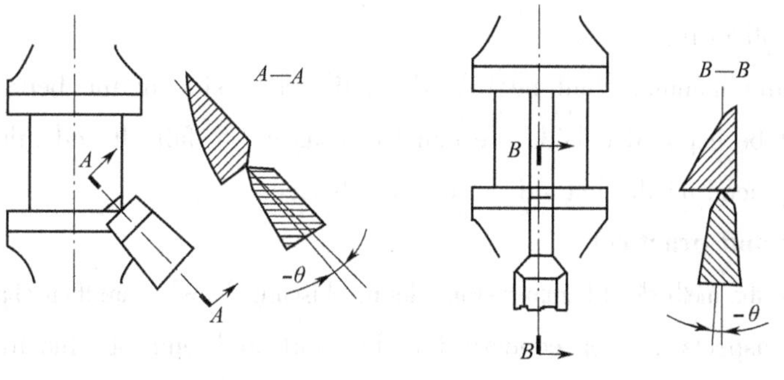

(a) (b)

Fig. 1-5-10 Methods to Start Chiseling

(a) Start chiseling from a bevel (b) Start chiseling from the front side

(2) A chiseling depth of 1 mm is advisable. When the chiseling allowance is greater than 2 mm, the chiseling can be completed in several times.

(3) For the end of chiseling, further chiseling will crack the edge of the workpiece every time it is chiseled to about 10 mm from the edge. Chiseling shall be ended in time, and the remaining part shall be chiseled off from the opposite direction by turning the chisel around, as shown in Fig. 1-5-11.

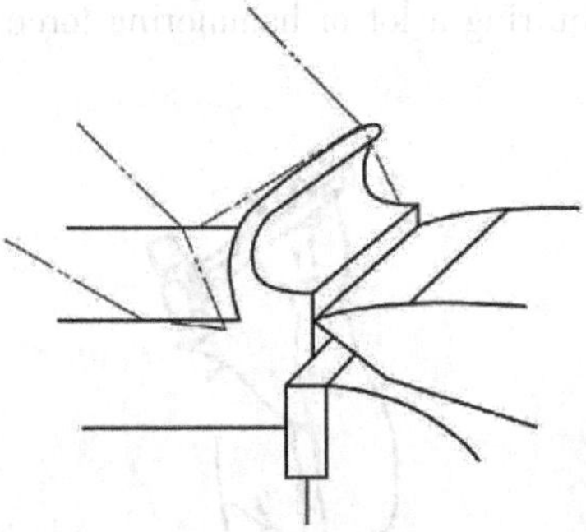

Fig. 1-5-11 Turning around and End of Chiseling

6) Safety precautions for chiseling

(1) The workpiece must be clamped in the middle of the bench vice, and the workpiece should protrude generally 10-15 mm from the vice jaw, and a wooden pad should be placed below.

(2) If the wooden handle of the hammer is found to be loose or damaged, it shall be replaced or fastened immediately. The wooden handle shall be free of oil to avoid slipping out during use.

(3) When there are burrs that can be easily seen obviously on the chisel head, they shall be ground off in time.

(4) The hand hammer shall be placed on the right side of the bench vice, and the handle shall not be exposed outside the bench, so as not to fall off and injure the feet. The chisel shall be placed on the left side of the bench vice.

[Thinking and practice]

(1) What are the methods of hammer swing during chiseling ? What are their characteristics ?

(2) Which aspects are of concern for the start and end of chiseling during plane chiseling ?

(3) What is the content of chisel grinding ?

Task VI Cuboid filing

I. Drawings and technical requirements

As shown in Fig. 1-6-1, The blank size of the workpiece is 120 mm×30 mm×30 mm and the material is 45 steel. Scoring table see Table 1-6-1.

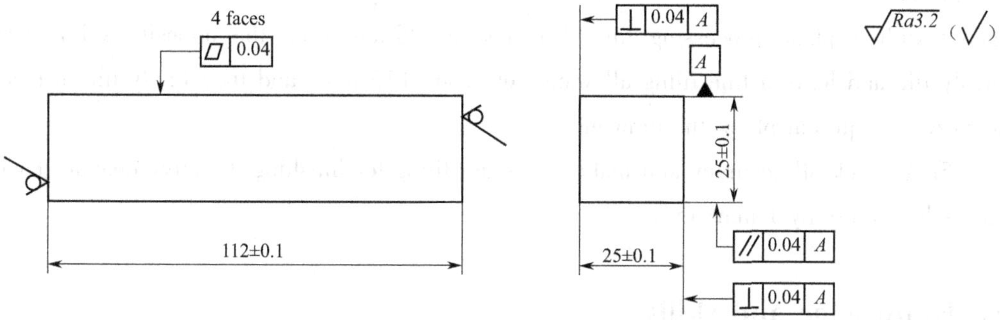

Fig. 1-6-1 Cuboid Filing

Table 1-6-1 Scoring Table

Technical requirements	Assigned score	Measured results	Score
length: 112±0.1mm	12		
width: 25±0.1mm (2 places)	12		
Verticality 0.04 mm (2 places)	8		
Parallelism 0.04 mm (1 place)	4		
Flatness 0.04 mm (4 places)	16		
Surface roughness *Ra*3.2 μm (4 places)	16		
Neat filing pattern and even chamfering (4 places)	16		
Correct filing posture	10		
Safe and civilized production	6		

II. Analysis of processing

(1) Rough and fine filing of reference plane A: Carry out rough filing with the 300 mm thick plate file and carry out fine filing with the 250 mm thin plate file. It is required to meet the requirements of flatness 0.04 mm and surface roughness *Ra*3.2 μm. (The surface roughness shall be checked visually by comparison with the sample block).

(2) The opposite side of the roughly and finely filed reference plane A shall be marked with a plane processing line distanced at 25 mm with a vernier height scale. Rough filing shall be carried out first, leaving a fine filing allowance of about 0.15 mm, and then the carry out fine filing to meet the drawing requirements.

(3) Any adjacent side of the roughly and finely filed reference plane A shall be marked with a plane processing line with a 90° square and a scriber, and then filed to meet the relevant requirements of the drawing (verticality is checked with a 90° square).

(4) The other adjacent side of the roughly and finely filed reference plane A shall be marked with a plane processing line distanced at 25 mm from the opposite side, then roughly file and leave a fine filing allowance of about 0.15 mm, and then finely file to meet the relevant requirements of the drawing.

(5) Recheck all accuracy and make necessary filing for finishing. Finally, chamfer each sharp edge evenly by 1 mm×45°.

III. Knowledge and skills

Filing is a method of cutting the workpiece surface with a file. The filing accuracy can reach 0.01 mm and the surface roughness can reach *Ra*0.8 μm. Filing is widely applied in different occasions: filing of planes, curved surfaces, angular surfaces, grooves, and surfaces with complex shapes.

1. Filing tool

The file is made of carbon tool steel T12 and T13 or T12A and T13A, heat-treated and quenched, and the hardness of the cutting part reaches HRC 62-67. See Fig. 1-6-2 for the structure of the file.

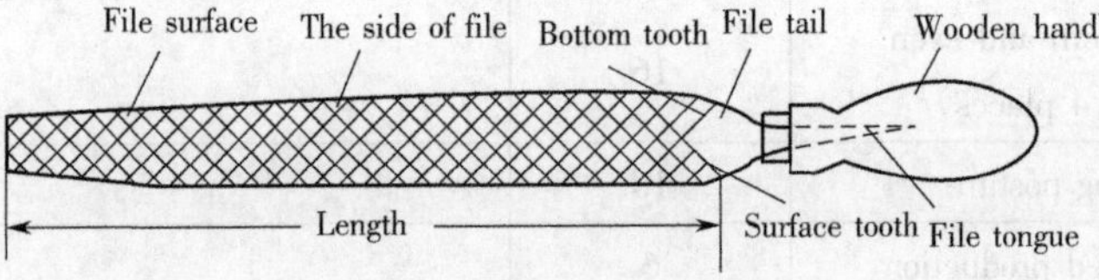

Fig. 1-6-2 Structure of File

1) File teeth and tooth pattern

The file teeth are the teeth on the file surface used for cutting, including milled teeth and chopped teeth. The tooth pattern is the arrangement of file teeth, including single-tooth pattern and double-tooth pattern, as shown in Fig. 1-6-3.

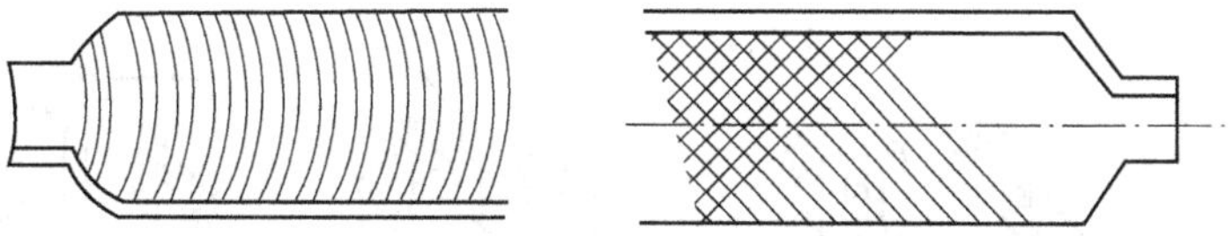

(a) (b)

Fig. 1-6-3 File Tooth Pattern

(a) Single-toothed pattern (b) Double tooth pattern

The file teeth of a single-toothed file are mostly milled teeth, and the teeth are inclined to the axis at a certain angle. The single-toothed file is suitable for filing soft non-ferrous metals. The file teeth of a double-toothed file are mostly chopped teeth: The firstly chopped tooth pattern is the bottom tooth pattern, with a shallow tooth depth, and the subsequently chopped tooth pattern is the surface tooth pattern, with a deeper depth. The surface tooth pattern is cross-arranged over the bottom tooth pattern, playing the role of filings separation and breaking, saving the filing force, and it is suitable for filing hard metals.

2) Types of file

(1) According to the coarseness of the file teeth, the files are divided into coarse files, medium files, finishing files, and dead smooth files. Generally, the number of coarse file teeth is 5.5-14/10 mm, and it is suitable for roughing, as well as the processing of malleable metals; the number of medium file teeth is 8-20/10 mm, and it is suitable for semi-finishing; the number of the finishing file teeth is 11-28/10 mm, and it is suitable for finishing; the number of dead smooth file teeth is 32-56/10 mm, and it is suitable for surface roughness improvement and size refinement.

(2) Depending on the shape of the file profile, files can be divided into plate files (flat or mill files), square files, triangular files, half-round files, and round files.

3) Selection of file

The file shall be selected according to the material of the workpiece and the quality and shape of the workpiece surface. See Fig. 1-6-4 for detailed selection.

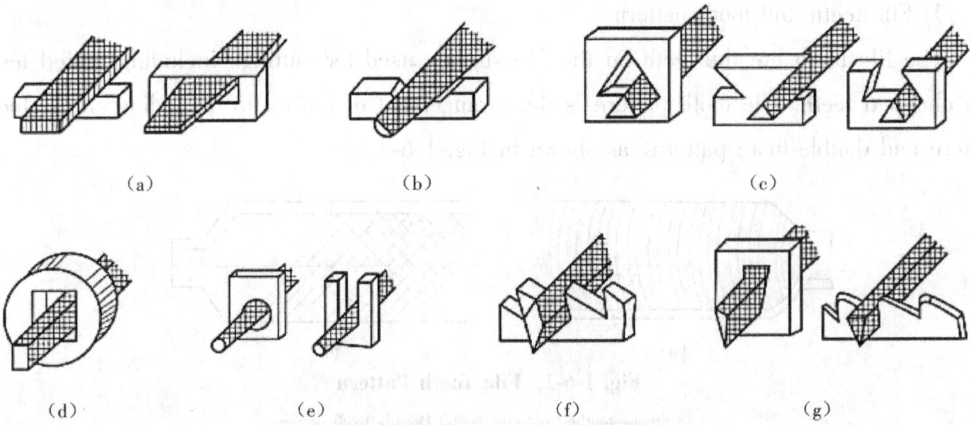

Fig. 1-6-4 Selection of File

(a) Plate file (b) Half-round file (c) Triangular file (d) Square file (e) Round file (f) Diamond file (g) Knife edge file

2. Filing

1) Clamping the workpiece

The workpiece must be firmly clamped in the middle of the bench vice jaws. The surface to be filed should be slightly higher than the jaws, but not too high. When clamping the machined surface, place copper or aluminum sheets between the jaws and the workpiece.

2) Holding method of file

The holding method of the file varies with the size and use of the file. For a large file, hold the file handle with the right hand, with the handle end against the palm at the root of the thumb, the thumb on the upper part of the file handle, and the rest four fingers holding the file handle from bottom to top; press the root muscle of the left hand thumb on the file head, with the thumb naturally straightened and the rest four fingers bent to the center of the palm, and pinch the front end of the file with the middle finger and the ring finger. Push the file with the right hand and determine the direction of pushing. The left hand assist the right hand in keeping the file balanced. See Fig. 1-6-5 for the holding method of a large plate file.

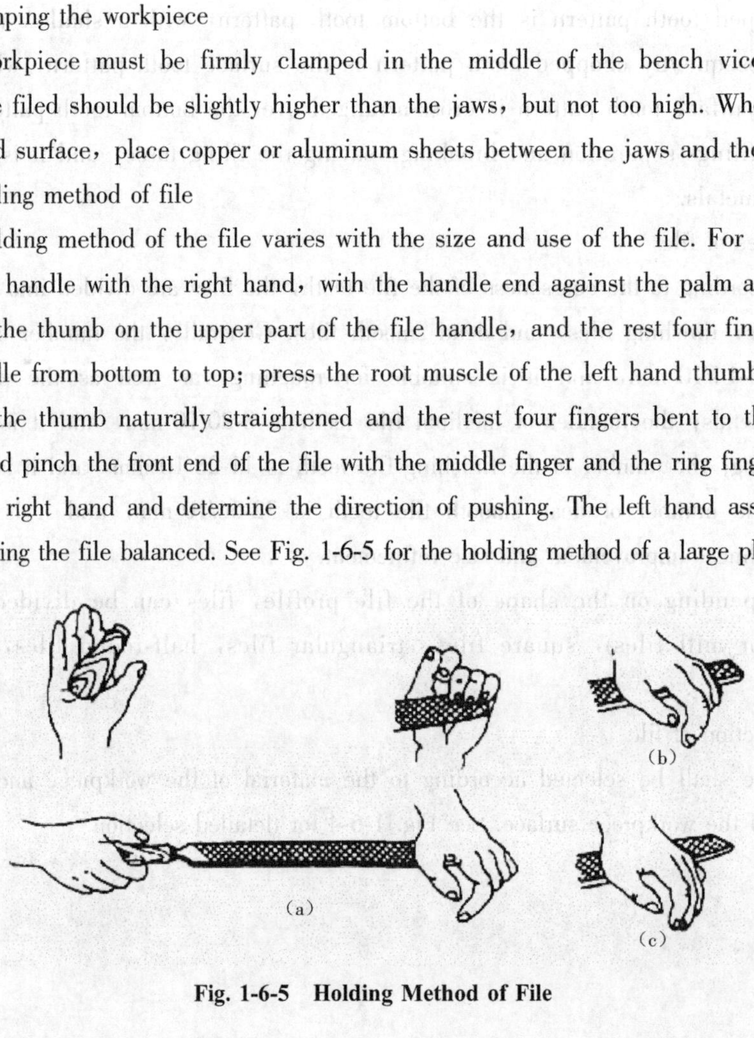

Fig. 1-6-5 Holding Method of File

3) Posture and movement of filing

The posture and movement of filing vary with different filing forces, as the correct filing posture can reduce fatigue and improve the quality and efficiency of filing. During rough filing, the filing force is large, so the posture should facilitate the stability of the body, and the movement should facilitate the application of the pushing force. During fine filing, as the filing force is small, the filing posture should be natural and the movement amplitude should be smaller to ensure the stability of the file movement and make the quality of the filing surface easily controlled, as shown in Fig. 1-6-6.

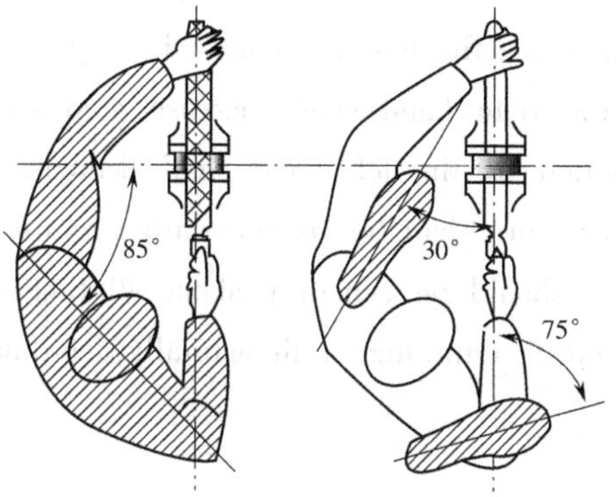

Fig. 1-6-6 Standing Position and Posture during Filing

Stand naturally during filing, the center of gravity of the body shall fall on the left foot, the right knee shall be straightened, and the left knee shall switch between the bent and straight states with the reciprocating motion of the filing. Move the body and arms during the forward movement of the file. At the beginning, the body tilts forward by about 10° and the right elbow contracts backward as far as possible; at the first 1/3 of travel, the body leans forward to about 15°, with the left knee bent slightly; when the filing process is carried to 2/3, the right elbow pushes the file forward and the body tilts to about 18°; when the filing process is carried to the last 1/3 of travel, the right elbow continues to advance and the body returns to about 15° due to the counter-acting force during filing; after the filing travel is completed, the hands and body return to the original posture, and the file is returned horizontally, as shown in Fig. 1-6-7.

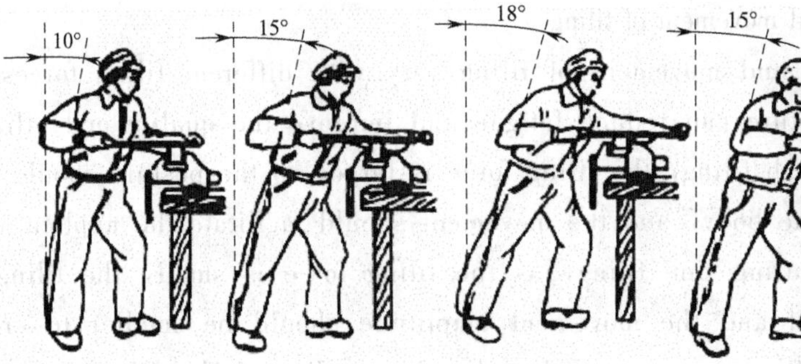

Fig. 1-6-7 Filing Posture

4) Filing force applied by both hands and filing speed

To file a straight plane, the file movement must be kept in a straight line. Therefore, the pressure applied by the right hand shall increase gradually with the file's pushing process, the pressure applied by the left hand shall decrease gradually with the file's pushing process, and no pressure shall be applied during return to reduce the wear of the file teeth. The filing speed should be generally about 40 times/min, slightly slow during pushing and slightly fast during returning, with naturally coordinated movement, as shown in Fig. 1-6-8.

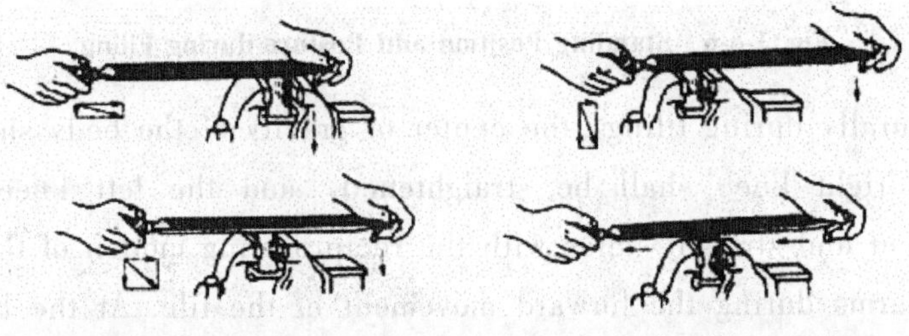

Fig. 1-6-8 Force Applied by Both Hands during Plane Filing

3. Methods of filing

1) Forward filing method

Forward filing is the most common method of filing. The movement direction of the file is always consistent with the clamping direction of the workpiece. This method is often used for planes with a small area. Forward filing can result in neat and consistent filing marks which are relatively beautiful and commonly used in fine filing. See Fig. 1-6-9(a).

2) Cross filing

The movement direction of the file is 30°-40° from the clamping direction of the workpiece, and the first-time filing crosses with the second-time filing. Since the filing marks are crossed, it is easy to judge the unevenness of the filing surface and to file the surface flat. The cross filing method is quick to remove filings and is suitable for rough filing of planes, as shown in Fig. 1-6-9(b).

3) Push–filing method

The file moves perpendicular to the file body. As it is easy to balance during push-filing and the filing amount is small, it is easy to obtain a relatively flat processed surface and a relatively low value of surface roughness. Push-filing is generally suitable for processing narrow and long surfaces. Since the force of the arm cannot be exerted during push-filing, the filing efficiency is low, while push-filing is suitable for surface roughness improvement and size refinement. See Fig. 1-6-9(c).

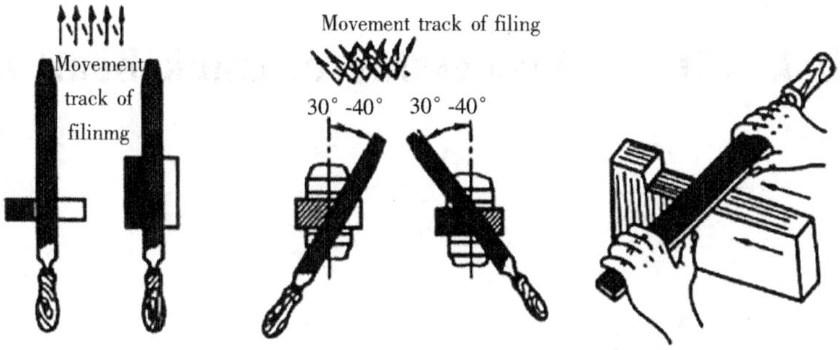

(a) (b) (c)

Fig. 1-6-9 Methods of Plane Filing

(a) Forwarding filing (b) Cross filing (c) Push-filing

4. Precautions of filing

(1) The file must be handled to avoid puncturing the wrist. The loose file handle shall be tightened before use, and the file body or file handle that has cracked or the file with no hoop shall not be used.

(2) Do not blow filings with your mouth or remove them by hand. After the file is blocked, the filings shall be removed with a wire brush along the direction of the file tooth pattern.

(3) Hard leather or sticky sand on castings, flashes or burrs on forgings, etc. shall be ground off with the grinding wheel and then filed.

(4) It is not allowed to touch the filed surface by hand when during filing, as there are oil stains on the hand and it is easy to slip when filing is carried out on the stained surface.

(5) To prevent the file from breaking and hurting people, the file shall not be used to strike the workpiece.

(6) When placing the file, do not expose it out of the workbench surface to prevent the file from falling and hurting the operator's feet; do not stack the file with another file or stack the file with a measuring tool.

[Thinking and practice]

(1) What are the types of files ? To which occasions do they apply ?

(2) How to select the file reasonably ?

(3) Briefly describe the standard posture of filing.

(4) What are the common methods of plane filing? To which occasion does each method apply?

(5) Why not apply pressure in the return process of filing?

(6) What is the approximate filing speed?

Task VII Processing of duck beak hammer

I. Drawings and technical requirements

As shown in Fig. 1-7-1, The blank size of the workpiece is 120 mm×25 mm×25 mm and the material is 45 steel. Scoring table see Table 1-7-1.

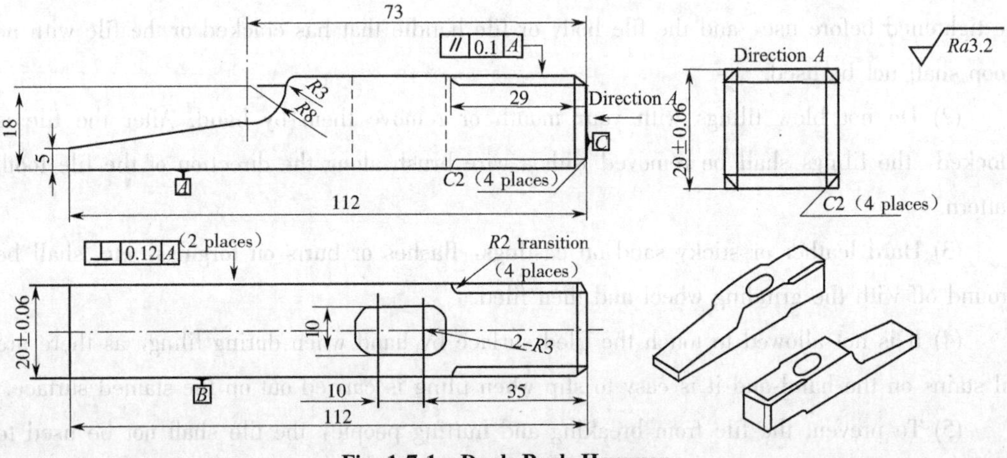

Fig. 1-7-1 Duck Beak Hammer

Table 1-7-1 Scoring Table

Technical requirements	Assigned score	Measured results	Score
Workpiece length: 112 mm	8		
20±0.06 mm (2 places)	10		
Width of slotted hole: 10 mm	10		

Continue Table

Technical requirements	Assigned score	Measured results	Score
Parallelism: 0.1 mm (2 places)	12		
Verticality: 0.12 mm (2 places)	12		
Smooth connection of *R*3 mm and *R*8 mm oblique planes	10		
Straightness of slope: 0.05	10		
*R*3 mm arc smooth connection (4 places)	8		
Chamfer size: *C*2(8 places)	8		
Surface roughness *Ra* 3.2 μm	2		
Safe and civilized production	10		

II. Analysis of processing

(1) Check the incoming material dimensions for compliance with the drawings.

(2) File a 20 mm×20 mm cuboid as required by the drawings.

(3) File one end surface with the long surface as the reference to achieve basic verticality, with surface roughness *Ra*3.2 μm.

(4) Drill ϕ5 mm holes at *R*8 mm according to the drawing, saw off the excess part with a hand saw according to the processing line in the drawing, and leave a filing allowance.

(5) Roughly file the *R*8 mm inner arc with a half-round file according to the line, and roughly file the inclined surface and the *R*3 mm arc with a plate file to the scribed line. Then finely file the inclined surface with a thin plate file, finely file the *R*8 mm inner arc surface with a half-round file, and then finely file the *R*3 mm outer arc surface with a thin plate file. Finally, use a thin plate file and a half-round file for push-filing and finishing to achieve smooth, clean, and neat-patterned connection of all surfaces.

(6) File the 5 mm flat head and ensure that the workpiece is 112 mm long.

(7) Mark the control processing line and the inspection line for drilling according to the drawing, and drill holes with a ϕ9.8 mm drill bit.

(8) File through two holes with a round file and file the slotted hole as required by the drawing.

(9) Mark the processing line based on the long face and the end face. It is required that both sides should be marked at the same time, and a 4-*C*2 chamfer processing line should be marked according to dimensions in the drawing.

(10) File the 4-$C2$ chamfer to meet the requirements: First, roughly file the $R2$ mm arc with a round file, then use thick and thin plate files respectively to roughly and finely file the 2 mm×2 mm chamfer, then finely process the $R2$ mm arc with a round file, and finally straighten the file tooth pattern with the push-filing method and polish the workpiece with an abrasive cloth;

(11) Chamfer the edges of the four corners by $C2$.

(12) Smooth each processing surface with an abrasive cloth and submit the workpiece for inspection.

(13) After it passes the dimensional inspection, process the slotted hole into an arc bell mouth, file the 20 mm×20 mm end face into a slightly convex arc surface, and finally harden both ends of the workpiece.

III. Knowledge and skills

Drilling is a method of processing holes on solid materials with a drill bit. When drilling on the drilling machine, the rotation of the bit is the main motion, and the axial movement of the bit is the feed motion.

1. Twist drill

A standard twist drill is a common tool for drilling and is generally made of high-speed steel.

1) Structure of drill bit

The twist drill consists of the shank, neck, and working part. The cutting part plays the main cutting role in drilling. The guiding part refers to the part between the cutting part and the neck, and it plays a guiding role during drilling and also plays the role of chip removal and smoothing the hole wall. There are two types of twist drill shanks: straight shank and taper shank. Generally, the straight shank is for drills with a diameter less than 13 mm, and the taper shank is for drills with a diameter greater than 13 mm. The specifications, materials and trademark of the bit are engraved in the neck. See Fig. 1-7-2.

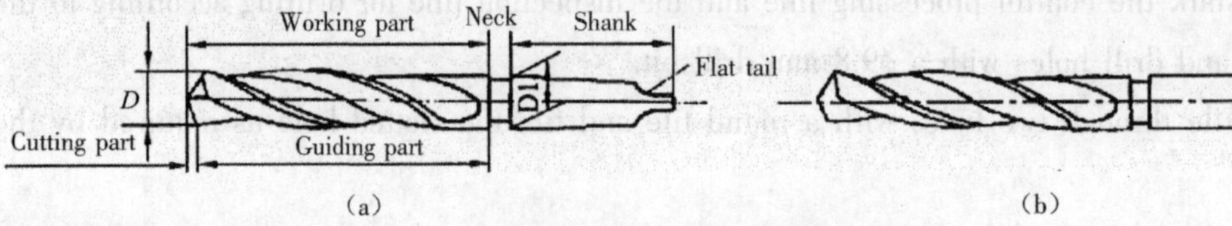

Fig. 1-7-2 Twist Drill

(a) Taper shank (b) Straight shank

2) Cutting angle of twist drill (see Fig. 1-7-3)

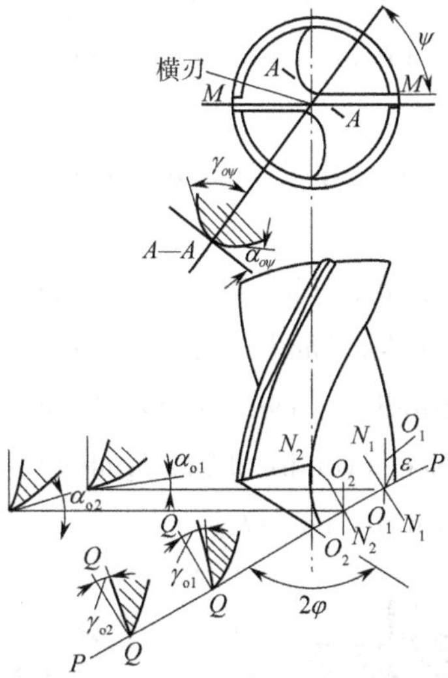

Fig. 1-7-3 Cutting Angle of Standard Twist Drill

(1) Vertex angle 2φ. The vertex angle is generally 100°-140°, and the vertex angle of a standard twist drill is 118°±2°.

(2) Anterior angle γ_0. The anterior angle affects the sharpness and strength of the cutting edge. The anterior angle of each point on the main cutting edge of the twist drill is different, the maximum at the outer edge is about 30°, and the angle gets smaller with the distance smaller to the center; it becomes a negative value within the $D/3$ range of the center, and the minimum near the chisel edge is about −30°. The larger the helix angle is, the larger the anterior angle will be, and less cutting force will be required.

(3) Relief angle α_0. In consideration of the relief angle of the workpiece, the closer the position is to the center, the greater amount of grinding will be required, and the smaller the relief angle will be; the larger the friction between the flank face of the drill bit and the cutting surface of the workpiece is, the higher strength of the cutting edge will be required. Therefore, the relief angle for drilling of hard material should be appropriately smaller, and the relief angle for drilling of soft material should be slightly larger. For the convenience of measurement, the relief angle is measured in a cylinder.

(4) Chisel edge angle ψ. The chisel edge angle of a standard twist drill is 50°-55°.

(5) Edge. The edge is the angled relief of the twist drill. In order to guide and reduce friction, the end relief angle of the edge is zero but with a slight back taper.

3) Grinding of a standard twist drill

Due to the many disadvantages of the standard twist drill, its cutting performance is usually improved by grinding. Generally, the drill bit is selectively ground according to the specific requirements of drilling. Generally, it is only necessary to grind the two flank faces of the twist drill, and simultaneously grind the vertex angle, relief angle, and chisel edge angle. The technical requirements for grinding are high. The operating steps are as follows:

(1) Check whether the grinding wheel surface is flat, and correct the grinding wheel in case of unevenness or runout.

(2) Hold the front end of the drill bit with the right hand as the supporting point, hold the shank of the drill bit with the left hand, place flat the main cutting edge of the drill bit above the central plane of the grinding wheel, so that the included angle between the axis of the drill bit and the grinding wheel cylindrical bus is about 1/2 of the vertex angle, and the drill tail tilts downward, as shown in Fig. 1-7-4.

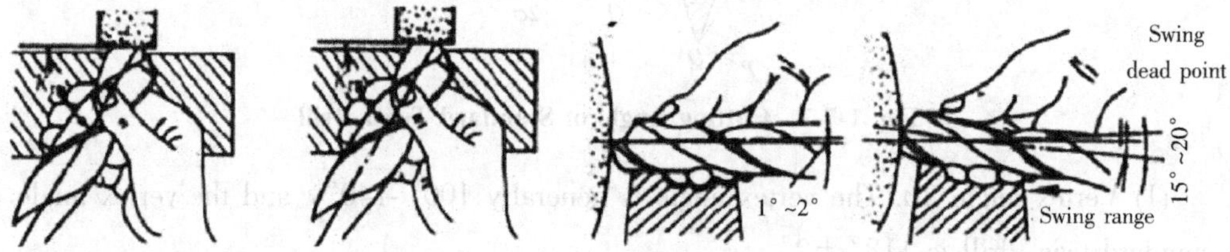

Fig. 1-7-4 Grinding of Twist Drill

(3) During grinding, take the front supporting point of the drill bit as the center of the circle, hold the shank with the left hand to slowly swing up and down and rotate it slightly, and at the same time grind out the main cutting edge and the flank face. When swinging upward, do not exceed the horizontal line, when swinging downward, ensure the amplitude of swing is appropriate to prevent wearing off the other main cutting edge, as shown in Fig. 1-7-5.

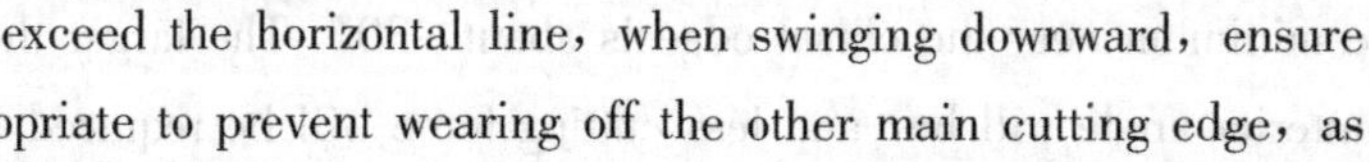

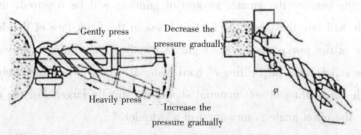

Fig. 1-7-5 Grinding of Main Cutting Edge and Flank Face of Twist Drill

(4) Turn the bit 180° and grind the other main cutting edge and the flank face in the same way. The two cutting edges shall be frequently ground alternately until the requirements are met.

(5) Grind the chisel edge as required. The chisel edge shall be shortened so that the anterior angle at the core bit becomes large.

(6) After grinding of the twist drill, it is usually subject to visual inspection. Vertically erect the drill bit at a position equal to the eye height, and visually check the two edges in a bright background; turn the drill bit 180° and then observe and compare repeatedly until the two edges are basically symmetrical. If any deviation is found during use, grinding is required again.

(7) The sharpened twist drill can be checked with an angle meter. Put one side of the angle meter against the edge of the twist drill and the other side on the edge of the twist drill to measure its edge length and angle; then turn the twist drill 180° and check the other main cutting edge in the same way, as shown in Fig. 1-7-6.

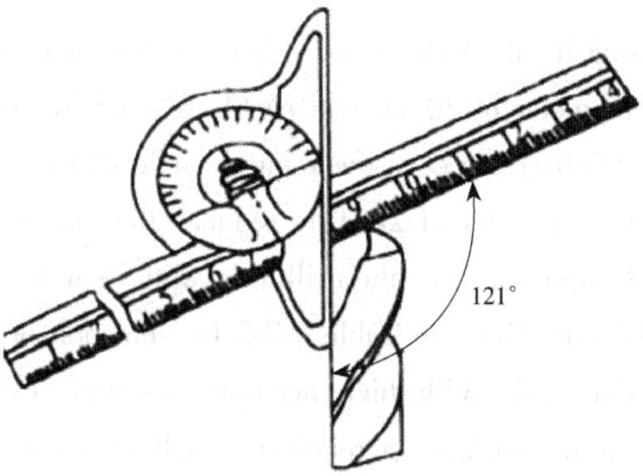

Fig. 1-7-6 Inspection of Main Cutting Edge of Twist Drill

4) Selection of twist drill

The diameter selection of the twist drill is based on two aspects:

(1) For hole processing with low accuracy requirements, use the twist drill directly for drilling (with a diameter equal to the hole diameter).

(2) For hole processing with higher accuracy requirements, sufficient allowance for bearizing and reaming must be reserved. As for the length of the twist drill, the guide part is slightly longer than the hole depth in general.

2. Common drilling machines

(1) Common drilling machines include bench drilling machines, vertical drilling machines, and radial drilling machines.

(2) Precautions for drilling machine operation.

① Strictly abide by the operating procedure of the drilling machine, and do not operate with gloves or cotton yarn.

② When testing is required during drilling, be sure to stop drilling for testing.

③ During operating of the drilling machine, operators wearing long hair shall wear a work cap; the workpiece, fixture, and tool must be clamped firmly and securely.

④ It is strictly prohibited to touch iron chips by hand or blow iron chips by mouth.

⑤ When drilling deep holes or cast iron workpieces, the cuttings shall be removed frequently; when drilling through holes, the bottom of the workpiece shall be cushioned with a backing plate or equally high sizing blocks to avoid damage to the workbench or flat-nose pliers.

3. Selection of cutting amount and cutting fluid

1) Selection of cutting amount

It is the general term of cutting speed, back cutting depth, and feed rate. Reasonable selection of the cutting amount can improve drilling accuracy and production efficiency, and prevent machine tool overload or damage. The selection method of the cutting amount is as follows:

(1) Selection of cutting depth. Holes with a diameter less than 30 mm can be drilled in one time; holes with a diameter of 30~80 mm can be drilled in two times. The toe hole is first drilled with a bit (0.5~0.7) *D* (*D* is the required hole diameter), and then the hole is enlarged with a bit with a diameter of *D*. This reduces the cutting depth and axial force, protects the machine tool, and improves the drilling quality as well.

(2) Selection of feed rate. Refer to Table 1-7-2 for selection of the feed rate of a high-speed steel standard twist drill. With high accuracy requirements for the hole and low roughness requirements for the surface, a relatively small feed rate shall be adopted; when the hole is deep and the bit is long, with poor rigidity and strength, a relatively small feed rate shall also be adopted.

Table 1-7-2 Feed Rate of High-speed Steel Standard Twist Drill

Diameter of drill bit /mm	<3	>3-6	>6-12	>12-25	>25
Feed rate/ (mm/r)	0.025-0.05	>0.05-0.10	>0.10-0.18	>0.18-0.38	>0.38-0.62

(3) Selection of drilling speed. When the diameter and feed rate of the drill bit are determined, the drilling speed shall be reasonably determined according to the service life of the drill bit. This is generally based on experience. See Table 1-7-3 for reference. When the hole depth is large, a relatively small cutting speed shall be chosen.

Table 1-7-3 Cutting Speed of High-speed Steel Drill Bit

Workpiece material	Cutting speed/ (m/min)
Cast iron	14-22
Carbon steel	16-24
Brass or bronze	30-60

2) Selection of cutting fluid

In order to facilitate heat dissipation and cooling of the drill bit, reduce the friction between the drill bit and the workpiece and cuttings during drilling, and eliminate the built-up edge attached to the surface of the drill bit and workpiece, thus reducing the cutting resistance, improving the service life of the drill bit and improving the quality of the surface of the machined hole, sufficient cutting fluid shall be filled during drilling. For drilling holes on steel parts, 3%-5% emulsion may be used; for drilling holes on cast iron, it is generally accepted that no emulsion is filled or the 5%-8% emulsion is continuously filled. See Table 1-7-4 for cutting fluids selected for drilling of various materials.

Table 1-7-4 Cutting Fluids Selected for Drilling of Various Materials

Workpiece material	Cutting fluids
Steels of various structures	3%-5% emulsion; 7% vulcanized emulsion
Stainless steel, heat resistant steel	3% soap plus 2% linseed oil water solution; vulcanized cutting oil
Pure copper, brass, bronze	5%-8% emulsion
Cast iron	None; 5%-8% emulsion; kerosene
Aluminum alloy	None; 5%-8% emulsion; kerosene; mixture of kerosene and rapeseed oil
Organic glass	5%-8% emulsion; kerosene

4. Method of drilling

1) Scribing the workpiece

Scribe the centerline of the hole according to the position and dimensional requirements of the hole, and punch the center point of hole. Mark the circumferential line of the hole according to the hole size. For larger holes, several checking circles and checking boxes of different sizes shall also be marked to facilitate checking and correcting the drilling position during drilling, as shown in Fig. 1-7-7.

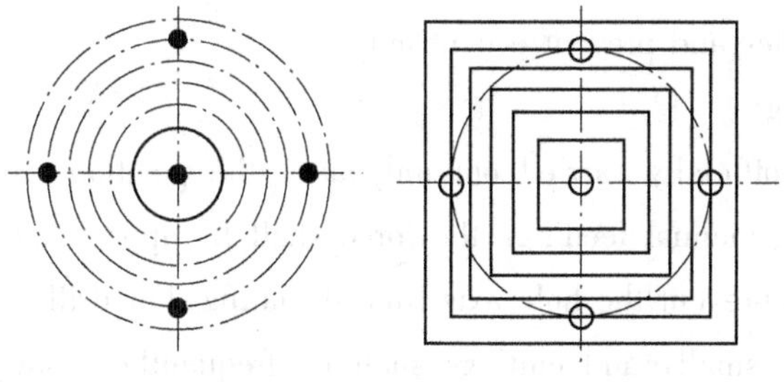

(a) (b)

Fig. 1-7-7 Checking Main Cutting Edge of Twist Drill

(a) Checking circles (b) Checking boxes

2) Bit clamping

(1) Clamping of the straight-shank drill bit: Clamp the straight-shank drill bit with a drill chuck. First, extend the shank of the drill bit (not less than 15 mm) into the three jaws

of the drill chuck, then rotate the jacket with the drill key, so that the ring nut drives the three jaws to move, making clamping and loosening movements, as shown in Fig. 1-7-8(a).

(2) Clamping of taper-shank drill bit: The shank of the taper-shank drill bit is directly connected to the taper hole of the spindle of the drilling machine; the taper hole and the taper shank shall be wiped clean; the flat tail of the drill bit shank shall be aligned with the slotted hole on the spindle, and clamping shall be carried out in one time using the impact force, as shown in Fig. 1-7-8(c). When the drill bit taper shank is smaller than the spindle taper hole, a transition sleeve connection is required, as shown in Fig. 1-7-8(b). See Fig. 1-7-8(d) for the tilting pad iron for removal of the drill bit on the taper sleeve and the spindle of the drilling machine.

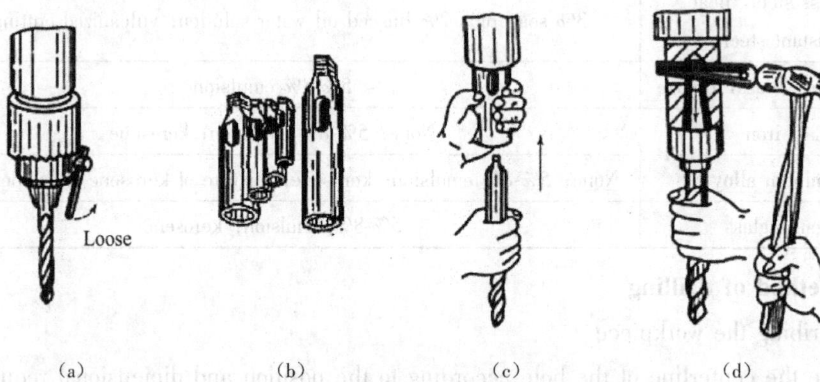

Fig. 1-7-8 Bit Clamping

(a) Install bit to drill chuck (b) Clamp bit with tapered sleeve (c) Tapered sleeve (d) Remove bit with tilting pad iron

3) Start of drilling

At the start of drilling, align the bit with the center hole to drill a shallow pit, observe whether the centering is accurate and continuously correct to make the shallow pit concentric with the checking circles and prevent eccentricity.

4) Manual feeding

Drilling can be officially carried out only after the position required for the start of drilling is met. During manual feeding, the force shall be appropriate so as not to bend or deform the drill bit to prevent the hole axis from deviating. For drilling small or deep holes, the feed rate shall be small, and cuttings shall be frequently removed to prevent cuttings from blocking and breaking the drill bit. When the hole is about to be drilled through, the feed force must be reduced to avoid accidents caused by a sudden excessive feed rate, breaking the bit or causing the workpiece to rotate with the bit.

5. Precautions for drilling

(1) When the hole is not required to be drilled through, the stop block, depth scale, or other control measures on the drilling machine shall be adjusted according to the drilling

depth to avoid excessive or shallow drilling, and cuttings shall be removed.

(2) When the hole is required to be drilled through, and when the hole is about to be drilled through, the feed force shall be reduced to avoid over-cutting, affecting the machining quality, breaking the twist drill or causing the workpiece to rotate with the twist drill, thus leading to accidents.

(3) When drilling deep holes, when the drilling depth reaches 3 times the diameter of the twist drill, the bit shall be withdrawn for cuttings removal; cooling and lubrication shall be carried out.

(4) For drilling small holes less than 1 mm, the cutting speed can be selected at 2000~3000 r/min and above, the feed force is small and stable and should not be too large, and the feed rate should not be too high, so as to prevent bending and sliding of the twist drill; the twist drill should be frequently withdrawn for cuttings removal, and the cutting fluid should be filled.

(5) For large holes over 30 mm, drilling is generally carried out in two times: In the first time, use the twist drill with 0.6-0.8 times the required diameter; in the second time, use the twist drill with the required diameter.

(6) When drilling holes on a bevel, initially drill a toe hole with a center drill, or mill a small plane at the drilling position with a milling cutter, or use a drill bushing for guidance.

6. Precautions for processing

On the basis of mastering the method of drilling, attention shall also be paid to the following processing links:

1) Sawing and filing a cuboid

Selection principles of filing reference:

(1) Select the processed largest flat surface as the filing reference.

(2) Select the surface with good quality as the filing reference.

(3) Select the scribing reference and measurement reference as the filing reference.

(4) Select the surface with the highest processing accuracy as the file reference.

The filing of each surface of the cuboid workpiece must be carried out in a certain sequence to quickly and accurately meet the requirements of dimensions and relative position accuracy. The general principles are as follows:

(1) Select the plane with the maximum area and relatively good surface quality as the reference plane for filing to meet the specified flatness requirement.

(2) File large planes at first and then small planes to control small planes with reference to large planes. In addition, compared with the filing of small planes, it is easier to control the filing of large planes, and it is more difficult to file small planes than large

planes.

(3) File the parallel planes at first and then vertical planes, i.e. ensure the perpendicularity of the relevant planes after the specified parallelism requirement is met. On the one hand, this is convenient for dimensional control; on the other hand, while the verticality is guaranteed, the measurement and comparison of parallelism and verticality errors can be carried out to reduce the accumulated errors.

2) Processing inclined planes and arcs

Stereoscopic scribing:

(1) The workpiece has obvious stereoscopic scribing characteristics.

(2) Stereoscopic scribing is to scribe on different surfaces (usually perpendicular to each other) of the part.

(3) Prerequisites for stereoscopic scribing: There shall be three or more scribing references; most sides of stereoscopic scribing are perpendicular to each other.

Selection of half-round file:

The half-round file has different specifications, and the selection principles are similar to those of a flat file, and the appropriate half-round file is selected according to the arc size.

3) Processing of slotted hole

During scribing, the centerline of the circle shall be marked with a height vernier caliper and the center point of hole shall be punched.

During drilling, a shallow pit shall be drilled with a $\phi 2.5$ drill bit aligned with the center drill to observe whether the centering is accurate, and if not, it shall be corrected. The purpose is to make the shallow pit concentric with the checking circles, and then a shallow pit shall be drilled with a $\phi 10$ drill bit aligned with the pilot hole to observe whether the centering is accurate, and if not, it shall be corrected. Drilling can be officially carried out by pulling the handle only after the position required for the start of drilling is met.

Then the filing process shall be carried out. The two tangent circles shall be filed through with a round file. During filing, the two semi-circles shall be protected to prevent the round file from damaging the two semi-circles that have been drilled. Then the raised parts shall be filed off with a square file, and the front and rear eight points of tangency of the slotted hole shall be reserved to ensure the smooth connection between the arc and the plane. Thus, the processing of the slotted hole is carried out.

4) Chamfering

(1) Scribing. Mark the processing limit of the 29 mm chamfer on four large surfaces with a height vernier caliper.

(2) Filing *R*2 mm arc. Before processing the longer edges on the cuboid, clamp the opposite two edges at an angle of 45°, and process the *R*2 mm arc with a round file, with

the starting point of the arc at the processing limit of 29 mm. Process the four edges into four $R2$ mm arcs.

(3) Filing plane. On the four edges of the large surface of the cuboid, file the edges of the narrow plane along the direction of the vertical edges with a flat file, and this plane is smoothly connected to the processed $R2$ mm above. For chamfering of the four edges, the same processing technology shall be adopted.

(4) Filing four edges of reference plane *C*. Clamp the workpiece onto the bench vice, and make sure that the filing edge is at an included angle of 45° during clamping. Use a flat file to file the machined narrow plane along the direction of the vertical edge, and the width of the plane is equal to the width of the four previously filed edges.

(5) Filing four vertexes. Adjust the clamping position of the workpiece, clamp the filing plane of the four vertexes of the workpiece reference plane *C* to the angle required for filing, and file four regular triangles with a flat file.

[Thinking and practice]

(1) Briefly describe the structure of the twist drill.

(2) How to select the cutting amount during drilling ?

(3) What are the common clamping methods for workpieces during drilling ?

Item 2

Turning

[Item description]

Common turning is one of the most widely used methods in mechanical cutting, and is the main method for processing shafts, discs, and sleeves. The internal and external surfaces of various cursors can be processed by turning, such as internal and external cylindrical surfaces, taper surfaces, grooves, and turning threads.

[Learning objective]

(1) Understand the basic knowledge of turning and the process characteristics and scope of turning.

(2) Preliminarily understand the model, structure, and transmission system of the lathe, be familiar with the components and functions of the lathe, master the main operating methods of the horizontal lathe and be able to operate correctly.

(3) Get familiar with the general structure, characteristics, clamping methods, and applications of common accessories for lathes, and master the characteristics of clamping methods for shafts.

(4) Master the components, installation, and edge grinding of ordinary turning tools, understand the main angles and functions of turning tools, and be able to use common turning tools and measuring tools correctly.

(5) Master method of radial turning, cylindrical turning, taper face turning, thread turning and groove turning, and measuring method, be familiar with the range of dimensional accuracy and surface roughness that can be achieved by turning, be able to independently process general parts, and have certain operating skills and turning process knowledge.

Task I General knowledge of lathe work

I. Function and typical job contents of lathe work

Turning is the most basic cutting method in machining, which can be used to complete many machining work. Generally, the rotating surfaces with tolerance grade below IT8 and surface roughness *Ra* value above 1.6 μ m can be machined on lathe.

Lathe movement includes workpiece rotating and tool movement. The rotating movement of the workpiece is the main motion, and the movement of the tool is called feed motion. The main work contents on ordinary lathe include: cylindrical turning, radial turning, grooving and cutting, drilling of center hole, drilling, boring, reaming, internal and external taper faces turning, various threads turning , form surface turning, knurling and spring winding, etc. (as shown in Fig. 2-1-1).

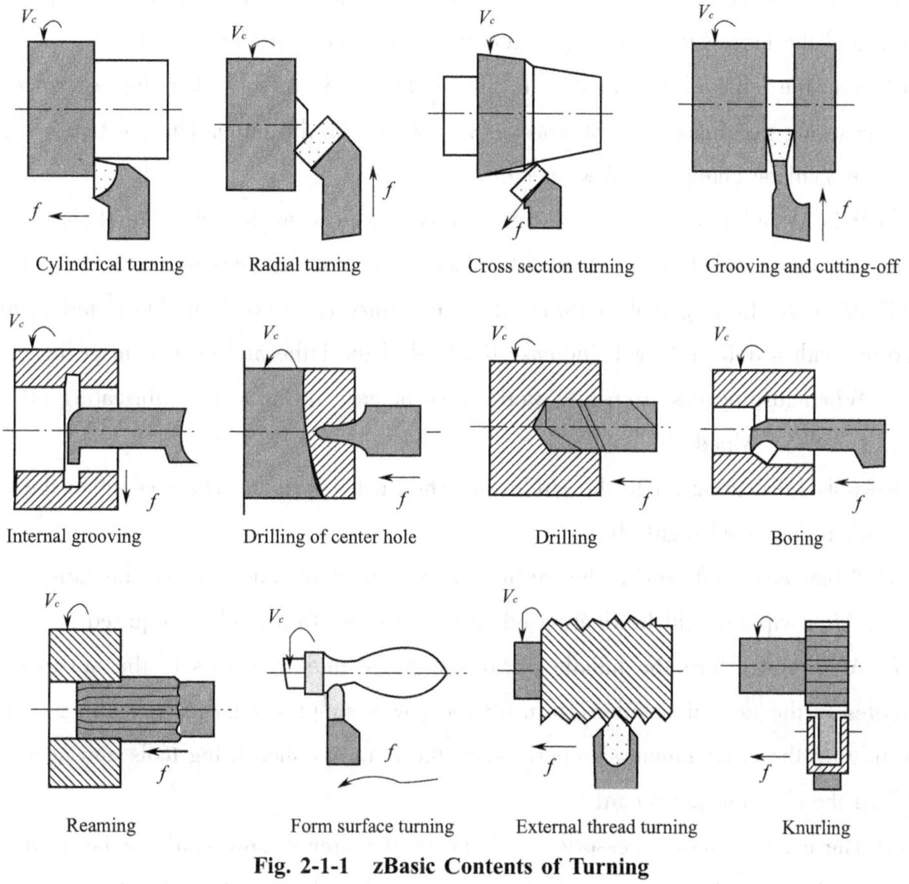

Fig. 2-1-1 zBasic Contents of Turning

II. Civilized production and safe operation technology for lathe work

1. Civilized production

Civilized production is a very important part of factory management, which directly affects the quality of products and the service life of equipment, tools, fixtures and measuring tools. Therefore, as students in vocational colleges, the good habit of civilized production shall be developed from the beginning of learning basic operation skills. Thus, students are required to do the following.

(1) Before operating the lathe, check whether the mechanisms of each part of the lathe are in good condition, and whether the positions of each transmission handle, speed change handles, limit stop, stop iron are correct to prevent damage to the machine tool due to sudden impact during operation. In addition, check whether the safety protection devices are complete. After operating it, the spindle shall idle at low speed for 5min, so that the lubricating oil can be spread to the place where the lathe needs lubrication (especially in winter), and the lathe shall not be put into operation until it works normally.

(2) Stop the lathe when adjusting the machine speed, travel, clamping workpieces and tools, measuring the dimensions of workpieces and wiping the lathe. The position of the feed box handle shall be changed at low speed.

(3) It is not allowed to strike or correct the workpiece on the chuck and the guide rail of lathe bed body, and it is not allowed to place tools or workpieces on the bed surface.

(4) After the turning tool is worn, it is necessary to grind it in time, and continuing the turning with a dull knife will increase the load of the lathe and even damage it.

(5) When turning cast iron and workpieces of gas cutting , the lubricating oil on the guide rail shall be wiped off.

(6)Apply lubricating oil to the guide rail when using coolant. The coolant in the cooling pump shall be replaced regularly.

(7) When going off work, the cuttings and coolant on and around the lathe shall be removed. After wiping, add lubricating oil to the parts for filling oil as required.

(8) After work, turn the handle of large carriage until it moves to the tailstock, push all handles in the neutral position, turn off the power supply, remove iron filings, wipe the lathe, tidy up the environment, properly keep the fixtures, measuring tools and accessories, and fill in the shift change record.

(9) The machine tool accessories shall be used correctly and shall not be used beyond load or specification. If any abnormal phenomena are found, stop immediately for inspection.

(10) Tools shall be uniformly placed on the tool cabinet beside the lathe instead of being placed randomly.

(11) Take care of the tools and measuring tools, wipe them after use to keep them clean, apply oil, put them in the box and return them to the tool room in time.

2. Operators shall pay attention to the reasonable placement of tools, fixtures, measuring tools and drawings

(1) The tools, fixtures, measuring tools and workpieces used in working shall be as

close to and concentrated around the operator as possible. When arranging objects, put the one held by the right hand on the right side, and the one held by the left hand on the left side. Put the commonly used ones closer, and put the infrequently used ones farther. Objects shall be placed in a fixed position, and put back after use.

(2) Toolboxes shall be arranged in a classified manner and kept clean and tidy.

(3) Drawings and operation cards shall be placed in places convenient for reading and attention shall be paid to keeping them clean and complete.

(4) Blanks, semi-finished products and finished products shall be separated and neatly arranged in order to be placed or taken.

(5) The operating environment shall always be neat and clean.

3. Safe operation technology

(1) Before operating the lathe, carefully check whether all parts of the lathe are in good condition and whether the handles are in correct positions. After running the lathe, the spindle shall idle at low speed for 5 min, and the lathe shall not be put into operation until it works normally.

(2) When the spindle needs to change speed at work, stop the machine for speed change.

(3) Wear work clothes and sleeves at work. Female students shall wear work hats.

(4) Rings or other trinkets are not allowed at work.

(5) The head shall not be too close to the workpiece during working, and protective glasses must be worn when cutting at high speed.

(6) Don't wear gloves at work.

(7) Don't stop the rotating chuck by hand.

(8) When the lathe is rotating, the workpiece is not allowed to be measured, and the surface of the workpiece is not allowed to be touched by hand.

(9) Cuttings shall be removed with special hooks instead of hands.

(10) After the workpiece is clamped, remove the chuck wrench at hand. Material rack or baffle shall be used when the bar extends too long from the rear end of the spindle.

(11) After each shift, the main power supply of the machine tool shall be turned off.

4. Accident emergency treatment

(1) Stop the machine immediately and turn off the power supply.

(2) Protect the site.

(3) Report to teachers and related personnel in time, so as to analyze the causes and draw lessons from it.

[Thinking and practice]

(1) What are the typical job contents of lathe work ?

(2) What are the safe operation technologies of lathe work？

(3) What aspects must be done in the civilized production of lathe work？

Task II Operation for lathe work

I. Structure and transmission relationship of lathe

The following is an example of CA6140 horizontal lathes commonly used in practice.

1. The main structure of lathe

Among all lathes，ordinary lathes have a wide processing range，and can be applied to machine repair workshops，or single piece or small batch production. Taking CA6140 horizontal lathe as an example，the basic structure of lathe will be introduced.

CA6140 ordinary horizontal lathe can process basic lathe with the maximum swing diameter of 400 mm，featuring good performance，simple operation，neat and aesthetic appearance，etc. The form and main parts of the horizontal lathe are shown in Fig. 2-2-1.

(1) Spindle box：It is used to hold the spindle and the spindle speed change mechanism. The spindle will rotate at different speed because of the action of speed change mechanism in spindle box whose power is from motor’s motion force transmitted through V-belt. Then the spindle will transmit the motion force to feed box by driving the change gear to rotate through the drive gear.

(2) Feed box：it is a speed change mechanism for feed motion，which can be adjusted to achieve required feed or pitch and change the feed speed.

(3) Apron：It is the control box of lathe feed motion，which can change the rotating movement from feed rod into the longitudinal or lateral movement in a straight line required by turning tool，or operate the split nut to make the tool post directly driven by the lead screw to turn threads.

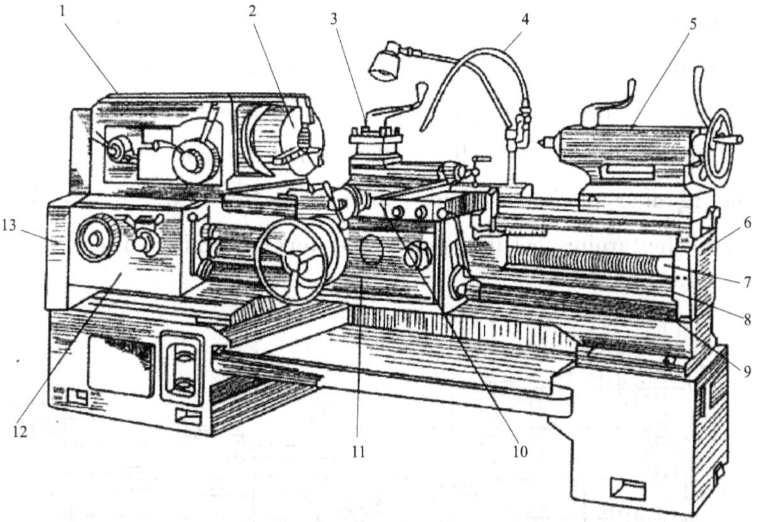

Fig. 2-2-1 CA6140 Horizontal Lathe

1—Spindle box; 2—Chuck; 3—Tool post; 4—Cutting fluid pipe; 5—Tailstock; 6—Bed; 7—Long lead screw; 8—Feed rod; 9—Control rod; 10—Sliding plate; 11—Apron; 12—Feed box; 13—Change gear box

(4) Sliding plate: it consists of saddle, middle sliding plate and small sliding plate. The saddle is used for longitudinal turning; The middle sliding plate is used for transverse turning and cutting depth control; The small sliding plate is used for longitudinal turning of short workpieces or taper turning.

(5) Tool post: It is used to hold the turning tool for longitudinal, transverse or inclined feed motion.

(6) Tailstock: it is installed on the guide rail of the lathe bed. A center can be installed in the tailstock sleeve to support the workpiece, or a drill bit and reamer can be installed to drill and ream the workpiece.

(7) Lathe bed: it is the basic component of lathe. It can be used to connect the main components and ensure the correct relative position between the components. The guide rail on the bed is used to guide the tool post and tailstock to move correctly relative to the spindle box.

(8) Lead screw: It is used for turning threads, and the turning tool can move in an accurate straight line according to the required transmission ratio through the pallet.

(9) Feed rod: It transmits the motion of the feed box to the apron, so that the turning tool can make feed motion in a straight line at the required speed. It is used for automatic feeding.

2. Transmission system of lathe

The transmission schematic diagram of CA6140 horizontal lathe is shown in Fig. 2-2-2. The main motion brings motion force to the spindle 5 from motor 1 through drive belt 2. The spindle will rotate at different speeds because of the action of the speed change mechanism 4, and then the workpiece will be driven to rotate through the chuck 6 (or fixture). For the feed motion, the spindle box outputs the rotating motion force to the change gear box 3, and then drive the apron: 9, saddle 10, sliding plate 8 and the tool post 7 by the lead screw 11 or the feed rod 12 after the speed change through the feed box 13, so as to control the movement track of the turning tool to complete the work of turning various surfaces.

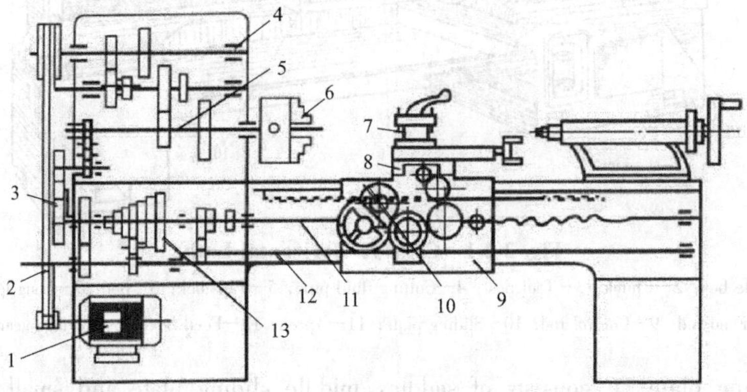

Fig. 2-2-2 Transmission Schematic Diagram of CA6140 Horizontal Lathe

1—Motor; 2—Drive belt; 3—Gear box; 4—Speed change mechanism; 5—Spindle; 6—Chuck; 7—Tool post; 8—Sliding plate; 9—Apron; 10—Saddle; 11—Lead screw; 12—Feed rod; 13—Feed box

The CA6140 horizontal lathe has 3 main motions, it means there are 3 drive chains. That is, main motion drive chain, thread turning drive chain and longitudinal and transverse feed motion drive chain. The transmission relationship between the main components of CA6140 horizontal lathe is shown in Fig. 2-2-3.

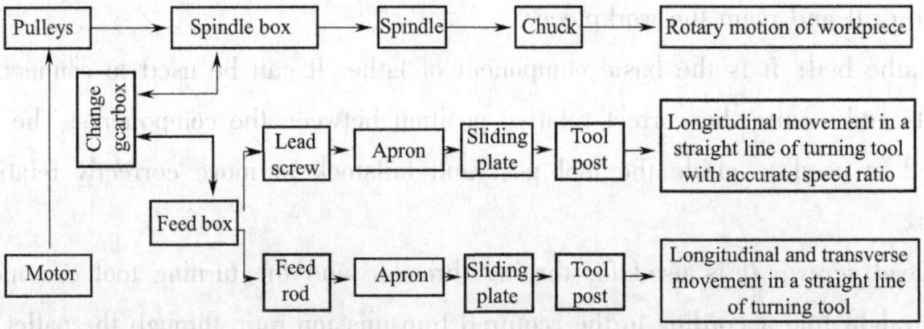

Fig. 2-2-3 Transmission Diagram of CA6140 Horizontal Lathe

II. Basic operation of lathe

1. Starting operation of lathe

(1) Check whether the speed change handles of the lathe are in the neutral position,

whether the clutch is in the correct position, and whether the control rod is in the stop state. After confirmation, close the main switch of lathe power supply.

(2) Press the green start button on the saddle to start the motor.

(3) Lift up the control rod handle on the right side of the apron, and the spindle rotates forward; When the control rod handle returns to the middle position, the spindle stops rotating; When the control rod is pressed downward, the spindle rotates reversely.

(4) The conversion of forward and reverse rotation of the spindle shall be carried out after the spindle stops rotating.

(5) Press the red stop button on the saddle to stop the motor.

2. Operation of changing spindle speed and feed speed

(1) Change the lathe spindle speed. There is a control handle to change the speed outside the spindle box of the horizontal lathe. Different rotating speeds can be achieved by changing the position of the handle. The spindle box indicates various rotating speeds with a nameplate and graphically indicates the position of each handle. The required spindle rotating speed can be achieved by changing the handle position as indicated on the nameplate during operation.

The speed change of the spindle box is controlled by changing the position of the two handles (spindle transmission handle) on the right side of the front of the spindle box. The front handle has 6 gears, each of which has 4 levels of speed, controlled by the rear handle, so the spindle has 24 levels of speed, as shown in Fig. 2-2-4.

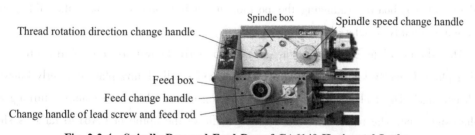

Fig. 2-2-4 Spindle Box and Feed Box of CA6140 Horizontal Lathe

The handle on the left side of the front of the spindle box (thread rotation direction change handle) is used to change the thread rotation direction from left to right and increase the screw pitch. There are four gears, namely right-handed thread, left-handed thread, right-handed thread of larger pitch and left-handed thread of larger pitch.

To ensure safety, the operation of changing the speed change handle of the spindle shall be carried out on the premise that the motor of the machine tool has been turned off. When changing the spindle speed, the force of turning the handle shall not be too large. If it is found that the handle can't turn or turn in place, the main reason is that the gears in the spindle box cannot engage. To solve the problem, the chuck can be turned manually to change the circumferential position of the gear, and then the handle can be turned.

(2) Change the lathe feed speed. There is a handle (feed change handle) on the front left side of the feed box of CA6140 lathe. The handle can rotate 360° after it is pulled outward. There are 8 uniformly distributed numbers on the circumferential line, which shall be adjusted according to the nameplate indication. After the handwheel is turned to the desired position, the feed rate or pitch can be changed by pushing it back.

There are two handles on the right, i.e. front and rear. The front handles are a lead screw and a feed rod change handle. The rear handle has four gears, which cooperate with the feed change handle to adjust the pitch or feed rate. When adjusting the pitch or feed rate, the specific position of handwheel and handle can be determined by looking up the allocation table on the oil cover of the feed box.

To ensure safety, the operation of changing the position of the handle of the feed box shall be carried out on the premise that the motor of the machine tool has been turned off. If it is found that the handle can't turn, the chuck can be turned by hand. Note: When rotating the chuck, the spindle speed shall be adjusted at the high speed position, as the low speed position is generally difficult to rotate by hand. After the adjustment is finished, adjust the spindle speed back to the initial state to avoid misoperation when starting the spindle again.

3. Manual operation of sliding plate

The use and working position of the control handles outside the apron are generally marked with signboards. Changing the position of each handle can make the sliding plate move longitudinally or laterally.

The sliding plate is divided into saddle (formerly known as large carriage), middle sliding plate (formerly known as middle carriage) and small sliding plate (formerly known as small carriage). The tool post is on the small sliding plate, which can clamp 4 turning tools at the same time. The position of the control handle of the apron of CA6140 lathe is shown in Fig. 2-2-5.

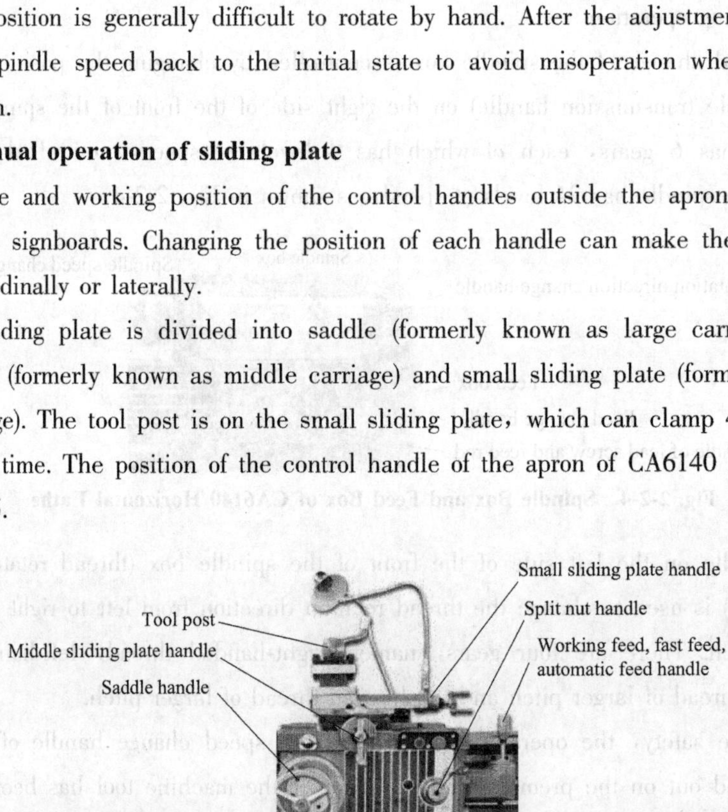

Fig. 2-2-5 Position of Control Handle of CA6140 Lathe Apron

(1) The lateral movement of the saddle is controlled by the saddle handle. Turn the

saddle handle to make the saddle move laterally. The dial on the handwheel indicates the movement distance of the saddle. The movement distance is 1 mm for each interval of rotation of the scale. Tighten the embossed screw to lock the scale ring; if the screw is loosened, turn the dial ring by hand to adjust the zero position.

(2) The lateral movement of the middle sliding plate is controlled by the middle sliding plate handle. When the handle is turned clockwise, the sliding plate moves forward (that is, lateral feed); When turning the handwheel counterclockwise, it moves to the operator (that is, lateral tool retraction). The dial on the handwheel shaft is evenly divided into 100 divisions, and the handwheel turns 1 division, moving 0.05 mm horizontally.

(3) The small sliding plate can move longitudinally for a short distance under the control of the small sliding plate handle. When the small sliding plate handle rotates clockwise, the small sliding plate moves to the left, and when the handle rotates counterclockwise, the small sliding plate moves to the right. The dial on the wheel shaft of the small sliding plate is evenly divided into 100 divisions, and the handwheel moves 0.05mm longitudinally or obliquely for each division.

The dividing plate of the small sliding plate can deflect the required angle within 90° clockwise or counterclockwise when the tool post needs to be fed obliquely to turn the short taper. During adjustment, loosen the lock nut, turn the small sliding plate to the required angle position, and then lock the nut to fix the small sliding plate.

4. Operation for sliding plate power feed

(1) The longitudinal and transverse power feed and rapid movement of CA6140 lathe are controlled by the single handle. The automatic feed handle is located on the right side of the apron, and can be pulled longitudinally and transversely along the cross groove. The pulling direction of the handle is consistent with the moving direction of the tool post, which is simple and convenient to operate. When the handle is in the center of the cross groove, the feed motion will be stopped. There is a fast feed button on the top of the automatic feed handle. Press this button, the rapid movement motor will work, and the saddle or the middle sliding plate handle will move quickly longitudinally or laterally in the direction of pulling. Release the button, and the rapid movement motor will stop rotating and the fast movement will be stopped.

(2) There is a split nut control handle on the right side of the front of the apron, which is used to control the movement connection between the apron and the lead screw.

When turning non-threaded surfaces, the split nut handle is located above. When turning the thread, pull down the split nut handle clockwise to make the split nut close and engage with the lead screw, and transmit the motion force of the lead screw to the apron, so that the apron and the saddle can be longitudinally fed according to a predetermined pitch.

After the thread processing is completed, the split nut handle shall be pulled back to its original position immediately.

5. Tailstock operation

The operation of lathe tailstock mainly includes moving tailstock and tailstock sleeve. The structure of lathe tailstock is shown in Fig. 2-2-6. The tailstock can move back and forth along the bed guide rail to support different lengths of work. The taper hole of the tailstock sleeve can be used to install the center and drill bit, and the sleeve can move forward and backward. The method is as follows.

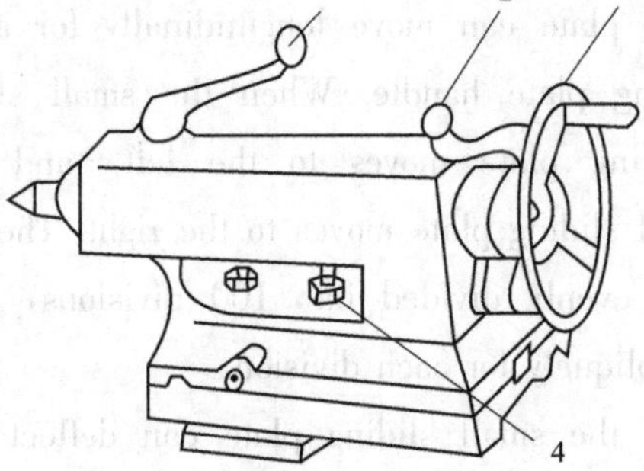

Fig. 2-2-6 Lathe Tailstock

1—Sleeve pressing handle; 2—Tailstock pressing handle; 3—Handwheel; 4- Lock nut

1) Movement and locking of tailstock

Loosen the pressing handle of the tailstock, so that the pressure plate at the bottom of the tailstock is separated from the bed guide rail. Push the tailstock by hand to make the tailstock move along the bed guide rail. Tighten the tailstock pressing handle, so that the pressure plate at the bottom of the tailstock is tightly pressed on the bed guide rail, and lock the tailstock. If the lathe tailstock is fixed on a certain position of the guide rail, the lock nut can be directly tightened with a wrench.

2) Movement and locking of tailstock sleeve

Turn the handwheel to move the sleeve forward and backward. Note that the sleeve shall not be extended too long to avoid affecting the rigidity and prevent the sleeve from extending to the limit so that the lead screw in the sleeve is disengaged from the nut.

When it is necessary to fix the tailstock sleeve, tighten the sleeve pressing handle directly and lock the tailstock sleeve.

III. Lubrication and maintenance of lathe

In order to ensure the normal operation of the lathe, reduce wear and prolong service life, all friction parts of the lathe shall be lubricated, and daily maintenance shall be paid attention to.

1. Lubrication mode of lathe

(1) Pouring oil for lubrication. It is usually used for exposed sliding surfaces, such as bed guide rail surface and sliding plate guide rail surface, etc., which are wiped clean with cotton yarn and lubricated with oiler by pouring oil. As shown in Fig. 2-2-7.

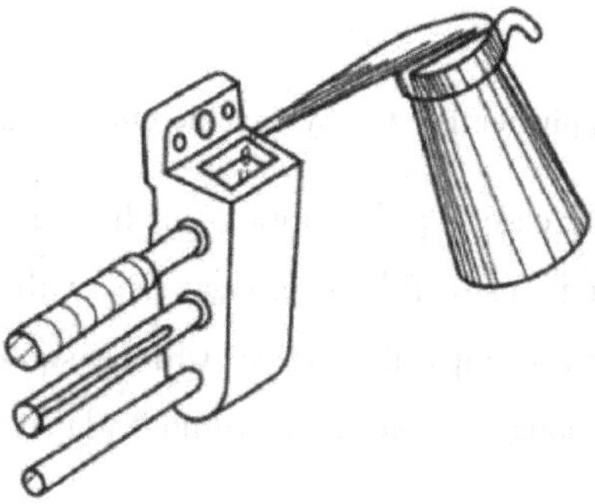

Fig. 2-2-7 Pouring Oil for Lubrication

(2) Splashing oil for lubrication. It is usually used for the sealed box, such as the spindle box of the lathe, which splashes the lubricating oil into the oil groove through the rotation of the gear and then delivers it to each bearing through a copper pipe for lubrication.

(3) Delivering oil through oil wick for lubrication. It is usually used for the oil pool of the lathe feed box and the apron, which slowly delivers the oil to the required lubrication place using the properties of oil absorption and penetration of hairline, as shown in Fig. 2-2-8.

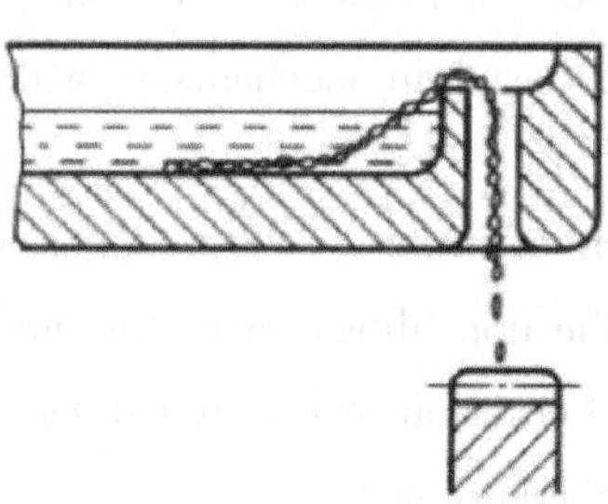

Fig. 2-2-8 Delivering Oil through Oil Wick for Lubrication

(4) Supplementing oil by marble oil cup for lubrication. It is usually used at the bearing where the handle of tailstock and carriage is turned. When filling oil, press down the marble with the nozzle and inject lubricating oil. The advantage of marble oil cup is that it can prevent dust and chips, as shown in Fig. 2-2-9.

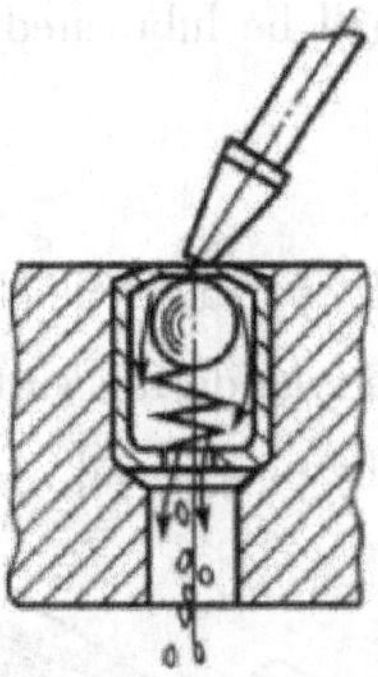

Fig. 2-2-9 Supplementing Oil by Marble Oil Cup for Lubrication

(5) Supplementing oil by grease cup for lubrication. It is usually used for the countershaft of the change gearbox. When in use, fill the grease cup with industrial grease first. When tightening the cover of the grease cup, the grease will be squeezed into the bearing sleeve. It has the characteristics of long storage time without daily refueling, as shown in Fig. 2-2-10.

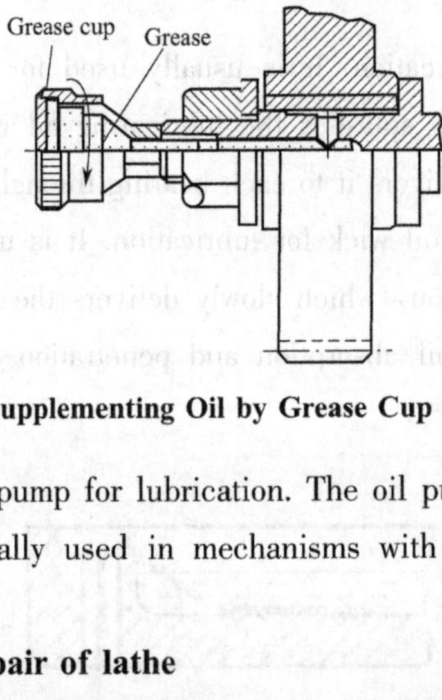

Fig. 2-2-10 Supplementing Oil by Grease Cup for Lubrication

(6) Delivering oil by oil pump for lubrication. The oil pump in lathe is used to provide lubricating oil, which is usually used in mechanisms with high rotating speed and large demand for lubricating oil.

2. Maintenance and repair of lathe

(1) After the work, clean the iron filings, wipe the guide rail surface of the lathe. It is required that the lathe shall be free of oil stain, iron filings and lubricated with oil to make the lathe appearance clean and the site neat.

(2) Every week, the guide rail surface of the bed, the guide rail surface and the transmission parts of the medium and small sliding plates are required to be clean, lubricated, with unblocked oil hole and clear oil pointer and window. The felt of the machine tool shall be cleaned and the lathe surface shall be clean and the site shall be neat.

(3) After the maintenance, the tailstock and apron shall be moved to the tail of the lathe.

[Thinking and practice]

(1) Briefly describe the main structure of the lathe ?

(2) Briefly describe the basic process of manual operation of the lathe sliding plate.

(3) What are the lubrication methods of lathe ?

(4) What are the contents of routine maintenance of the lathe ?

Task III Turning tool and edge grinding

I. Types of common turning tools on lathes

1. Turning tools are divided into: radial turning tools, cylindrical turning tools, grooving tools, thread turning tools, boring tools and formed surface turning tools according to the application; it is divided into turning tools of 45°, 90° and 75° according to the basic angle (tool cutting edge angle) (as shown in Fig. 2-3-1).

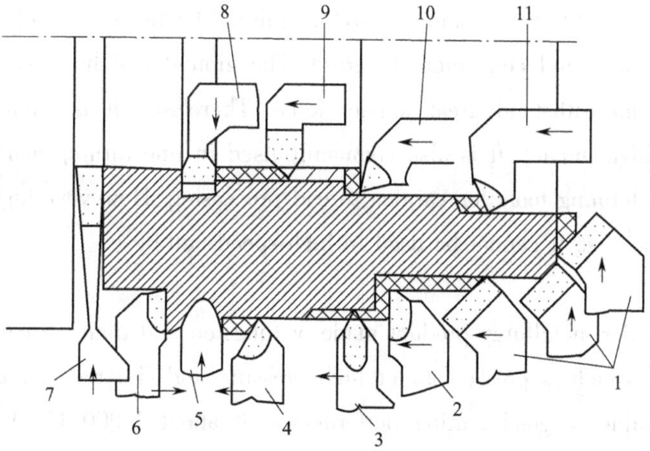

Fig. 2-3-1 Types of commonly used turning tools and their functions

1—45° elbow turning tool; 2—90° cylindrical turning tool (right-hand side tool); 3—External threading tool; 4—75° cylindrical turning tool; 5—Formed surface turning tool; 6—90° cylindrical turning tool (left-hand side tool); 7—Cut-off tools; 8—Inner groove turning tool; 9—Internal threading tool; 10—Blind hole turning tool; 11—Through hole turning tool

(1) 75° cylindrical turning tool is mainly used for turning cylindrical part of workpiece,

which has better impact resistance than 90° cylindrical turning tool and is generally used for rough turning.

(2) 45° elbow turning tool is mainly used for radial, cylindrical and chamfering turnings of workpiece.

(3) 90° cylindrical turning tool (right-hand side tool) is mainly used for turning the cylindrical face, step and end face of workpiece.

(4) The cut-off tool is mainly used to cut off the workpiece or cut grooves on the workpiece.

(5) Formed surface turning tool is used to turn round corners and grooves at steps, or turn various special profile workpieces.

(6) The external threading tool is mainly used for turning various threads.

(7) The bore turning tool is used to turn the bore of the workpiece.

2. According to the material of the cutting part of the tool

According to the material of the cutting part of the tool, it is divided into hard alloy turning tool, high-speed steel turning tool, ceramic turning tool, diamond turning tool and coated turning tool. At present, the materials of commonly used turning tool are high-speed steel and hard alloy.

1) High-speed steel

Commonly used high-speed steel grades include W18Cr4V and W6Mo5Cr4V2 (the number after each chemical element refers to the percentage of the element contained in the material, such as W18, which means 18% tungsten). It is a tool steel with many alloy elements such as tungsten, chromium and vanadium. High-speed steel turning tools are simple to manufacture and convenient to grind. The ground tool has sharp edge with good toughness, and can withstand great impact force. Therefore, it is often used to process workpieces with high impact. It is also commonly used in fine turning tools, rough and fine turning tools and forming tools for various thread processing. However, high-speed steel has poor heat resistance, so it cannot be used in high-speed cutting.

2) Hard alloy

It is the powder metallurgy product made of tungsten and titanium carbide powder plus cobalt as binder, which is pressed under high pressure and sintered at high temperature. It has good red hardness, good cutting performance at about 1 000 ℃, high hardness and good wear resistance. Therefore, its cutting speed can be several times or even dozens times higher than that of high-speed steel tools, and it can be used for difficult-to-machine materials and hardened materials.

Compared with high-speed steel, hard alloys have many advantages, but there are also some disadvantages, mainly poor toughness, fear of impact and less sharp cutting edge than high-speed steel. However, these shortcomings can be improved by choosing a reasonable tool angle, so tools with hard alloy are widely used.

II. Geometrical angle and function of turning tools

1. Composition of turning tool

The turning tool consists of a tool body and a tool handle. As shown in Fig. 2-3-2, the tool body is used for turning, also known as cutting part, and the handle is used to clamp the turning tool.

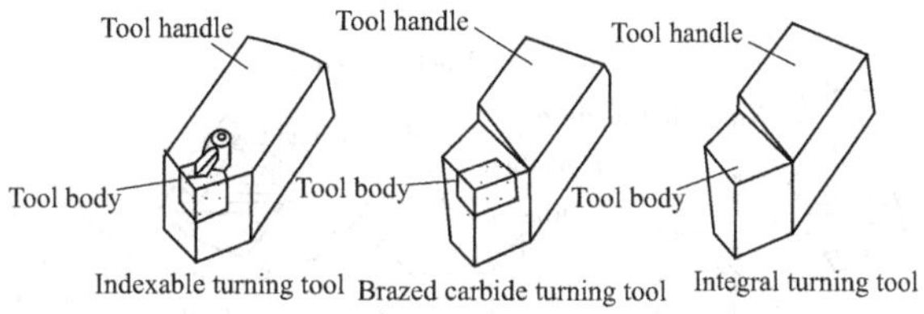

Fig. 2-3-2 Composition of Turning Tool

The body is of the wedge shape, as shown in Fig. 2-3-3, which consists of the following tool surfaces and edges.

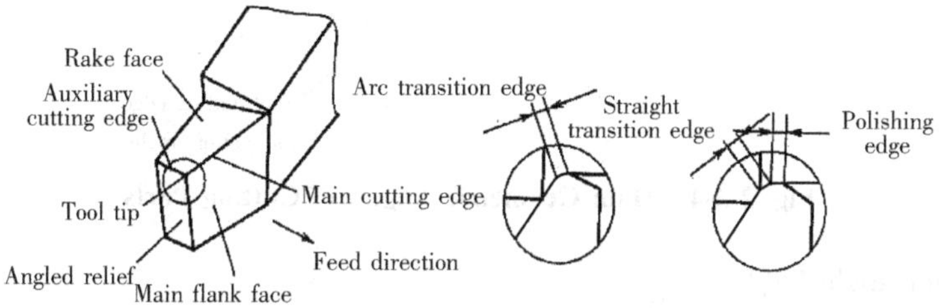

Fig. 2-3-3 Composition of Tool Body

(1) Rake face: the surface on which cuttings flow.

(2) Flank face: it is divided into main flank face and angled relief. Main flank face: the flank face corresponding to the transition surface of the workpiece. Angled relief: The flank face corresponding to the machined surface of the workpiece.

(3) Main cutting edge: the intersection of rake face and main flank face, which is used for the main cutting work.

(4) Auxiliary cutting edge: The intersection of rake face and angled relief, which is used for secondary cutting work.

(5) Tool tip: A small cutting edge where the main cutting edge and the auxiliary cutting edge intersect is called the tool tip. In fact, the tool tip is not very sharp when grinding, and it is always ground into the arc or straight transition edge to increase its

strength.

(6) Polishing edge: A small straight cutting edge near the auxiliary cutting edge is called a polishing edge. When loading the tool, the polishing edge must be parallel to the feed direction, and the length of the polishing edge must be greater than the feed rate, so as to have the polishing effect.

2. The cutting angle of the tool and its function

There are six independent basic angles in the cutting part of turning tool, which are: tool cutting edge angle, tool minor cutting edge angle, anterior angle, main relief angle, end relief angle and tool cutting edge inclination angle; There are also two derived angles: tool included angle and wedge angle. As shown in Fig. 2-3-4.

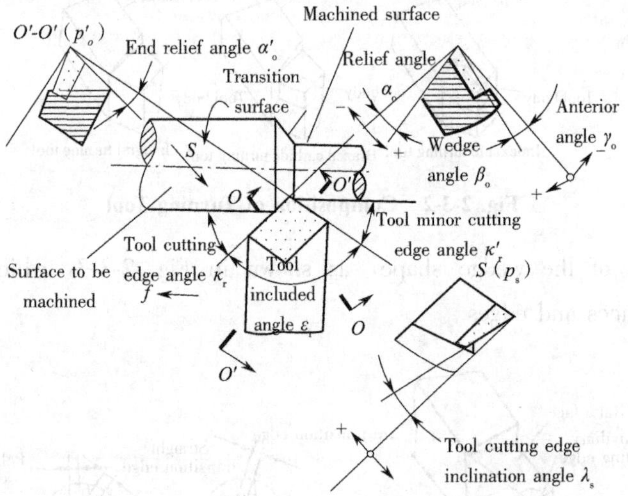

Fig. 2-3-4 Main Geometric Angles of Cutting Parts

1) Anterior angle (γ_0)

It is the angle between rake face and base face. It affects the sharpness and strength of the cutting edge, cutting deformation and cutting force (whether cutting is labor-saving or not, and the difficulty of cuttings discharge).

The selection of the anterior angle is based on the following principles:

(1) When turning plastic metal materials, a larger anterior angle is desirable; When turning brittle metal materials, a smaller anterior angle is desirable.

(2) For rough machining, especially when turning castings and forgings with hard skin, in order to ensure the cutting edge has enough strength, a smaller anterior angle shall be taken: For finishing, in order to refine and make the cutting edge sharp through the surface roughness, the anterior angle shall be larger.

(3) For poor strength and toughness of turning tool material, the anterior angle shall be smaller; On the contrary, the anterior angle shall be larger.

2) Relief angle (α_0)

Main relief angle: the included angle between the main flank face and the main cutting plane. Reduce the friction between the main flank face of the turning tool and the transition surface of the workpiece.

End relief angle: the included angle between the angled relief and the tool minor cutting edge plane. Reduce the friction between the angled relief and the machined surface of the workpiece.

The selection of the relief angle is based on the following principles:

(1) During rough machining, a smaller relief angle shall be taken (hard alloy turning tool: α_0 = 5°-7°; High speed steel turning tool: α_0 = 6°-8°).

(2) When finishing, a larger relief angle (hard alloy turning tool: α_0 = 8°-10°; High speed steel turning tool: α_0 = 8°-12°).

For hard workpiece material, the relief angle shall be smaller; For soft workpiece material, the relief angle shall be larger.

3) Tool cutting edge angle (κ_r)

It is the angle between the projection of the main cutting edge on the base surface and the feed direction. It changes the force on and thermal conductivity of the main cutting edge and affects the cutting thickness.

The shape of the workpiece shall be considered first in selecting the tool cutting edge angle. For processing workpiece of shaft with steps, the tool cutting edge angle must be equal to or slightly greater than 90°; For processing workpiece by cutting in the middle, the angle is generally 45°-60°; For processing shaft workpiece with poor rigidity, the angle shall be larger in order to reduce the radial force during cutting.

4) Tool minor cutting edge angle (κ_r')

It is the angle between the projection of the auxiliary cutting edge on the base surface and the opposite of feeding direction. It reduces the friction between the auxiliary cutting edge and the machined surface of the workpiece. Reducing tool minor cutting edge angle can reduce the residual area of cutting, so the surface roughness of the workpiece can be reduced. On the contrary, if the tool minor cutting edge angle is too large, the tool included angle ε_r will decrease, which will affect the strength of the tool bit. Generally, it is 6°-8°.

5) Tool included angle (ε_r)

It is the included angle between the projections of the major and minor cutting edges on the base surface. It will affect the strength and heat dissipation performance of the tip.

6) Tool cutting edge inclination angle (λ_s)

It is the angle between the main cutting edge and the base surface. The main function of the inclination angle is to control the discharge direction of cuttings. When the inclination angle is negative, it can enhance the strength of the tool bit and protect the tool tip when impacted.

The selection of the tool cutting edge inclination angle is based on the following principles:

(1) The inclination angle shall be positive during finishing and negative during rough machining.

(2) For intermittent cutting with larger impact load, the inclination angle with larger negative value shall be taken.

(3) When machining high hardness materials, negative value of the inclination angle shall be taken to improve the tool strength.

(4) Generally, the tool cutting edge inclination angle shall be zero for forming turning, negative for rough machining and positive for fine turning.

3. Grinding of turning tool

1) Selection of grinding wheel

The material of commonly used grinding wheel includes white alumina and grey-green silicon carbide.

(1) Alumina grinding wheel: It is suitable for grinding cutting tools such as high-speed steel turning tools, carbon tool steel turning tools and the shank of hard alloy turning tools.

(2) Silicon carbide grinding wheel: it is suitable for grinding the blade part of hard alloy carbide.

2) Grinding steps and methods of turning tools

Generally, the grinding of turning tools has the following steps, but it shall be changed appropriately according to different turning tools.

(1) Roughly grind the main flank face, and grind out the tool cutting edge angle and main relief angle at the same time, as shown in Fig. 2-3-5(a).

(2) Roughly grind the angled relief, and grind out the tool minor cutting edge angle and end relief angle at the same time, as shown in Fig. 2-3-5(b).

(3) Roughly grind the rake face, and grind out the anterior angle and tool cutting edge inclination angle at the same time, as shown in Fig. 2-3-5(c).

(4) Grind the rake face.

(5) Grind the main flank face and the angled relief.

(6) Grind the tool tip arc, as shown in Fig. 2-3-5(d).

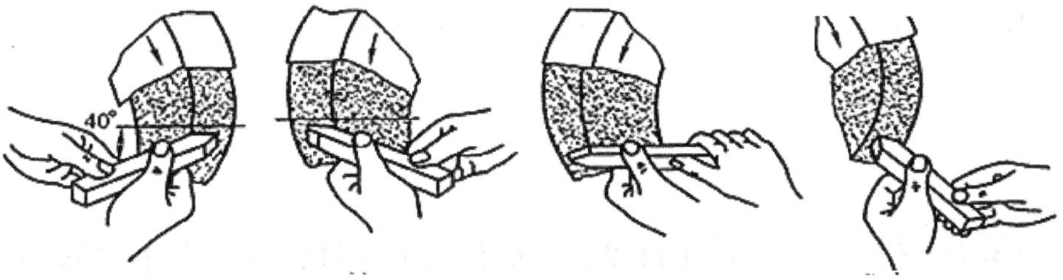

(a) (b) (c) (d)

Fig. 2-3-5 Grinding of Turning Tool

(a) roughly grind the mainflank face (b) roughly grind the angled relief
(c) roughly grind the rake face (d) grind the tool tip arc

4. Angle detection of turning tool

(1) Visual inspection: observe whether the turning tool angle meets the requirements, whether the cutting edge is sharp, and whether there are cracks and other defects on the surface.

(2) Measure with a template.

(3) Measure with protractor.

Measure the angle with the turning tool protractor if accurate angle is required for turning.

5. Safety precautions for grinding turning tool

(1) Before grinding, carefully check the protective facilities of grinder.

(2) Wear protective glasses when grinding the turning tool.

(3) Stand on the side of the grinding wheel when grinding to prevent pieces caused by grinding wheel broken from flying out and hurting people.

(4) Hold the turning tool tightly with both hands when grinding the turning tool, and do not exert too much force, to prevent slipping and hurting the hands.

(5) For grinding hard alloy turning tools, it is not allowed to put the bits into water for cooling to prevent the blades from breaking. For grinding high-speed steel turning tools, water shall be used for cooling at any time to prevent the turning tools from overheating and reducing annealing hardness.

(6) The grinding surface of the grinding wheel must be polished frequently so that there is no obvious runout of the grinding wheel.

(7) Turn off the power supply of grinder after grinding.

[Thinking and practice]

(1) What kinds of turning tools can be classified according to their uses ?

(2) Briefly describe the geometric angle and function of turning tool.

(3) What is the selection method of grinding wheel for cutting edge of turning tool ?

(4) Briefly describe the steps and methods of cutting edge grinding.

(5) What is the method of turning tool angle detection ?

Task IV Manual feed turning of step shaft

I. Drawings and technical requirements

As shown in Fig. 2-4-1, The blank dimension of the workpiece shall be ϕ45 mm×80 mm, the material shall be 45 steel, and the workpiece processing does not require cutting. Scoring table see Table 2-4-1.

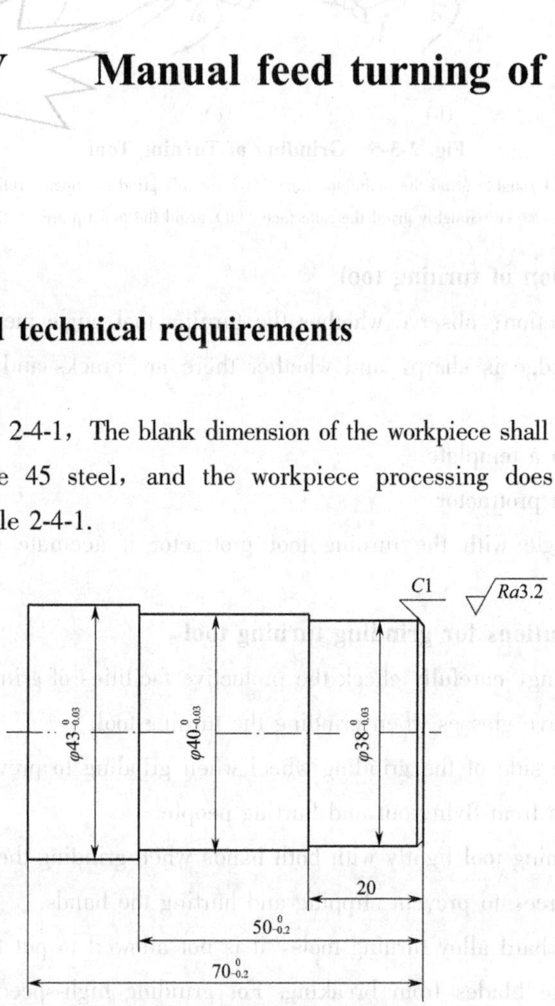

Fig. 2-4-1 Step Shaft Processing Ⅰ

Table 2-4-1 Scoring Table

Technical requirements	Assigned score	Measured results	Score
ϕ $38^{0}_{-0.03}$ mm	15		
ϕ $40^{0}_{-0.03}$ mm	15		
ϕ $43^{0}_{-0.03}$ mm	15		
20 mm	10		
$50^{0}_{-0.2}$ mm	10		
$70^{0}_{-0.2}$ mm	10		

Continue Table

Technical requirements	Assigned score	Measured results	Score
Chamfering *C*1 (1 place)	5		
Surface roughness *Ra*3.2 μm	15		
Safe and civilized operation	5		

II. Analysis of processing

Refer to Table 2-4-2 for the process arrangement of each profile of the part.

Table 2-4-2 Process Card for Ordinary Lathe Machining

Part drawing No.	Fig. 2-4-1	Process card for ordinary lathe machining		Machine model	CA6140	
Part name	Step shaft	Machine tool number				
Table of turning tools				Table of measuring tools		
Name of turning tools			Turning tool parameters	Measuring tool name	Specification/mm	
90° cylindrical turning tool			YT15	Vernier caliper Micrometer	0-150/0.02 25-50/0.01	
Working procedure	Process content	Cutting amount				Processing property
		S/ (r/min)	*F*/ (mm/r)	a_p/mm		
Ordinary lathe	Cylindrical turning and radial turning for determining reference	400	Manual	Manual		Manual
1	Roughly turn the outer circle of ϕ43 mm, length: 70 mm	400	Manual	Manual		Manual
2	Fine turning of the outer circle of ϕ43 mm, length: 70 mm	600	Manual	Manual		Manual
3	Roughly turn the outer circle of ϕ40 mm, length: 50 mm	400	Manual	Manual		Manual
4	Fine turning of the outer circle of ϕ40 mm, length: 50 mm	600	Manual	Manual		Manual
5	Roughly turn the outer circle of ϕ38 mm, length: 20 mm	400	Manual	Manual		Manual
6	Fine turning of the outer circle of ϕ38 mm, length: 20 mm	600	Manual	Manual		Manual
7	Chamfering *C*1	600	Manual	Manual		Manual

III. Knowledge and skills

1. Clamping methods and requirements for turning tools

(1) When clamping the turning tool, the tip shall be aligned to the center line of the workpiece, as shown in Fig. 2-4-2.

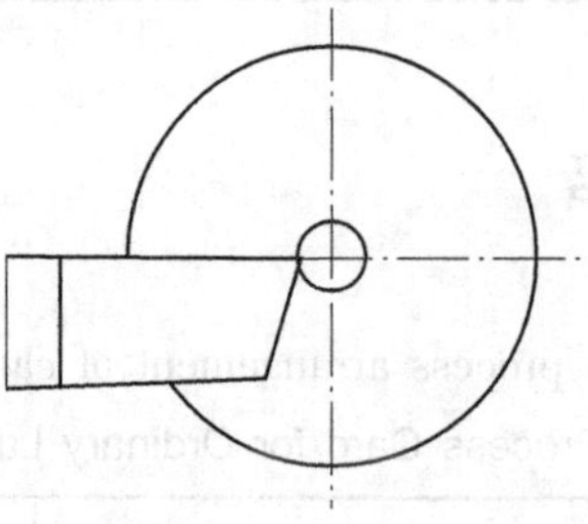

Fig. 2-4-2 The Tip shall be Aligned to the Center line of the Workpiece

① Align the center to a steel ruler according to the value of the lathe center height.

② Level the tool tip and the center, with the center aligned. As shown in Fig. 2-4-3.

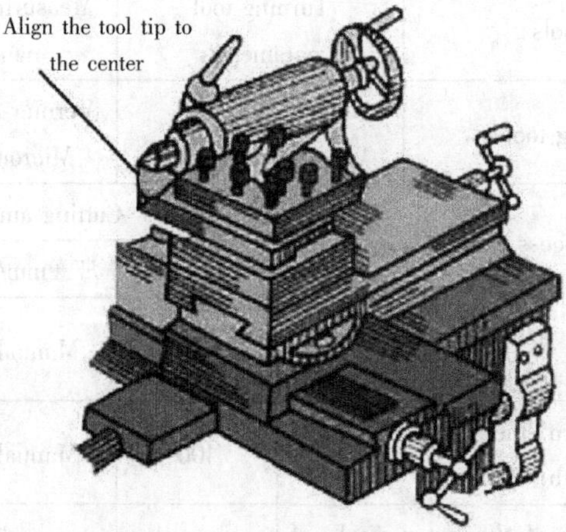
Align the tool tip to
the center

Fig. 2-4-3 The Tip shall be Aligned to the Center line of the Center

(2) The extension length of the turning tool shall be as short as possible. If the extension is too long and the rigidity is poor, vibration may easily occur. The extension length is about 1-1.5 times of the thickness of the tool arbor.

(3) The center line of the tool arbor shall be perpendicular to the surface of the workpiece, otherwise the values of the tool cutting edge angle and the tool minor cutting edge angle will be changed. As shown in Fig. 2-4-4.

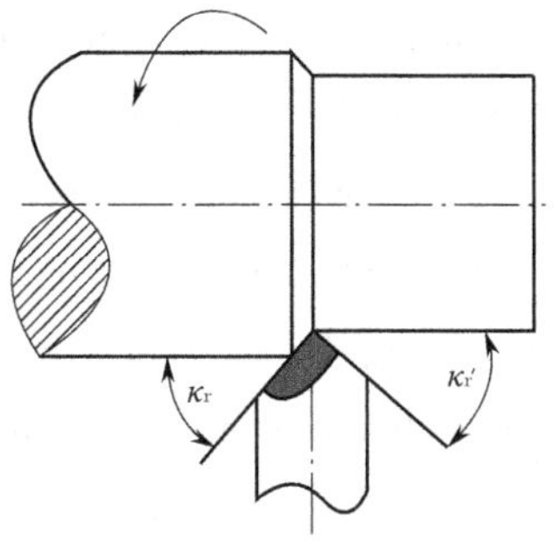

Fig. 2-4-4 The Center Line of the Tool Arbor shall be Perpendicular to the Surface of the Workpiece

(4) The turning tool shall be clamped and fastened to the tool post with at least two screws.

(5) For turning step workpieces, a 90° cylindrical turning tool is usually used. The clamping of turning tools shall be distinguished according to the allowance in rough and fine turning. For large allowance, it is advisable to clamp the turning tool with the tool cutting edge angle of less than 90° (generally 85° to 90°) in order to increase the tool feed and reduce the tool tip pressure, as shown in Fig. 2-4-5(a). For fine turning, the tool cutting edge angle shall be more than 90° (generally about 93°) to ensure that the step plane is perpendicular to the axis line, as shown in Fig. 2-4-5(b).

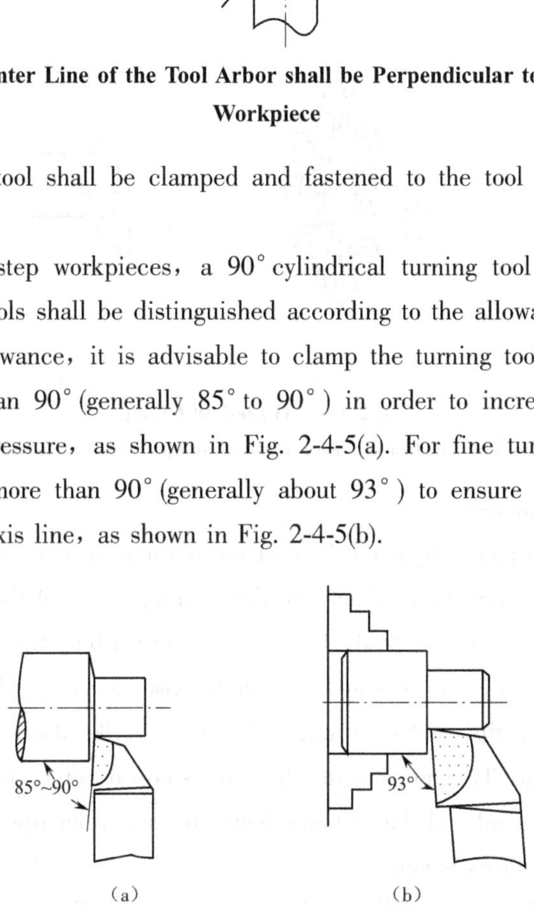

Fig. 2-4-5 Clamping of Turning Tool

2. Clamping of workpiece

Before turning, the workpiece shall be installed on the lathe. According to the shape, dimension, machining accuracy and quantity of shaft workpieces, the clamping methods shall be different to ensure the machining quality and production efficiency.

1) Scroll chuck

The scroll chuck is shown in Fig. 2-4-6(a) and the structure is shown in Fig. 2-4-6(b). When the small bevel gear is rotated by the chuck wrench, the big bevel gear also rotates. Under the action of the plane thread on the back of the large bevel gear, the three claws move to the center or withdraw at the same time to clamp or loosen the workpiece. It is characterized by good neutrality and automatic centering accuracy of 0.05-0.15 mm; Workpieces with smaller diameters can be clamped, as shown in Fig. 2-4-6(c). When clamping cylindrical workpieces with larger diameters, three counter claws can be used, as shown in Fig. 2-4-6(d). However, the scroll chuck is generally only suitable for light-weight workpieces because of its low clamping force. When clamping heavy workpieces, it is advisable to use four jaw independent chuck or other special fixtures.

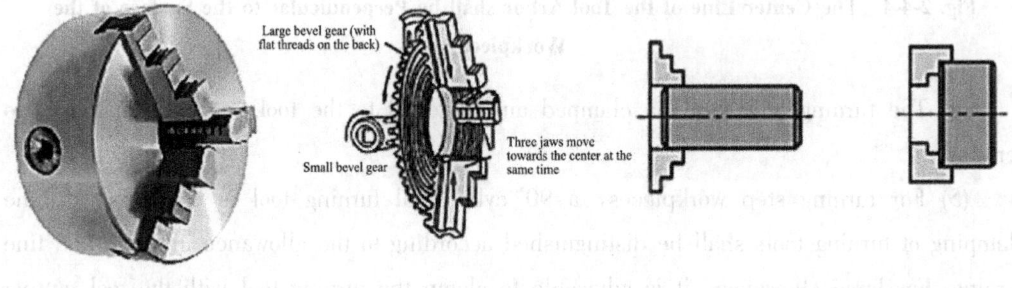

(a) (b) (c) (d)

Fig. 2-4-6 Three-jaw Chuck

(a) Three-jaw chuck (b) Structure (c) Clamping the bar with positive jaw (d) Clamping the heavy bar with inverse jaw

2) Workpiece alignment

The so-called workpiece alignment is to clamp the machined workpiece on the chuck so that the center of the workpiece is aligned to the rotating center of the lathe spindle, which is called alignment. Alignment method: First, make the scribing tip close to the cylindrical surface of the workpiece, turn the chuck by hand, observe the gap between the workpiece surface and the scribing tip, and then adjust the jaw or strike the workpiece with a copper bar based on the gap. The adjustment shall be stopped after several times when the workpiece is rotated around and the distance between the scriber tips and the surface of the workpiece of different places is equal.

(1) Shaft parts alignment: cylindrical part *A* and part *B* of workpiece of shaft parts usually be aligned, as shown in Fig. 2-4-7(a). The method is to align the cylindrical part *A* first and then the cylindrical part B. When align the cylindrical part *A*, the jaw shall be adjusted; When align the cylindrical part B, the copper rod shall be used for striking.

(2) Alignment of disk parts: Usually, the cylindrical part and plane of disc parts shall be aligned, as shown in Fig. 2-4-7(b). For alignment of cylindrical part *A*, the adjustment shall be carried out by moving jaws; For alignment of plane at B, the copper rod shall be used for striking.

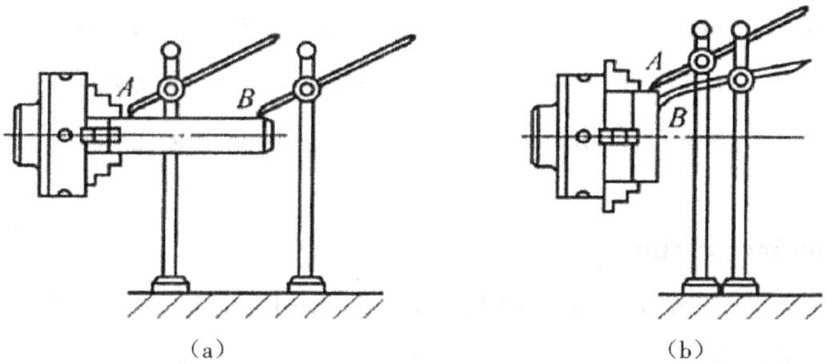

Fig. 2-4-7 Alignment

(a) Shaft parts (b) Disc parts

3. Principle and application of dial

1) The principle of dial

Take the dial of the middle sliding plate as an example: The dial plate is installed on the lead screw of the middle sliding plate. When the handle rotates for a round with the dial plate, the lead screw also rotates for a round. At this time, the nut fixed on the middle sliding plate drives the middle sliding plate and the turning tool to move a lead. If the lead of the transverse feed lead screw is 5mm and the dial is divided into 100 divisions, and when the feed lead screw is turned for a round, the middle sliding plate will move 5 mm. When the dial turns one grid, the movement of the middle sliding plate is: 5 ÷ 100 = 0.05 mm.

2) Cautions in use of dial

(1) How to eliminate the idle travel movement of middle sliding plate.

① Cause: There is a gap between the screw and the nut, resulting in idle travel movement. That is, the dial rotates but the sliding plate does not move.

② Elimination method: When using it, you must slowly turn to the required number of grids, if you have turned a few more grids accidentally (or when the feed is too large, you can' t simply make it back for a few grids, because the result is that only a few grids, instead of the sliding plate, are back due to the gap), you must make all idle travel movement back in the opposite direction, and then turn to the required number of grids, as shown in Fig. 2-4-8.

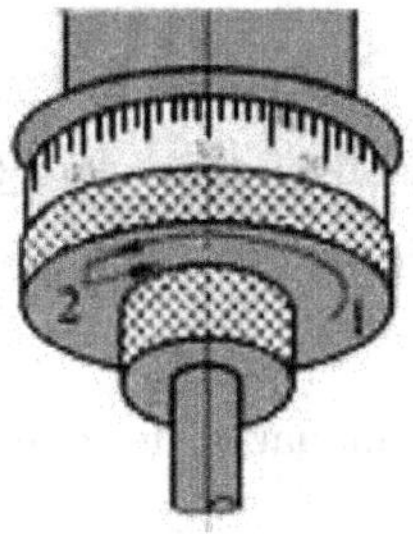

Fig. 2-4-8 The Method to Eliminate the Idle Travel Movement of Middle Sliding Plate

(2) Due to the rotation of the workpiece, after the turning tool feeds from the surface of the workpiece to the center, the cut part is twice the cutting depth. Therefore, it shall be noted that the cutting depth of the turning tool is half of the allowance when using the sliding plate dial.

4. Radial turning method

When turning a workpiece, the end face of the workpiece is often used as the reference for measuring the axial dimension, so it must be processed first. Generally, the roughness is not very high and the end face is flat. This method of turning the end face of the workpiece is called radial turning. The tool used for radial turning is related to the clamping method of the workpiece. When the chuck is clamped, 45° elbow turning tool and 90° turning tool are commonly used for radial turning. The strength and bit dissipation conditions of 45° turning tools are better than those of 90° turning tools, which are often used for turning the end face and chamfering of workpieces. The tip strength of 90° turning tool is poor, so it is often used for finishing. As shown in Fig. 2-4-9.

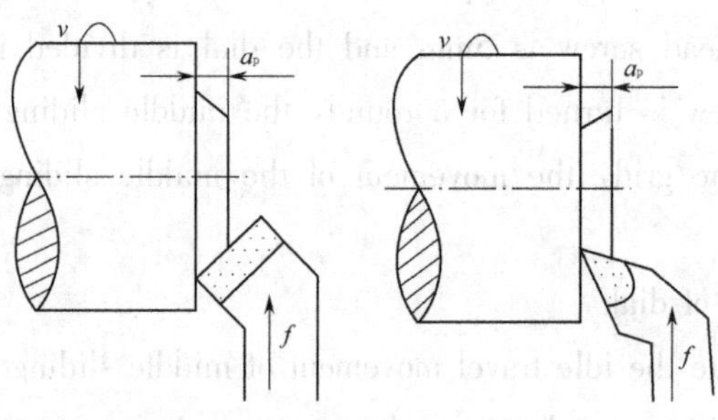

Fig. 2-4-9 Turning End Face

(a) 45° turning tool for radial turning (b) 90° turning tool for radial turning

Steps of radial turning.

(1) Clamp the workpiece on the chuck, align the cylindrical part and end face for clamping.

(2) Clamp the turning tool as required, and adjust the reasonable speed and feed rate.

(3) Turn the saddle and the middle sliding plate to make the tool tip 2-3 mm away from the end face to be machined.

(4) Lock the saddle on the bed.

(5) Turn the middle and the small sliding plate to make the tip contact with the end face of the workpiece, and return the middle sliding plate (the small sliding plate does not move).

(6) Set the small sliding plate graduation to zero or remember the graduation of small sliding plate.

(7) Adjust the cutting depth as required with the graduation of small sliding plate.

(8) Turn the handwheel of the middle sliding plate with both hands, keep the speed uniform, and retract the tool after turning transversely to the center of the end face.

(9) Turn in sequence until the end face of the lathe is flattened by turning or the requirements of the drawing are met.

Note: If the tool is not aligned to the rotation center of the workpiece when loading, there will be bulges or tool tip breakage when turning the end face to the center, as shown in Fig. 2-4-10.

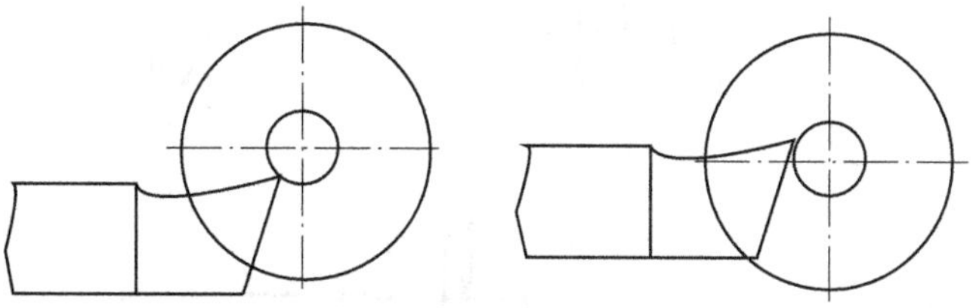

(a) (b)

Fig. 2-4-10 Tool Loading Height

(a) Too low tip is easy to be broken (b) Too high tip is not easy to cut

5. Cylindrical turning method

1) Rough turning steps

(1) Set clamp on the chuck, and clamp it tightly after alignment.

According to the turning length, clamp the workpiece on the chuck, and stretch out for an appropriate length. Clamp the workpiece slightly by hand, rotate the workpiece, strike the workpiece with an iron bar, and observe the runout of the workpiece when it rotates. After the runout is relatively small, it indicates that the workpiece has been aligned at this time, and then clamp the workpiece with the guide cylinder for increasing pressure. Or install a tool arbor on the tool post to make the workpiece rotate, turn the middle carriage to make the tool arbor contact with the workpiece, then slowly turn the middle carriage until the cylindrical surface of the workpiece is basically concentric with the rotation center of the workpiece, and finally clamp the workpiece. This method is suitable for the occasion where the workpiece is clamped once from rough machining to finishing.

(2) Clamp the turning tool and adjust the reasonable spindle speed and feed rate.

The purpose of rough turning is to cut off the excess metal layer as soon as possible, so that the workpiece is close to the final shape and dimension. The cutting depth shall be as large as possible to reduce the cutting time with full play to the performance of turning tools and lathes during rough turning.

(3) Turn related handle to send the turning tool to the end of the workpiece and start to operate the lathe with the tool 3-5 mm away from the end face of the workpiece.

(4) For trial cutting, control the cutting depth at 1-2mm and longitudinal feed at about

3mm through the graduation on the middle sliding plate. Then retract the turning tool in the longitudinal direction (keep still in the transverse direction) and measure the diameter of the workpiece after stopping operating the lathe, as shown in Fig. 2-4-11.

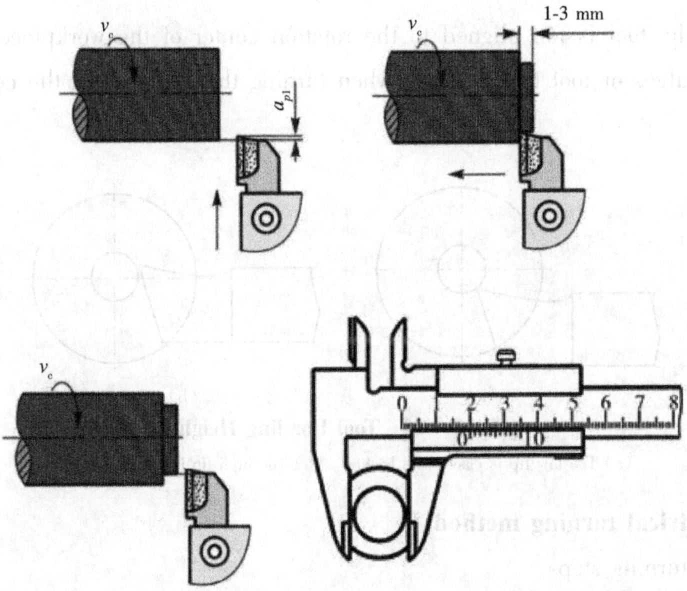

Fig. 2-4-11 Trial Cutting Steps

(5) Adjust the cutting depth according to the measured diameter dimension, leaving the allowance for fine turning: Allowance for high-speed steel fine turning with low speed: 0.1-0.2 mm; allowance for hard alloy fine turning: 0.4-1.0 mm. Carry out the longitudinal feed after the back cutting depth is adjusted. Turn the handwheel of the saddle with both hands at uniform speed until the required cutting depth is achieved. Then turn the middle sliding plate handle to withdraw the turning tool, and the saddle will quickly move back to its original position, as shown in Fig. 2-4-12.

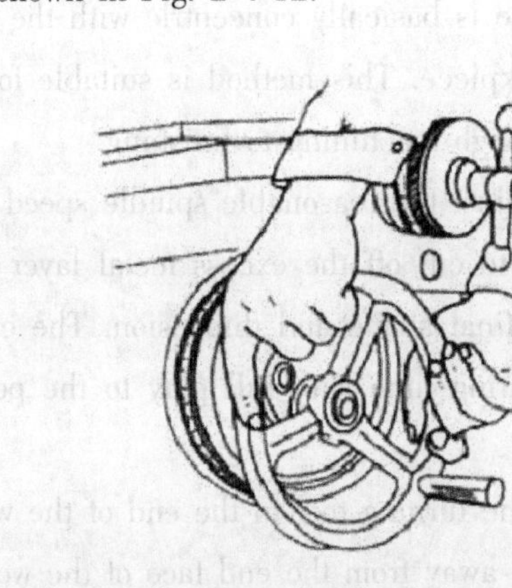

Fig. 2-4-12 Turn the Handwheel with Both Hands

2) Fine turning steps

(1) Clamp the fine turning tool as required, and adjust the reasonable speed and feed rate.

(2) Align the tool tip for cylindrical turning to the end of the workpiece and keep the cutting depth at 0.1-0.2 mm and trial cutting length at 3 mm through the dial on the middle sliding plate.

(3) Measure the outer diameter of the workpiece after stopping operating the lathe.

(4) Adjust the cutting depth according to the measured outer diameter dimension and the drawing dimensions.

(5) Turn the handwheel of the saddle with both hands at uniform speed and carry out the turning longitudinally until the required cutting depth is achieved.

(6) Check the dimensions after stopping operating the lathe and remove the workpiece after the drawing requirements are met.

3) Control of the cylindrical length and dimensions

Measure the distance from the tool tip to the end face of the workpiece with a steel ruler, caliper or template as cylindrical length required, and make a turning mark with tool tip (called scribing method). The turning shall then be carried out according to the mark. After turning, measure again until related requirements are met. As shown in Fig. 2-4-13.

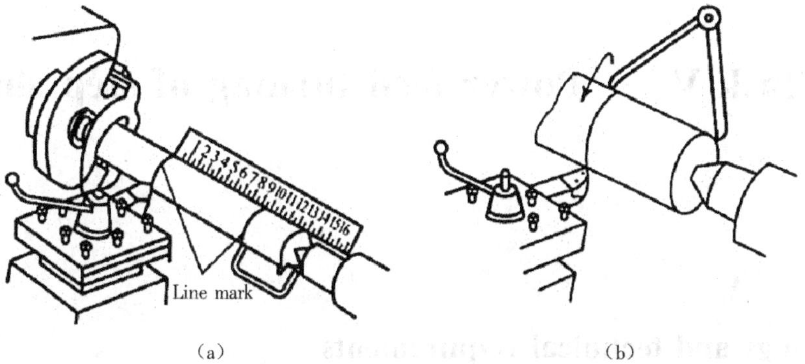

Fig. 2-4-13 Determination of Turning Length by Scribing Lines

(a) Scribe lines with a steel ruler and a template (b) Scribe lines on the workpiece with an inner caliper

6. Quality analysis for radial and cylindrical turning of shaft parts

(1) The end face of the workpiece is uneven, with bump, or little bulge on the center. The reason is that the grinding or installation of the turning tool is incorrect, the tool tip is not aligned to the center of the workpiece, the feed depth is too large, the lathe has clearance, and the carriage moves.

(2) The workpiece end surface has poor roughness. The reason is that the turning tool is not sharp, and the handle is rotated unevenly or too fast for feed by hand.

(3) The diameter or length of the workpiece is incorrect. It is mainly caused by misreading the dimension because of carelessness; error in dial calculation or manipulation; inappropriate and incorrect measurement.

(4) The surface roughness of the workpiece does not meet the requirements. The grinding angle of the turning tool edge is wrong, the tool installation is incorrect or the tool is worn, the cutting amount are improperly selected, and the gaps of various parts of the lathe are too large.

(5) The outer diameter of the workpiece is tapered. It mainly caused by excessive depth of feed, worn tool, loose tool or sliding plate. And the lower reference line of the turntable is not aligned to the"0"line when turning with the small sliding plate. The machining allowance is not enough during fine turning.

[Thinking and practice]

(1) Briefly describe the method of radial turning by manual feed.

(2) What are the methods of tool installation and tool alignment.

(3) How to control the machining length with dial during cylindrical turning ?

(4) Briefly describe the basic process and method of cylindrical turning.

(5) What should be paid attention to during radial and cylindrical turning ?

Task V Power feed turning of step shaft

I. Drawings and technical requirements

As shown in Fig. 2-5-1, The blank dimension of the workpiece is $\phi 50$ mm×100 mm and the material is 45 steel. Scoring table see Table 2-5-1.

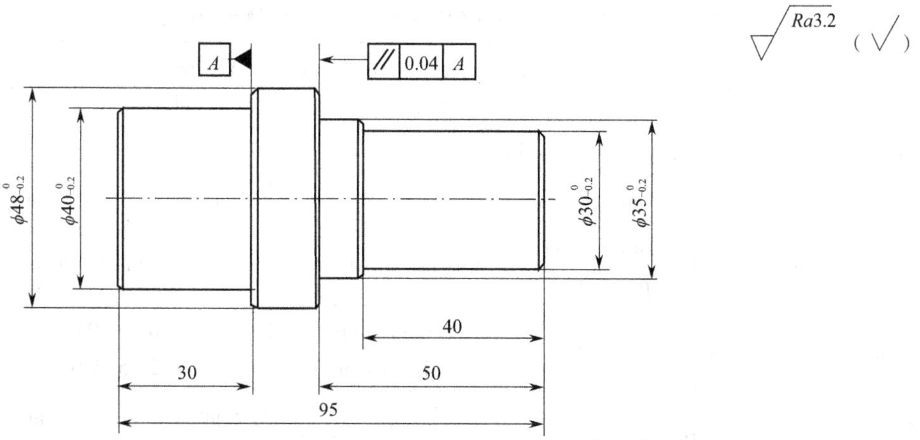

Fig. 2-5-1 Step Shaft Processing Ⅱ

Table 2-5-1 Scoring Table

Technical requirements	Assigned score	Measured results	Score
$\phi\ 30_{-0.2}^{\ 0}$ mm	10		
$\phi\ 35_{-0.2}^{\ 0}$ mm	10		
$\phi\ 40_{-0.2}^{\ 0}$ mm	10		
$\phi\ 48_{-0.2}^{\ 0}$ mm	10		
40 mm	5		
50 mm	5		
30 mm	5		
95 mm	10		
Parallelism: 0.04 mm	10		
Chamfering *C*1 (5 places)	3		
Surface roughness *Ra* 3.2 μm (4 places)	12		
Safe and civilized operation	10		

II. Analysis of processing

Refer to Table 2-5-2 for the process arrangement of each profile of the part.

Table 2-5-2 Process Card for Ordinary Lathe Machining

Part drawing No.	Fig.2-5-1	Process card for ordinary lathe machining		Machine model	CA6140
Part name	Step shaft	Machine tool number			
Table of turning tools				Table of measuring tools	
Name of turning tools	Turning tool parameters	Measuring tool name		Specification/mm	
90° cylindrical turning tool	YT15	Vernier caliper Micrometer		0-150/0.02 25-50/0.01	
Working procedure	Process content	Cutting amount			Machining Property
		S/ (r/min)	F/ (mm/r)	a_p/mm	
Ordinary lathe	Cylindrical turning and radial turning for determining reference	600		1	Manual
1	Roughly turn the outer circle of ϕ48 mm, the length shall meet the dimension requirements	600	0.2	1-2	Automatic
2	Roughly turn the outer circle of ϕ40 mm, the length shall meet the dimension requirements	600	0.2	1-2	Automatic
3	Fine turning of the outer circle of ϕ48 mm and ϕ40 mm, and chamfering *C*1	1 000	0.1	0.5-1	Automatic
4	Turn around to clamp ϕ40 mm outer circle and flatten the end face, ensure the total length	600		1	Manual
5	Roughly turn the outer circle of ϕ35 mm, the length shall meet the dimension requirements	600	0.2	1-2	Automatic
6	Roughly turn the outer circle of ϕ30 mm, the length shall meet the dimension requirements	600	0.2	1-2	Automatic
7	Fine turning of the outer circle of ϕ35 mm and ϕ30mm, and chamfering *C*1	1 000	0.1	0.5-1	Automatic

III. Knowledge and skills

1. Advantages of power feed turning

Labor-saving operation, uniform feed, low surface roughness after processing, etc.

2. Operation method of power feed turning of workpiece

For power feed, pull the power feed handle (the handle can be pulled along the cross groove, as shown in Fig. 2-5-2), whose direction is consistent with the moving direction of the tool; after turning to the specified position, retract the tool transversely, measure the workpiece, and the specified dimension of the workpiece will be achieved after multiple feeds. There is a button on the top of the power feed handle, which can be used to switch on the quick feed and retract.

Fig. 2-5-2 Power Feed

Special note: Pay attention to the fact that when the middle slide plate moves forward, the front of the tool post should not exceed the center of the chuck, so as to prevent the screw of the middle slide plate from disengaging from the nut. In case of reverse feed, the middle sliding plate shall be prevented from being damaged by collision with the dial when it retreats.

3. Alignment method of workpiece

Whenever the workpiece for feed connection is clamped, the alignment must be strictly required, otherwise it will cause the feed deviation and affect the processing quality. Generally, dial indicator is used to align the finished surface of the workpiece, as shown in Fig. 2-5-3.

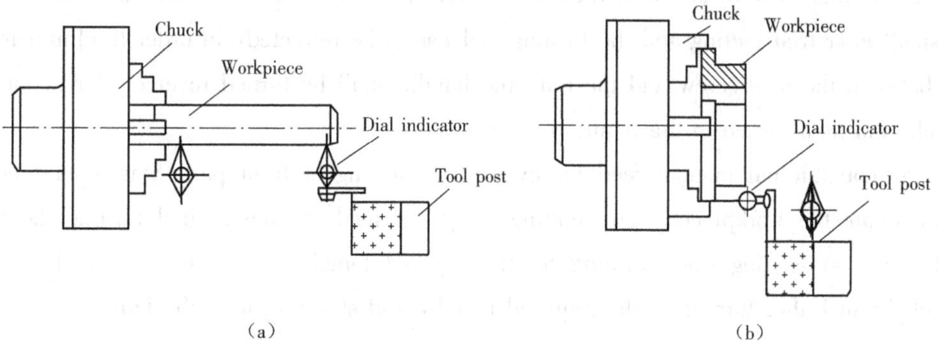

Fig. 2-5-3 Alignment of Dial Indicator

(a) Align the outer circle (b) Align the end face

Fix the dial indicator on the tool post, and press the pointer of the dial indicator on the outer circle or end face of the workpiece and make it rotate around the workpiece once.

If the slewing center of the workpiece coincides with the chuck center, the pointer will not move; If the slewing center of the workpiece does not coincide with the center of the chuck, and the workpiece surface is close to the pointer, the pointer turns right, and if the workpiece surface is far away from the pointer, the pointer turns left. The position of the jaw or workpiece shall be adjusted until the requirements are met. The centering accuracy of dial indicator alignment can reach 0.02-0.01 mm.

4. Basic steps for power feed turning of outer circle

(1) Select the spindle speed and feed rate, and adjust the position of related handles.

(2) For tool alignment, move the tool post so that the tip of the turning tool touches the surface of the workpiece, and start to operate the machine when the zero point is aligned.

(3) After the tool alignment, adjust the cutting depth with a dial. When adjusting the pre-cutting depth with the dial, you shall know the scale value of the dial of the middle sliding plate, that is, the horizontal cutting depth value of the turning tool every time it turns a small grid. Then calculate the number of grids to be turned according to the pre-cutting depth.

(4) For trial cutting, check whether the cutting depth is accurate, and the cutting shall be carried out transversely. In order to ensure the dimensional accuracy of the machining, the trial cutting method shall be used for turning. There are two cases where the trial cutting dimension is not qualified: if the dimension is too large, the back cutting depth ap shall be determined by cutting transversely again; if the dimension is too small, the turning tool shall be transversely withdrawn from a certain distance, and then the trial cutting shall be carried out. Repeat the steps (1) to (3) above until the dimension is qualified. The back cutting depth determined each time shall be less than half of the diameter allowance for each time.

If the dial handle has been turned excessively, or if it is found that the dimension is too small after trial cutting and the turning tool has to be retracted, in order to eliminate the gap between the lead screw and the nut, the handle shall be turned reversely for about half a cycle and then turned to the required scale value.

(5) Longitudinal manual feed for cylindrical turning of front part. Start operating the lathe, rotate the workpiece → try cutting → power feed → longitudinal turning the outer circle → stop feeding when turning to the required length → switch to manual feed → retract the tool after turning to the required length, and stop operating the lathe.

(6) Measure the dimension of the outer circle.

(7) Chamfering Select a 45° turning tool, and then move the saddle to the intersection of the outer circle and the end face of the workpiece for chamfering.

5. Precautions for power feed of fine turning of outer circle

For power feed of fine turning of step shaft, the feed handle shall be disconnected when the longitudinal turning with turning tool is carried out to the place 0.7 mm to 1 mm close to the step, and the manual feed of saddle shall be carried out until the tool is touched with the step and the small sliding plate shall be used to control the step length. For manual feed, the turning tool shall be rotated outward with sliding plate along the end face of the step at a uniform speed to ensure the length dimensional accuracy and surface roughness requirements of the step shaft will be met.

6. Turning method at feed connection place of workpiece

In order to ensure the quality of feed connection, it is usually required that turning length shall be longer when turning the first end of the workpiece; and the alignment distance between two points shall be larger during turning around and clamping; the turning tool shall be sharp during turning around and for fine turning, and the last cutting shall be of small allowance, otherwise, dents will easily occur on the workpiece.

[Thinking and practice]

(1) Briefly describe the machining method of power feed turning of outer circle.

(2) Briefly describe the alignment method of the workpiece.

(3) Briefly describe the turning process of the workpiece for feed connection.

Task VI Turning outer circle groove and cutting off

I. Drawings and technical requirements

As shown in Fig. 2-6-1, The blank dimension of the workpiece is $\phi 30$ mm×80 mm and the material is 45 steel. Scoring table see Table 2-6-1.

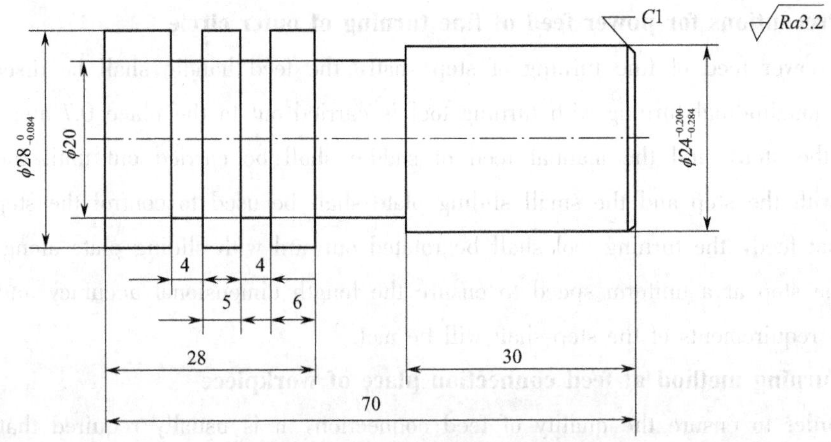

Fig. 2-6-1 Step Shaft Processing Ⅲ

Table 2-6-1 Scoring Table

Technical requirements	Assigned score	Measured results	Score
$\phi\ 24_{-0.284}^{-0.2}$ mm	12		
$\phi\ 28_{-0.084}^{0}$ mm	12		
Cylindrical turning of ϕ20 mm tool retraction groove	10		
Cylindrical turning of ϕ20 mm outer groove（2 places）	20		
Length without tolerance（7 places）	7		
Drilling of center hole	10		
Chamfering $C1$	3		
Surface roughness Ra 3.2 μm	16		
Safe and civilized operation	10		

II. Analysis of processing

Refer to Table 2-6-2 for the process arrangement of each profile of the part.

Table 2-6-2 Process Card for Ordinary Lathe Machining

Part drawing No.	Fig. 2-6-1	Process card for ordinary lathe machining	Machine model	CA6140
Part name	Step shaft	Machine tool number		
Table of turning tools			Table of measuring tools	

Continue Table

Name of turning tools	Turning tool parameters	Measuring tool name	Specification/mm
90° cylindrical turning tool and grooving tool Center drill	YT15, high speed steel turning tool B3 center drill	Vernier caliper Micrometer	0-150/0.02 0-25/0.01, 25-50/0.01

Working procedure	Process content	Cutting amount			Machining Property
		S/ (r/min)	F/ (mm/r)	a_p/mm	
1	Radial turning	350	0.2	0.5-1	Automatic
2	Drilling of center hole	700	—	—	Manual
3	Roughly turn the out circle to ϕ29 mm and ensure that it is 75 mm long	600	0.2	1-2	Automatic
4	Roughly turn the out circle to ϕ25 mm and ensure that it is 42 mm long	600	0.2	1-2	Automatic
5	Fine turning of the ϕ24 mm outer circle for required dimension and ensure that it is 42 mm long	800	0.1	0.5	Automatic
6	Fine turning of the ϕ28 mm outer circle for required dimension and ensure that it is 28 mm long	800	0.1	0.5	Automatic
7	Turn ϕ20 mm tool retraction groove and ensure the dimension meet related requirement	300	—	—	Manual
8	Turn seal groove and ensure the dimension meet related requirement	300	—	—	Manual
9	Chamfering and smoothing	300	—	—	Manual
10	Turn around, clamp the ϕ24 mm outer circle, and align the ϕ28 mm outer circle, turn the left end face and ensure that it is 28 mm long	350	0.2	0.5-1	Automatic

III. Knowledge and skills

1. Types and functions of grooves

For grooves provided on the part, it aims to facilitate subsequent processing, such as thread tool retraction groove turning and grinding the grinding undercut for processing; to ensure the accuracy of axial positioning of parts during assembly, such as shaft shoulder groove processing; to move or tighten at will, such as T-shaped grooves and dovetail grooves processing; to arrange different forms of lubricating grooves, seal grooves and dustproof grooves on the moving mating surfaces. The method of turning groove on the surface of workpiece is called grooving.

The structural forms of grooves include rectangular grooves, formed grooves, inclined grooves, end face grooves and so on. According to the groove position, it can be divided into outer groove and inner groove. As shown in Fig. 2-6-2, generally speaking, the dimension requirements of the groove are not high.

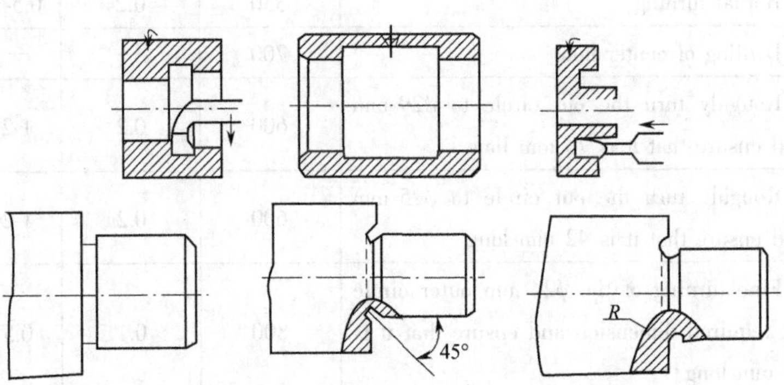

Fig. 2-6-2 Types of Grooves

2. Selection of cut-off tool (grooving tool)

1) High-speed steel cut-off tool

Refer to Fig. 2-6-3 for high-speed steel cut-off tool.

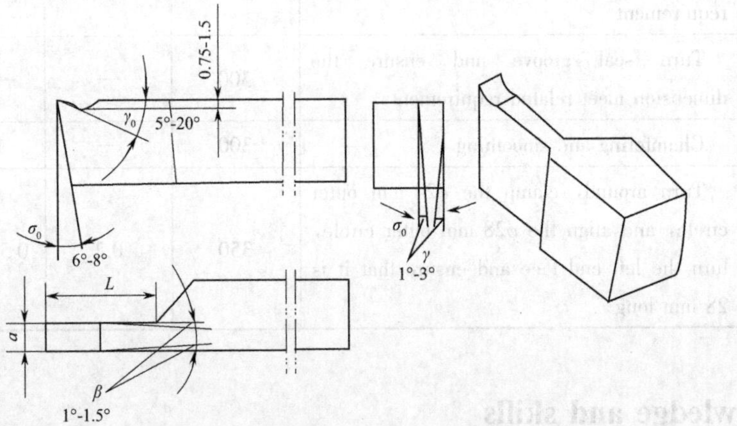

Fig. 2-6-3 High-speed Steel Cut-off Tool

(1) Anterior angle (γ_o): $\gamma_o = 20°\text{-}30°$ when cutting medium carbon steel material and $\gamma_o = 0°\text{-}10°$ when cutting cast iron material.

(2) Relief angle (α_o): it shall be larger when cutting plastic materials and smaller when cutting brittle materials, generally $\alpha_o = 6°\text{-}8°$.

(3) End relief angle (α_o'): The cut-off tool has two end relief angles $\alpha_o' = 1°\text{-}3°$, which are used to reduce the friction between the angled relief and the machined surface of the workpiece.

(4) Main deflection angle (κ_r): The cut-off tool mainly feeds horizontally, so $\kappa_r = 90°$. In order to prevent a small boss from being left outside the center of the end face of the workpiece during cutting, and to avoid flash during cutting hollow workpiece, the main cutting edge can be slightly ground.

(5) Tool minor cutting edge angle (κ_r'): The two tool minor cutting edge angles of the cut-off tool must be symmetrical, otherwise the flatness and the verticality of the section to the axis will be affected by the uneven cutting resistance on both sides. In order not to weaken the strength of the tool bit, it is generally taken as $\kappa_r' = 1°\text{-}1.5°$.

(6) Width of main cutting edge (a): if the main cutting edge is too wide, it will vibrate due to too much cutting force and waste materials at the same time; if it is too narrow, it will weaken the strength of the tool body.

(7) Length of tool body (L): too long a tool body can easily cause vibration and break the tool body.

(8) Chip-breaker groove: the chip-breaker groove of the cut-off tool shall not be ground too deep, generally 0.75-1.50 mm. If it is ground too deep, its tool bit strength is poor and it is easy to break; Moreover, the front part cannot be ground too low or be ground to the stepped shape, otherwise, it will bring problems including: difficulty in cutting and cuttings discharge, heavy cutting load and breaking the tool bit easily, as shown in Fig. 2-6-4.

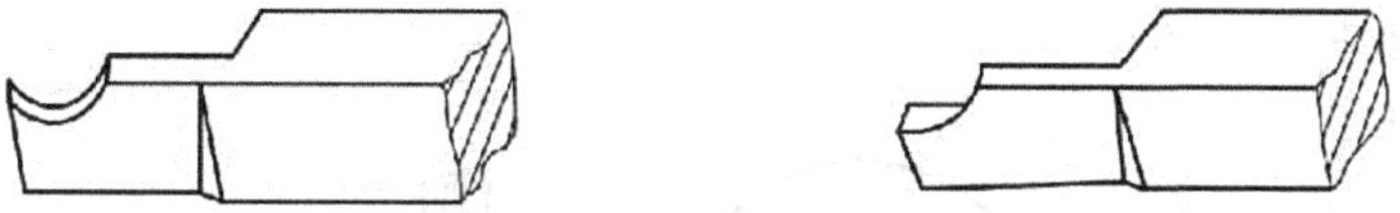

Fig. 2-6-4 Incorrect Grinding of Chip-breaker Groove

2) Hard alloy cut-off tool

When cutting the workpiece with hard alloy cut-off tool at high speed, the cuttings width and the groove width of the workpiece are equal, so it is easy to block the groove with cuttings. For smooth cuttings discharge, both sides of the main cutting edge can be chamfered or ground into herringbone shape, as shown in Fig. 2-6-5.

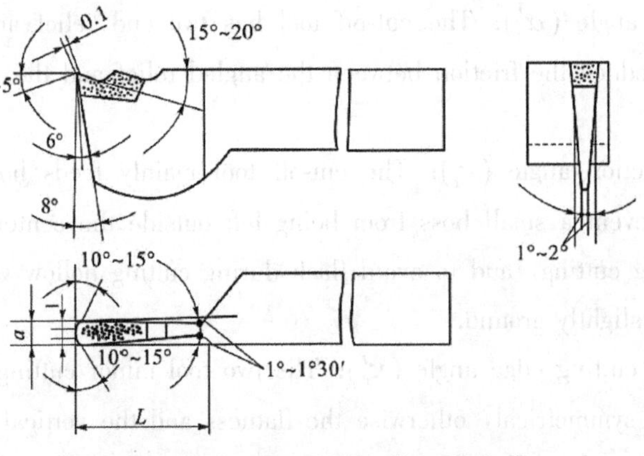

Fig. 2-6-5 Hard Alloy Cut-off Tool

When cutting off at high speed, it will generate a lot of heat. In order to prevent the blade from wielding off, sufficient cutting fluid shall be poured at the beginning of cutting. In order to increase the strength of the tool body, the lower part of the tool body is often made into a convex arc shape.

3) Reverse cutting tool

When cutting a workpiece with a larger diameter, it is easy to cause vibration due to the long tool bit and poor rigidity. At this time, the reverse cutting method can be adopted, that is, the workpiece is reversed and cut off with a reverse cutting tool, as shown in Fig. 2-6-6. In this way, when cutting, the direction of cutting force is consistent with the direction of gravity, which is not easy to cause vibration. In addition, when cutting in the reverse direction, the chips are discharged from below, which is not easy to block the workpiece groove.

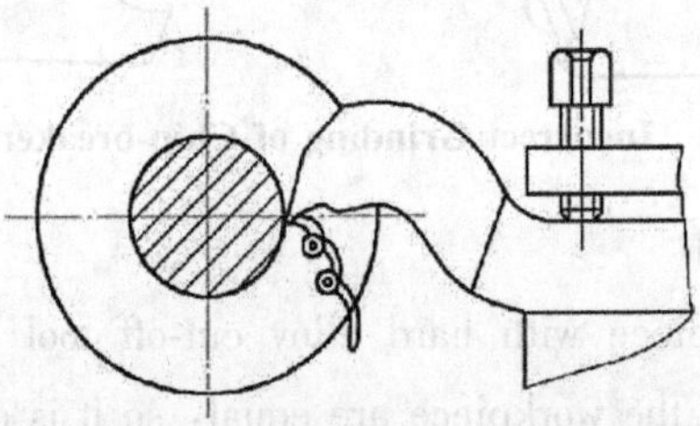

Fig. 2-6-6 Reverse Cutting Method and Reverse Cutting Tool

When using reverse cut-off, the connecting part between chuck and spindle must be equipped with safety device. At this time, the force on the tool post is upward, so the tool post should have enough rigidity.

4) Elastic cut-off tool

The blade body made of high-speed steel is clamped on the elastic handle; When the cutting amount is too large, the elastic handle will be deformed due to stress; Because the bending center of the handle is located above the handle, the tool bit will automatically give way to the tool, which can avoid breaking the tool due to scratches, as shown in Fig. 2-6-7.

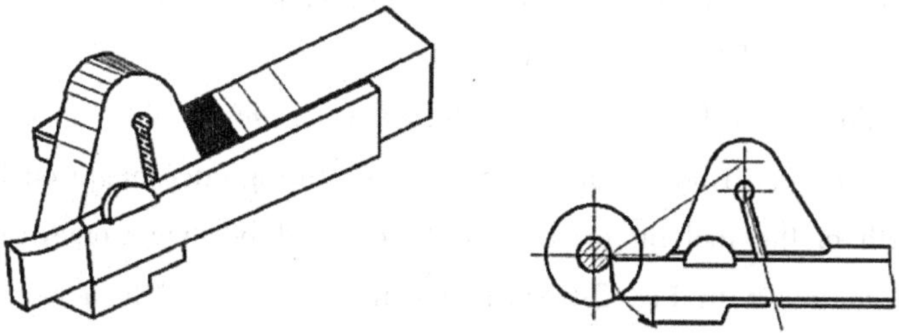

Fig. 2-6-7 Elastic Cut-off Tool

5) Grooving tool

The angle and shape of the outer groove turning tool are basically the same as that of the cut-off tool. When turning a narrow outer groove, the width of the main cutting edge of the grooving tool should be equal to the groove width, and the length of the tool body should be slightly larger than the groove depth.

3. Grinding of cut-off tool (grooving tool)

Cut-off tool is very similar to grooving tool, and their grinding is basically the same.

(1) The grinding of the cut-off tool mainly includes four basic steps, as shown in Fig. 2-6-8.

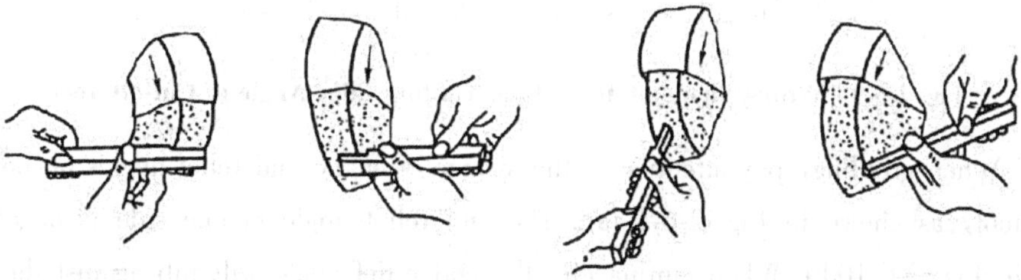

(a) (b) (c) (d)

Fig. 2-6-8 Grinding Steps of Cut-off Tool

(a) Grinding the left angled relief (b) Grinding the right angled relief (c) Grinding the main flank face (d) Grinding the rake face

① Grind the left angled relief. Hold the tool with both hands, with the rake face of the turning tool facing upward, as shown in Fig. 2-6-8(a), and simultaneously grind the end relief angle and tool minor cutting edge angle on the left side of the turning tool.

② Grind the right angled relief. Hold the tool with both hands, with the rake face of the turning tool facing upward, as shown in Fig. 2-6-8(b), and simultaneously grind the end relief angle and tool minor cutting edge angle on the right side of the turning tool.

③ Grind the main flank face. Hold the tool with both hands and grind the main relief angle of the turning tool at the same time, as shown in Fig. 2-6-8(c).

④ Grind the rake face. Hold the tool with both hands and grind the anterior angle of the turning tool at the same time, as shown in Fig. 2-6-8(d).

⑤ Grind the tool tip. Hold the tool with both hands, and grind the straight or arc transition edge at the tip of the knife respectively.

(2) Pay attention to the following points during tool grinding.

① When the tool is grinding, if the tool minor cutting edge angle on both sides is too large, the strength of the tool bit of the cut-off tool will become worse, which will easily cause breakage, as shown in Fig. 2-6-9(a); It can' t be ground to a negative value, for, doing so, straight feed method cannot be used for cutting off and phenomenon of tool clamping is easy to occur when cutting the groove, which will cause the two sides of the groove to be not perpendicular to the center of the workpiece, as shown in Fig. 2-6-9(b). Non-straight auxiliary cutting edge will cause cutting difficulties, as shown in Fig. 2-6-9(c); The left side of the turning tool shall not be worn away too much, otherwise the workpiece with high steps cannot be cut, as shown in Fig. 2-6-9(d).

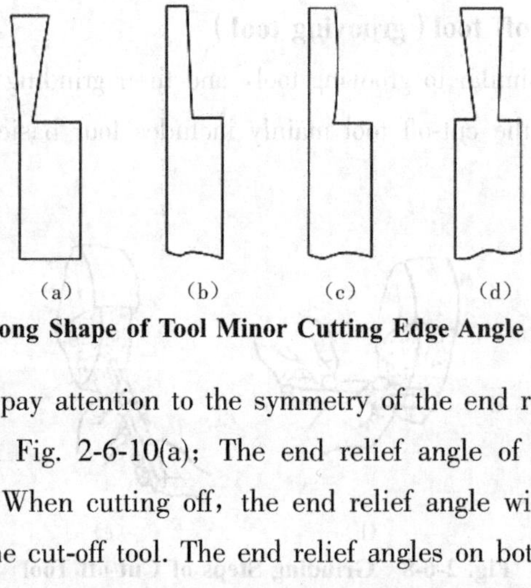

Fig. 2-6-9 Wrong Shape of Tool Minor Cutting Edge Angle of Cut-off Tool

② When grinding, pay attention to the symmetry of the end relief angles on both sides of the tool, as shown in Fig. 2-6-10(a); The end relief angle of one side is negative, as shown in Fig. 2-6-10(b). When cutting off, the end relief angle will rub against the side of the workpiece or break the cut-off tool. The end relief angles on both sides should not be too large, otherwise the strength of the tool bit will be weakened and easily broken, as shown in Fig. 2-6-10(c).

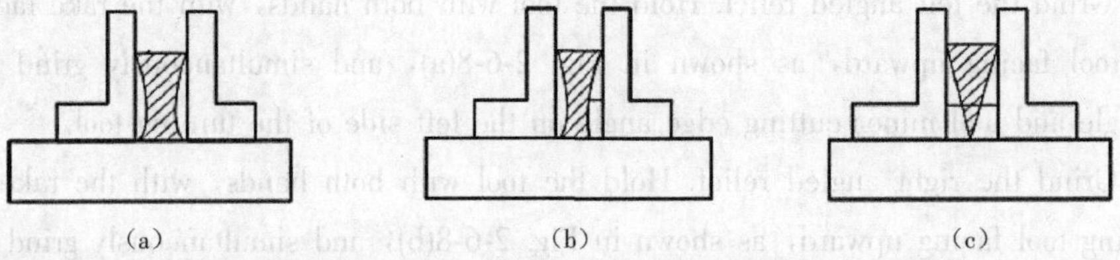

Fig. 2-6-10 Grinding Requirements for the End Relief Angle of the Cut-off Tool

(3) Safety precautions for grinding.

① Wear protective glasses when grinding the turning tool.

② When grinding the turning tool, don't use too much force to avoid slipping and hurting your hand.

③ The height of turning tool should be controlled in the horizontal center of grinding wheel, and the tool tip is slightly upward.

④ When grinding, the turning tool should move horizontally from left to right to avoid pits in the grinding wheel.

⑤ Grinding on the side of grinding wheel should be avoided.

⑥ The grinding surface of the grinding wheel must be polished frequently so that there is no obvious runout of the grinding wheel.

⑦ When grinding the turning tool, the operator should stand on the side of the grinding wheel.

⑧ When grinding high-speed steel turning tools, attention should be paid to cooling at any time to prevent annealing; When grinding hard alloy turning tools, the turning tools can't be cooled in water to prevent the blades from being broken, and at the same time, too much force can't be applied during grinding, otherwise the weld of the turning tool bit will easily fall off at high temperature.

⑨ When grinding the end relief angles on both sides, check with a metal straight edge or 90° square based on the bottom surface of turning tool.

4. Installation of cut-off tool (grooving tool)

Whether the tool clamping is correct or not has a direct impact on the machining quality.

(1) In order to increase the rigidity of cut-off tool and grooving tool, the turning tool should not be extended too long during installation.

(2) The center line of the main cutting edge of the grooving tool must be perpendicular to the axis of the workpiece to ensure that the two end relief angles are symmetrical; otherwise the grooving wall will not be straight.

(3) When installing the grooving tool, the main cutting edge of the tool must be parallel to the center line of the lathe spindle, otherwise the grooving tool will be damaged.

(4) When cutting a solid workpiece with a cut-off tool, the main cutting edge must be at the same height as the rotation center of the workpiece, otherwise the workpiece can't be turned to the center, and the turning tool is easy to be broken.

(5) The tool bit of the cut-off tool is narrower and longer, and the rigidity is worse. When cutting, the tool bit extends into the workpiece, which leads to poor heat dissipation, difficult chip removal, easy vibration and easy breakage of the tool bit. Therefore, when installing the workpiece, try to keep the cutting place close to the chuck to increase the rigidity of the workpiece.

5. Turning method of outer groove

(1) When turning a groove with low precision and narrow width, you can use a grooving tool with a tool width equal to the groove width to turn it out by using single straight feed method, as shown in Fig. 2-6-11(a).

(2) When turning a groove with precision requirements, double straight feed method is generally adopted, that is, when turning the groove for the first time, a fine turning allowance is left on both sides of the groove wall, and then fine turning is carried out according to the groove depth and groove width, as shown in Fig. 2-6-11(b).

(3) When turning a wide groove, multiple straight feed methods can be used, as shown in Fig. 2-6-11(c), and a certain fine turning allowance is left on both sides of the groove wall, and then fine turning is carried out according to the groove depth and width.

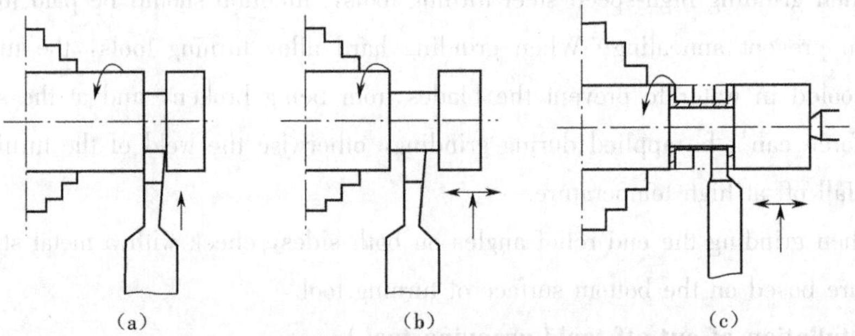

Fig. 2-6-11 Machining Method of Rectangular Groove

(4) When turning a narrow trapezoidal groove, it is generally done once with a forming tool, as shown in Fig. 2-6-12.

(5) A narrow arc groove is generally turned with a forming tool at one time.

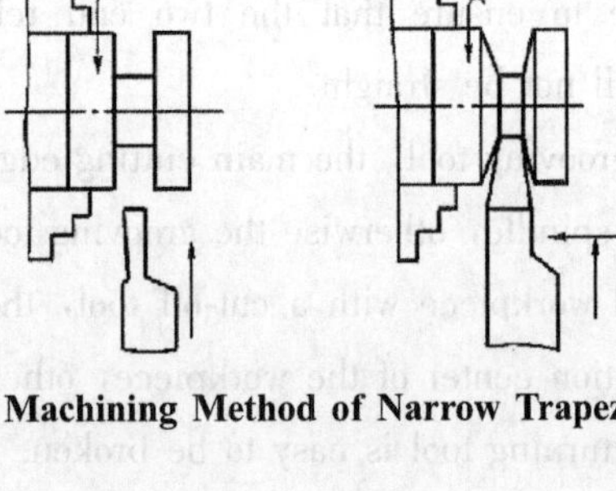

Fig. 2-6-12 Machining Method of Narrow Trapezoidal Groove

6. Cut-off method

The cut-off methods include straight feed method, reciprocating tool method and reverse cutting method, as shown in Fig, 2-6-13. Straight feed method is often used to cut brittle materials such as cast iron and bars with smaller diameters, and reciprocating tool method is often used to cut plastic materials such as steel and bars with larger diameters.

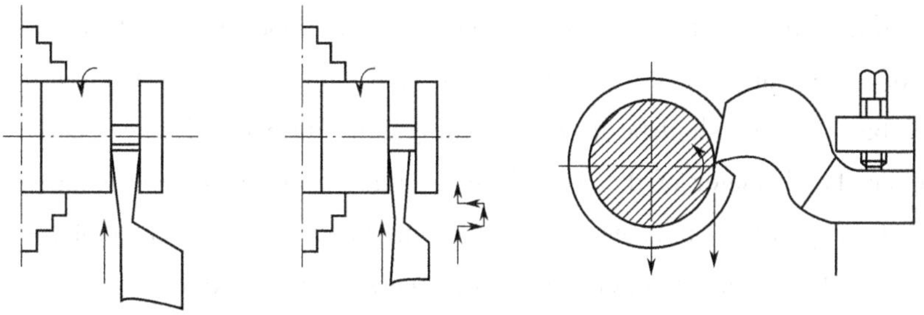

(a) (b) (c)

Fig. 2-6-13 Cut-off Method

(a) Straight feed method (b) reciprocating tool method (c) reverse cutting method

1) Straight feed method

Straight feed method refers to feed cutting in the direction perpendicular to the axis of the workpiece. This method has high efficiency, but it has high requirements for the grinding and installation of lathes and cut-off tools, otherwise it will easily cause the tool bit to break. The straight feed method is widely used as it is simple and saves materials.

2) Reciprocating tool method

The cut-off tool moves back and forth repeatedly in the axial direction of the workpiece, and feeds radially at both sides until the workpiece is cut off. This method is used in case of insufficient rigidity of the process system (system composed of lathes, tools, fixtures and workpieces).

3) Reverse cutting method

When the reverse cutting method is adopted, the spindle of the lathe reversely rotates, the lathe tool is installed reversely, and cutting and chip removal are smooth, but the lathe chuck must have a safety device.

7. Precautions for grooving and cutting-off operations

Although grooving and cutting-off are easy to operate, it is difficult to master well, especially cutting-off. If you are inattentive during operation, the tool bit will be broken. During operation, attention shall be paid to:

(1) The workpiece and tool post must be clamped firmly, and the tool post must be locked to prevent loosening. During cutting-off, the cutting-off position shall be close to the chuck to increase the rigidity of the workpiece and reduce the vibration during cutting.

(2) When the tool is installed, the tool tip must be aligned with the center of the workpiece. Too low tool position tends to break the tool, and a boss will be left at too high cutting-off position.

(3) The cutting speed should not be too quick or too slow. When the manual feed is adopted for cutting-off, the feed shall be uniform. The fit clearance between the spindle and each part of the tool post shall be small.

(4) Cutting fluid shall be used to promote lubrication and heat dissipation during cutting-off of steel parts. It is not allowed to add cutting fluid when cutting cast iron workpieces, but kerosene shall be used for cooling and lubrication if necessary.

(5) When the diameter of the workpiece is large or the workpiece is installed by one-clamping and one-jacking, it is not allowed to cut directly to the center of the workpiece, leaving 2-3 mm. After tool retraction, the workpiece shall be broken.

(6) The rake face shall not be ground too low or the chip breaker groove shall not be ground too deep in front of the cut-off tool, so as not to allow the tool bit to be easily broken due to poor chip removal.

(7) In case of reverse cutting-off, the chuck must have a safety device, and the compression nuts on both sides of the small sliding plate turntable shall also be locked; otherwise, the lathe is vulnerable to damage.

8. Measurement of grooves

(1) Grooves with low accuracy requirements. The width of the groove can be measured with a metal straight edge, and the diameter of the groove bottom can be measured with a metal straight edge and an external caliper, as shown in Fig. 2-6-14(a) and 2-6-14(b).

(2) Grooves with high accuracy requirements. Generally, the diameter of the groove bottom is measured with an outside micrometer as shown in Fig. 2-6-14(c); the width is measured with a template as shown in Fig. 2-6-14(d); the width is measured with a vernier caliper as shown in Fig. 2-6-14(e).

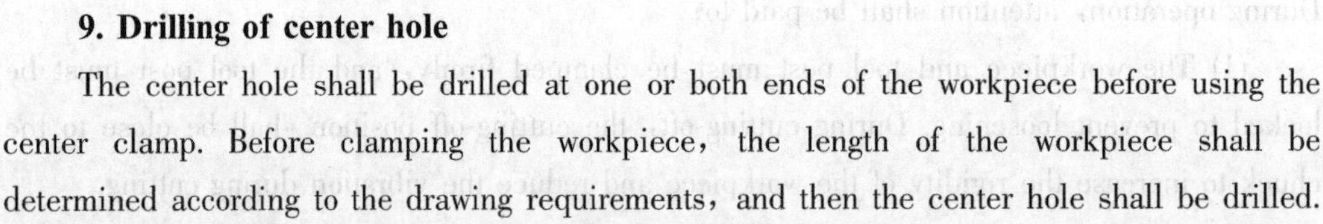
(a) (b) (c) (d) (e)

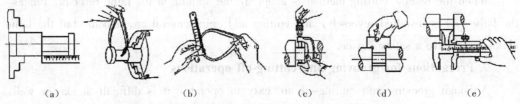

Fig. 2-6-14 Inspection and Measurement of Grooves

9. Drilling of center hole

The center hole shall be drilled at one or both ends of the workpiece before using the center clamp. Before clamping the workpiece, the length of the workpiece shall be determined according to the drawing requirements, and then the center hole shall be drilled. The shape of the center hole must comply with the requirements of the drawings and the standards.

According to the national standard GB/T145—2001, there are four types of center holes: Type A (without protective taper), Type B (with protective taper), Type C (with screw hole) and Type R (arc), as shown in Fig. 2-6-15. The cylindrical part of the center hole can be used to store grease, protect the tip, and make the tip fit well with the 60° taper hole.

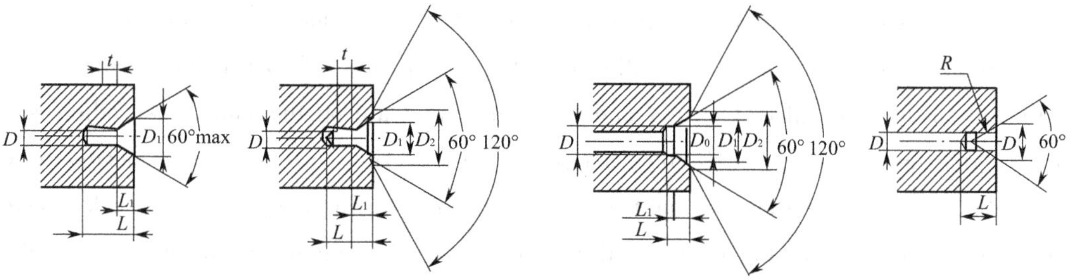

(a) (b) (c) (d)

Fig. 2-6-15 Shape of Center Hole

(a) Type A (b) Type B (c) Type C (d) Type R

The principle of selecting the center hole is as follows.

(1) For workpieces that do not require center holes or multiple clamping, the center holes without protective tapers will be selected (as shown in Fig. 2-6-15(a)).

(2) A 120° taper hole is added at the end of the center hole to protect the edge of the 60° taper hole from being damaged. The parts clamped for several times or the workpieces requiring the center hole to be retained are provided with center holes with protective tapers (as shown in Fig. 2-6-15(b)).

(3) The threads in the center hole are used for fastening connections between workpieces. If the machined part needs to fix other parts on the shaft head, the center hole with the threaded hole shall be selected (as shown in Fig. 2-6-15(c)).

(4) The arc of Type R center hole opening can reduce the friction with the tip to improve the positioning accuracy (as shown in Fig. 2-6-15(d)).

Before drilling the center hole with the center drill, the end face shall be leveled before drilling. During drilling, the spindle speed is quick; the tool shall be fed slowly and coolant shall be filled; in order to prevent the bit from breaking, the drilling cuttings must be removed in time.

10. Precautions for workpiece clamping

(1) When rough turning with center clamping workpieces is used, due to the large machining allowance, the live center shall be selected to avoid damaging the center hole when the cutting force is too large. During fine turning, the machining accuracy is mainly considered, the cutting allowance is small and the cutting force is also small. The dead center can be selected and greased to prevent burning loss due to overheating.

(2) The fit between the center hole of the workpiece and the tip shall be appropriate, and should not be too loose or too tight. If it is too loose, the workpiece cannot be centered and vibration often occurs during turning. If it is too tight, as the cutting temperature increases, the workpiece gradually elongates to make the workpiece tighter during cutting; if it is the dead center, the friction will increase and burning will occur; if it is the live

center, it is easy to damage the internal structure of the center due to excessive pressure.

(3) Before the workpiece is installed with the center, the tailstock center shall also be corrected so that the tailstock center and the front center are on the same axis; otherwise, the external taper turning will become a taper surface. During correction, it is necessary to move the tailstock to the chuck, hold the machined taper in the center of the chuck, make the center close to the taper tip, and visually check for alignment. If not, adjust the tailstock body laterally to make it meet the requirements. Then the workpiece is installed again, the diameter at both ends of the workpiece is measured after cutting on one end, and the lateral position of the tailstock is adjusted according to the diameter. If the diameter of the right end of the workpiece is large and the diameter of the left end is small, the tailstock shall be offset towards the operator; otherwise, the tailstock shall be offset in the opposite direction.

(4) The tailstock center sleeve shall protrude shorter to enhance the rigidity of the tailstock and reduce the vibration during cutting, and the tailstock screws shall be tightened.

[Thinking and practice]

(1) What are the types of cut-off tools ?

(2) How are cut-off tools installed ?

(3) Briefly describe the method of cutting-off and grooving.

(4) How to measure grooves ?

(5) What is the selection principle of center hole ?

Task VII External taper turning

I. Drawings and technical requirements

As shown in Fig. 2-7-1, The blank size of the workpiece is $\phi 50$ mm×70 mm and the material is 45 steel. Scoring table see Table 2-7-1.

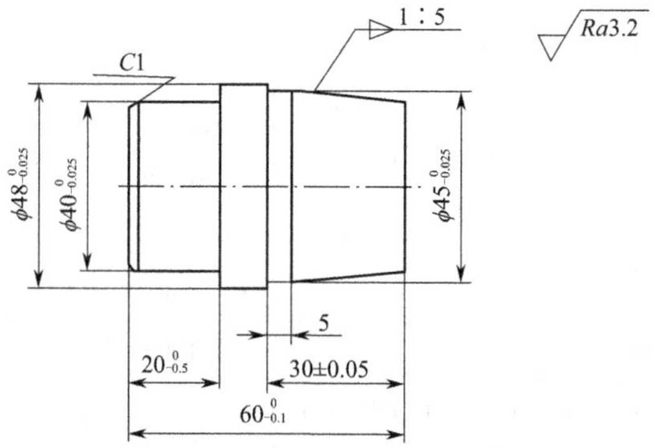

Fig. 2-7-1 External Taper Machining

Table 2-7-1 Scoring Table

Technical requirements	Assigned score	Measured results	Score
$\phi\ 48_{-0.025}^{\ 0}$ mm	10		
$\phi\ 40_{-0.025}^{\ 0}$ mm	10		
$\phi\ 45_{-0.025}^{\ 0}$ mm	10		
1 : 5 (5°42′ ±4′)	20		
30 mm±0.05 mm	6		
5 mm	5		
$60_{-0.1}^{\ 0}$ mm	5		
$20_{-0.05}^{\ 0}$ mm	5		
Chamfering *C*1	3		
Surface roughness *Ra*3.2 μm	16		
Safe and civilized operation	10		

II. Analysis of processing

Refer to Table 2-7-2 for the process arrangement of each profile of the part.

Table 2-7-2 Process Card for Ordinary Lathe Machining

<table>
<tr><td>Part drawing No.</td><td colspan="2">Fig. 2-7-1</td><td colspan="2" rowspan="2">Process card for ordinary lathe machining</td><td>Machine model</td><td>CA6140</td></tr>
<tr><td>Part name</td><td colspan="2">External taper</td><td>Machine tool number</td><td></td></tr>
<tr><td colspan="5">Table of turning tools</td><td colspan="2">Table of measuring tools</td></tr>
<tr><td colspan="2">Name of turning tools</td><td colspan="2">Turning tool parameters</td><td>Measuring tool name</td><td colspan="2">Specification/mm</td></tr>
</table>

Continue Table

90° external turning tool	YT15, high speed steel turning tool		Vernier caliper Micrometer	0-150/0.02 25-50/0.01		
Working procedure	Process content	Cutting amount			Machining Property
		S/ (r/min)	*F*/ (mm/r)	a_p/mm	
1	Radial turning	1 100	—	—	Manual
2	External taper rough turning to $\phi 46$ mm, with guaranteed length of 29 mm	450	0.33	2	Automatic
3	Turn around to clamp the turned step at $\phi 46$mm×29mm	—	—	—	Manual
4	Radial turning	1 100	—	—	Manual
5	External taper rough turning to $\phi 48$ mm, with guaranteed length of 30 mm, and allowance reserved	450	0.33	2	Automatic
6	External taper rough turning to $\phi 40$ mm, with guaranteed length of 20 mm, and allowance reserved	450	0.33	2	Automatic
7	External taper fine turning to $\phi 48_{-0.025}^{0}$ mm, with guaranteed length of $20_{-0.05}^{0}$ mm	1 100	0.08	0.2	Automatic
8	External taper fine turning to $\phi 48_{-0.025}^{0}$ mm, with guaranteed length of 30 mm	1 100	0.08	0.2	Automatic
9	Chamfering and smoothing	450	—	—	Manual
10	Turn around to fine turning end face, with guaranteed total length of $60_{-0.1}^{0}$ mm	1 100	—	—	Manual
11	External taper fine turning to $\phi 45_{-0.025}^{0}$ mm, with guaranteed length of 30 mm	1 100	0.08	0.2	Automatic
12	Rotate the 1 ∶ 5 taper of the small sliding plate rough turning and gradually align the 1 ∶ 5 taper, leaving an allowance of 0.3-0.5 mm for fine turning	560	—	0.5	Manual
13	Taper fine turning, with guaranteed length of 25 mm	1100	—	0.2	Manual
14	Chamfering	450	—	—	Manual

III. Knowledge and skills

1. Application of tapers

Taper surface mating is widely used in mechanical engineering, such as mating between center tail handle and tailstock sleeve, and mating between taper pin and taper hole. The taper surface fits closely and is easy to assemble and disassemble. After many times of disassembly, it can still ensure accurate centering effect, and the small taper mating surface can also transmit large torque. Therefore, taper shanks are used for large diameter of twist drill, as shown in Fig. 2-7-2.

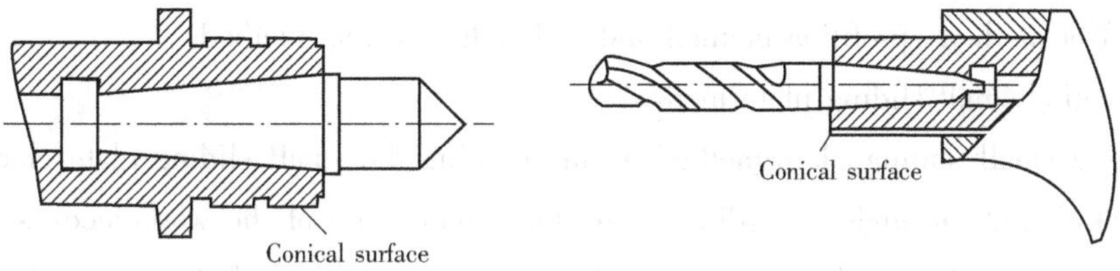

Fig. 2-7-2 Twist Drill with Taper Shank

2. The name of each part of the taper

As shown in Fig. 2–7–3.

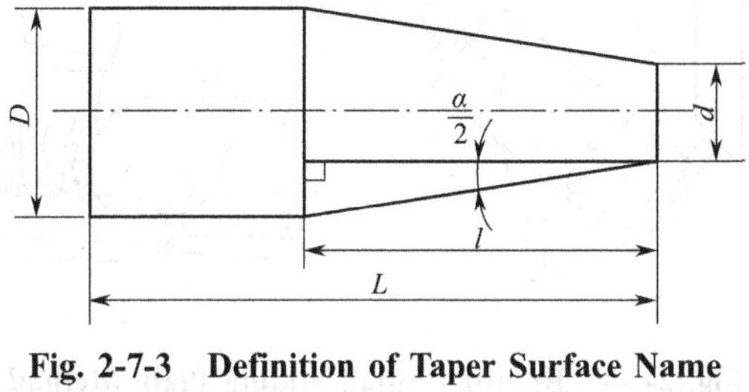

Fig. 2-7-3 Definition of Taper Surface Name

Where, D is the diameter of large end of the taper; d is the diameter of small end of the taper; L is the length of the part; l is the length of the taper, i.e. the axial distance (mm) between the maximum taper diameter and the minimum taper diameter; α is the taper angle.

The ratio of the difference between the diameters of small end and large end of the taper and the length of the taper is called taper C, i.e.

$$C = \frac{D-d}{l}$$

$$\tan\frac{\alpha}{2} = \frac{D-d}{2l}$$

In addition to the requirements for dimensional accuracy, shape accuracy and surface roughness, the taper surface shall be machined with taper requirements.

3. Selection and installation of taper surface tool

In addition to using the same tool as the external taper turning and inner hole, wide-edge turning tool and tapered reamer can also be used. When the turning tool is installed, the turning tool tip must be strictly aligned with the rotation center of the workpiece; otherwise, the taper plain line is not straight, affecting the machining quality.

4. Taper turning method

The taper plain line intersects with the axis to form a taper half-angle, so when the taper is turned, the correct taper surface can be turned out only when the movement track of the turning tool is parallel to the taper plain line.

Common methods for turning taper surfaces include rotating small sliding plate method, tailstock offset method, profiling method and wide-edge turning method.

1) Rotating small sliding plate method

Rotating small sliding plate method means rotating the small sliding plate clockwise or counterclockwise at an angle according to the taper half-angle of the workpiece, so that the movement track of the turning tool is parallel to the plain line of the required machined taper in the horizontal axis plane, and rotating the small sliding plate handle uniformly and continuously with both hands to manually feed taper surface turning, as shown in Fig. 2-7-4.

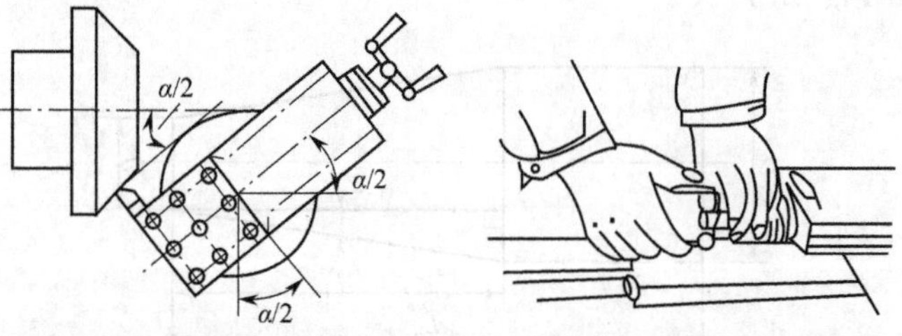

Fig. 2-7-4 Rotating Small Sliding Plate Method

(a) Rotation angle of small sliding plate (b) Rotation of small sliding plate handle alternately with both hands

(1) Rotating small sliding plate method.

① Loosen the two nuts on the turntable under the small sliding plate with a wrench.

② Determine the rotation direction of the small sliding plate counterclockwise and clockwise according to the external taper surface on the workpiece.

Turn the positive external taper surface (also called the forward turning), i.e. when the large end of the taper is close to the spindle and the small end is close to the tailstock, the small sliding plate shall rotate counterclockwise. Turn the reverse external taper surface (also called the reverse turning), i.e. when the small end of the taper is close to the spindle and the large end is close to the tailstock, the small sliding plate shall rotate clockwise.

③ Rotate the small sliding plate to the required position according to the taper half-angle of the workpiece, align the reference zero line of the small sliding plate with the taper half-angle marking line, and then lock the nut on the turntable.

④ If the taper half-angle of the workpiece is not an integer, the decimal part can be estimated by visual inspection, and after approximate alignment, it can be gradually aligned by trial run.

⑤ Check and adjust the fit clearance between the guide rail of the small sliding plate and the strip.

(2) Characteristics of rotating small sliding plate method.

① The taper surface with larger taper angle α can be turned.

② It is easy to operate and the internal and external taper surfaces of any taper angle can be machined. The taper surface and the taper hole can be turned and trimmed with a wide application range. When the taper surfaces of different taper angles are turned on the same workpiece, the angle shall be easily adjusted.

③ Only manual feeding is allowed, with high requirements for operators, high labor intensity, and difficult control of workpiece surface roughness values, so it is only suitable for single piece and small batch production.

④ The machining length is limited by the stroke of the small sliding plate, and only the shorter taper surface can be machined.

(3) Installation of tool in the rotating small sliding plate method.

The turning tool tip must be strictly aligned with the slewing center of the workpiece; otherwise, the turned taper plain line is not a straight line but a hyperbolic curve (the turning tool clamping method and the method of aligning the workpiece center are the same as those for turning end face).

(4) Procedures for turning external taper surface.

① Turn the taper part into a cylinder first according to the maximum taper diameter (plus 1 mm allowance) and the taper length.

② Move the medium and small sliding plates so that the turning tool tip is in exact contact with the external taper surface at the axle end, as shown in Fig. 2-7-5. Then withdraw the small sliding plate backward and adjust the scale of the middle sliding plate to zero position as the starting position of rough turning of external taper surface.

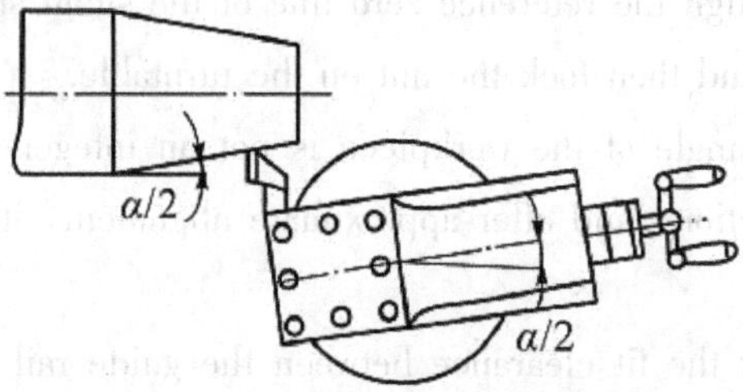

Fig. 2-7-5 Light Contact of Tool with Workpiece

③ Move the middle sliding plate, adjust the back cutting depth, start the lathe, rotate the small sliding plate handle alternately with both hands, and feed cutting at a uniform speed, as shown in Fig. 2-7-6. When turning to the terminal, withdraw the middle sliding plate and make the small sliding plate quickly back and reset.

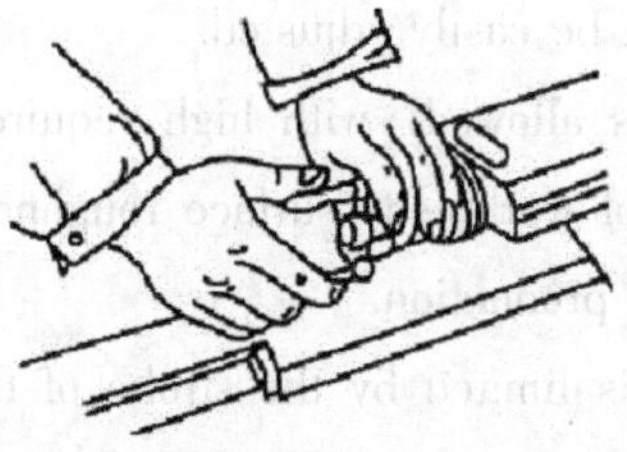

Fig. 2-7-6 Uniform Manual Feeding with Both Hands

④ Repeat step ③, adjust the back cutting depth, and manually feed the external taper surface of rough turning until the workpiece can be plugged into the socket gauge by about 1/2.

⑤ Detect the taper angle with a socket gauge, template or vernier universal angle meter and align the corner of the small sliding plate.

⑥ After alignment, continue rough turning of the external taper surface, leaving an allowance of 0.5-1.0 mm for fine turning.

⑦ After adjusting the corner of the small sliding plate accurately, fine turn the external taper surface to meet the requirements.

(5) Alignment of taper angle of external taper surface.

When the external taper surface is fine turned, the taper angle of the external taper surface must be aligned. Therefore, it is required to start alignment when the half of the external taper surface is fine turned, and the main methods for aligning the taper angle or taper are as follows:

① Light transmission test with angle template: As shown in Fig. 2-7-7, when checking with angle template, the angle of the small sliding plate is aligned mainly by the amount of light transmission, which is repeated several times until the requirements are met.

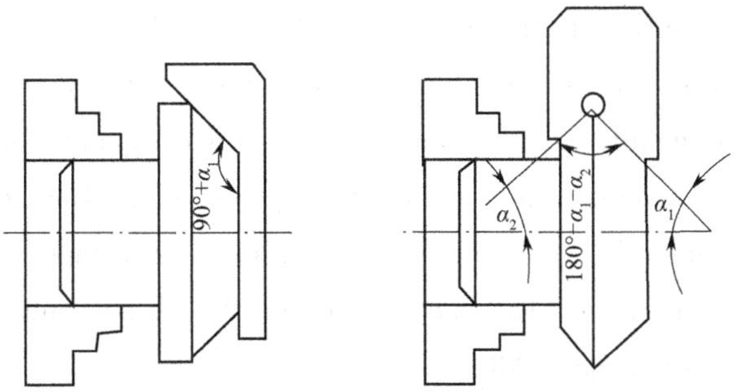

Fig. 2-7-7 Detection of Taper Angle with Angle Template

② Detection with a vernier universal angle meter: The vernier universal angle meter is adjusted to the angle to be measured so that the base ruler is against the end face through the center of the workpiece, and the knife straightedge is against the plain line on the external taper surface, and the taper half-angle is measured and adjusted as required, as shown in Fig. 2-7-8.

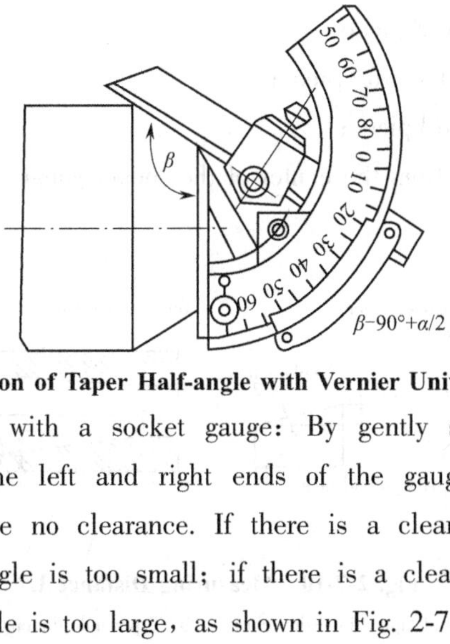

Fig. 2-7-8 Detection of Taper Half-angle with Vernier Universal Angle Meter

③ Coloring detection with a socket gauge: By gently sleeving the gauge on the workpiece, and holding the left and right ends of the gauge to swing up and down respectively, there shall be no clearance. If there is a clearance at the large end, it indicates that the taper angle is too small; if there is a clearance at the small end, it indicates that the taper angle is too large, as shown in Fig. 2-7-9. At this time, loosen the turntable nut, tap the small sliding plate with a copper hammer as required to make it rotate slightly, and then tighten the nut. Retest after commissioning until the requirements are met.

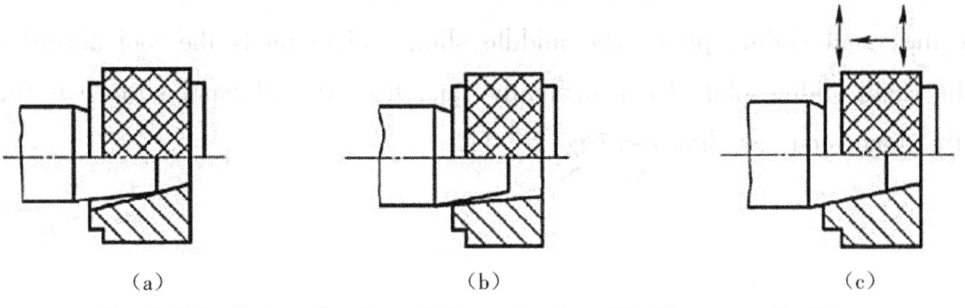

(a) (b) (c)

Fig. 2-7-9 Determination of Taper Angle with Clearance Position

(a) Too small taper angle (b) Too large taper angle (c) Suitable taper angle

(6) Control of geometric dimensions of external taper surface.

After the corner of the small sliding plate is adjusted accurately, fine turning of external taper surface mainly aims to improve the surface quality of the workpiece and control the dimensional accuracy of the external taper surface. Therefore, the turning tool must be sharp and wear-resistant during fine turning of external taper surface, and the feed shall be uniform and continuous. The back cutting depth can be controlled and adjusted by middle sliding plate and moving saddle.

① The back cutting depth is adjusted by middle sliding plate (calculation method). First measure the distance L (as shown in Fig. 2-7-10) from the small end face of the workpiece to the over-end interface of the socket gauge, and calculate the back cutting depth with the following formula:

$$a_p = L\tan\frac{\alpha}{2} = L\frac{C}{2}$$

Where: a_p is the back cutting depth (mm) when the center of the plug gauge or step is L from the end face of the workpiece;

α is the taper angle of the workpiece;

C is the taper of the workpiece;

L is the distance (mm) from the center of the socket gauge step to the small end face of the workpiece.

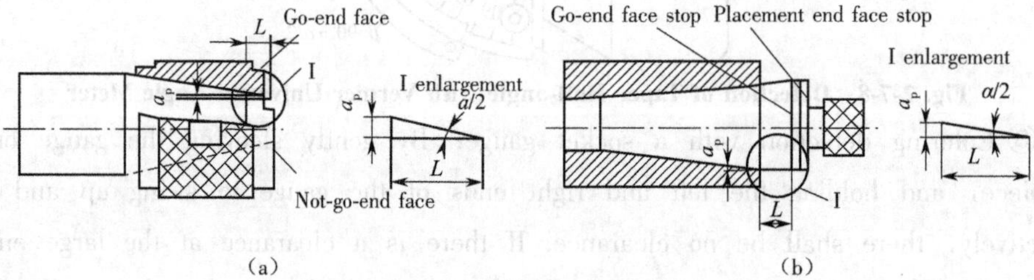

Fig. 2-7-10 Measuring Distance L

(a) External taper dimension check (b) Taper hole dimension check

Then move the medium and small sliding plates so that the tip of the tool touches the external taper surface of the small end of the taper of the workpiece gently and then withdraw the small sliding plate. The middle sliding plate feeds the tool according to the value, the small sliding plate feeds manually, and the external taper surface is fine turned to meet the dimension, as shown in Fig. 2-7-11.

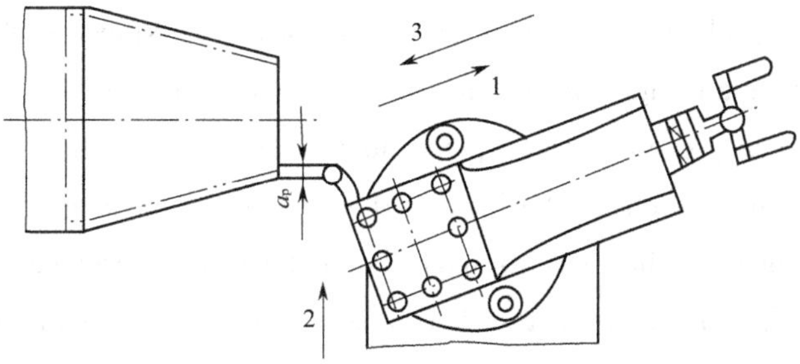

Fig. 2-7-11 Adjustment of Back Cutting Depth of Fine turning with Medium Sliding Plate a_p

② The back cutting depth is adjusted by moving saddle. According to the measured length a, make the turning tool tip touch the external taper surface of the small end of the taper of workpiece gently, withdraw the small sliding plate backward, and make the turning tool leave the end face of the workpiece for a distance along the axial direction (the clearance of the small sliding plate screw shall be eliminated before adjustment), as shown in Fig. 2-7-12. Then move the saddle so that the turning tool contacts the end face of the workpiece, as shown in Fig. 2-7-13. At this time, although the middle sliding plate is not moved, the turning tool has cut into a required back cutting depth a_p.

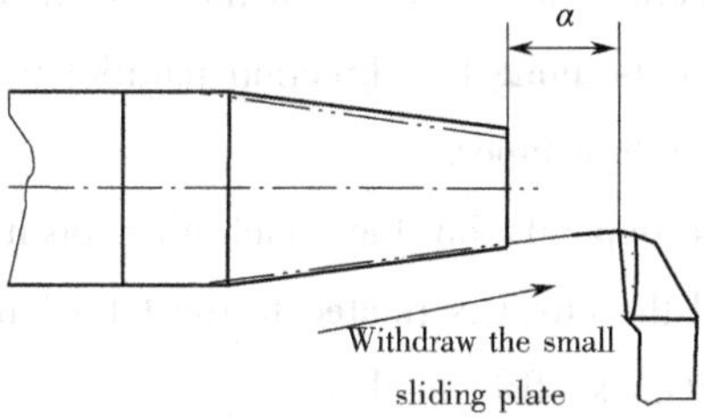

Fig. 2-7-12 Adjustment of Back Cutting Depth of Fine turning by Withdrawal of Small Sliding Plate

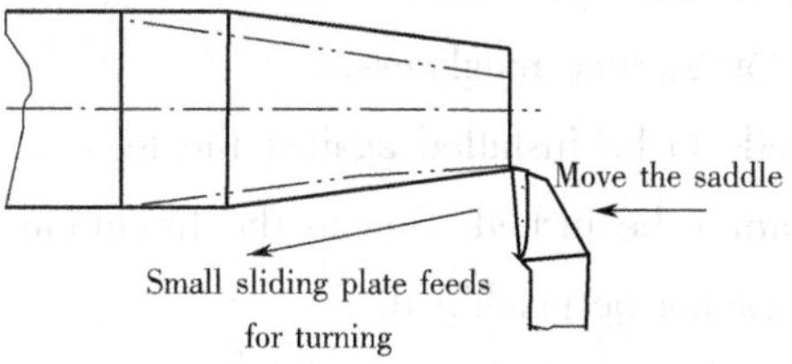

Fig. 2-7-13 Adjustment of Back Cutting Depth by Moving Saddle

(7) Attention shall be paid to the following contents when the taper surface of the sliding plate transfer method is used.

① The turning tool must be aligned with the rotating center of the workpiece; otherwise, hyperbolic error will occur.

② During manual feeding, the small sliding plate handle is held with both hands to move the small sliding plate evenly, and the surface of the workpiece shall be turned once.

③ During rough turning, the amount of feed should not be too large, and the taper shall be aligned first to prevent the workpiece from being scrapped due to small size.

④ When the taper is checked with a vernier universal angle meter and vernier angle gauge, the measuring edge shall pass through the center of the workpiece.

⑤ When check is made with a socket gauge, the surface roughness of the workpiece shall be small, the color shall be uniform and thin, and the rotation amount shall be generally within half a circle to avoid misjudgement.

⑥ When the small sliding plate is rotated to adjust the angle, the angle turned by the small sliding plate shall be greater than the taper half-angle, and then gradually aligned. When slightly adjusting the small sliding plate angle, just loosen the fastening nut slightly, press the left thumb against the small sliding plate and the middle sliding chassis, tap the small sliding plate slightly with a copper rod, and determine the fine adjustment according to the feeling of the finger.

2) Tailstock offset method

The sliding plate on the tailstock is laterally offset by a distance S, so that the angle between the two center connections after offset and the original two center centerlines is $\alpha/2$. When the saddle moves and cuts along the direction parallel to the spindle with the turning tool, the workpiece is turned into a taper.

The tailstock deflection depends on the machining position of the workpiece heads between the two centers, and the offset is related to the total length of the workpiece.

(1) Characteristics of tailstock offset method.

① It is applicable to workpieces with small taper, low accuracy requirements and long taper part.

② Longitudinal maneuverable feed turning can be adopted. The machined surface has uniform cutting marks and low surface roughness.

③ The workpiece needs to be installed against the top, so the whole taper cannot be turned or the taper hole cannot be turned. Due to the limitation of the tailstock offset, the workpiece with large taper cannot be machined.

④ Since the center is skewed in the central hole and cannot have good contact, the wear of the center and central hole is not uniform.

(2) Calculation of tailstock offset.

As shown in Fig. 2-7-14, the tailstock offset S can be approximately calculated by the following formula:

$$S = \frac{D-d}{2L_0}L = L\tan\frac{\alpha}{2} = \frac{L_0}{2}C$$

Where, S is the tailstock offset;

L_0 is the length of the taper part of the workpiece;

L is the total length of the workpiece;

D and d are the diameters of the large and small ends of the taper;

C is the taper of the workpiece.

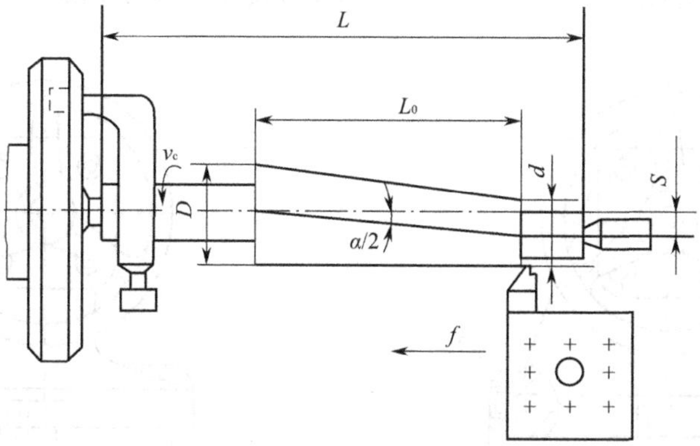

Fig. 2-7-14 Tailstock Offset

(3) Installation of workpieces.

① Adjust the position of the tailstock on the lathe guide rail so that the distance between the front and rear centers is equal to the total length of the workpiece and the extension length of the tailstock sleeve is less than 1/2 of the total length of the sleeve.

② Fill the central holes at both ends of the workpiece with grease first, then clamp the lathe dog at one end of the workpiece, and finally clamp the workpiece between the two centers, with a moderate degree of tightness (as shown in Fig. 2-7-15).

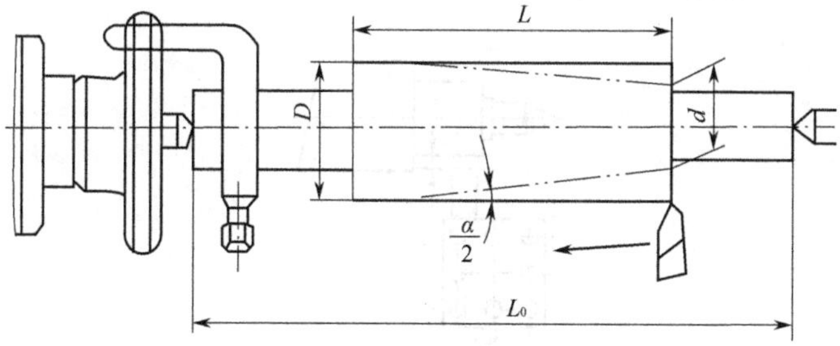

Fig. 2-7-15 Tailstock Offset

(4) Method of offsetting tailstock. The offset direction of the lathe end is determined by the taper direction of the workpiece. When the small end of the workpiece is close to the lathe end, the lathe end shall move inward; otherwise, the lathe end shall move outward. Align the front and rear centers (align the upper and lower zero lines of the tailstock), and then offset the upper position of the tailstock by the following four methods according to the

obtained offset S.

① Offset by the tailstock lower scale. Loosen the tailstock fastening nut and turn the screws 1 and 2 on both sides of the upper tailstock with a hexagon wrench to move the upper tailstock inward (operator direction) for distance S (as shown in Fig. 2-7-16). After alignment, tighten the tailstock fastening nut to prevent the offset S from changing during machining. This method can be used on a scaled lathe with a tailstock.

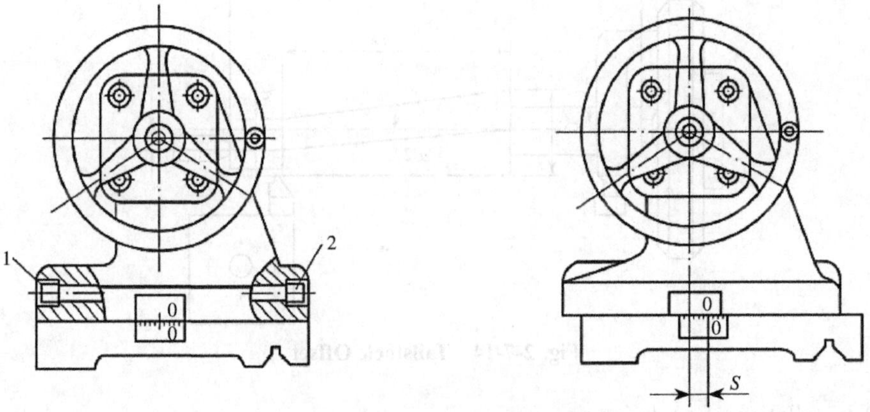

Fig. 2-7-16 Offset by Tailstock

② Offset by middle sliding plate scale. Clamp the flat end of the copper rod on the tool post, so that the flat end of the copper rod contacts the tailstock sleeve gently, record the scale of the middle sliding plate, and then move the middle sliding plate using distance *S* according to the tailstock offset (as shown in Fig. 2-7-17, move the upper position of the tailstock laterally) to make the tailstock sleeve contact the end face of the copper rod. At this time, the tailstock is laterally offset by distance *S*.

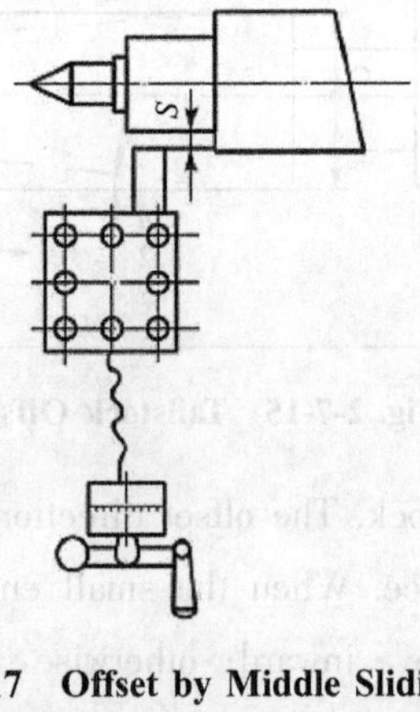

Fig. 2-7-17 Offset by Middle Sliding Plate

When moving the middle sliding plate, pay attention to removing the clearance between the middle sliding plate screw and the nut.

③ Offset by dial indicator. Fix the dial indicator on the tool post so that the measuring head of the dial indicator is in vertical contact with the tailstock sleeve and is at the same height as the center of the lathe. Adjust the dial indicator to zero position and then offset the tailstock. When the pointer on the dial indicator turns to offset value S, fix the tailstock (as shown in Fig. 2-7-18). This method can accurately adjust the offset.

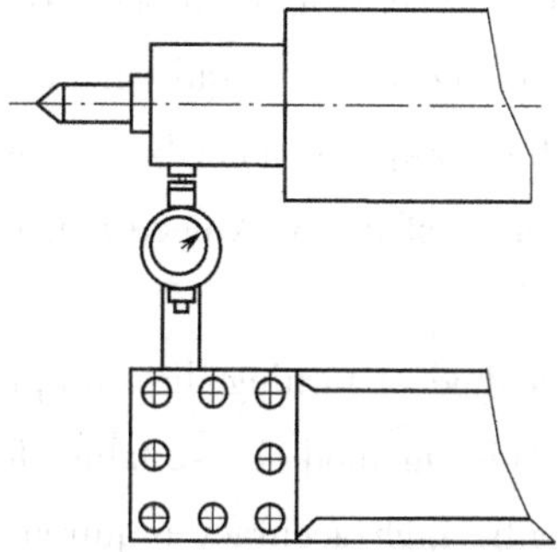

Fig. 2-7-18 Offset by Dial Indicator

④ Offset by taper measuring rod or standard sample. Install the taper measuring rod or standard sample between the two centers and fix a dial indicator on the tool post so that the measuring head of the dial indicator is in vertical contact with the taper surface of the measuring rod and is at the same height as the center of the lathe. When offsetting the tailstock, move the saddle longitudinally so that the readings of the dial indicator at both ends of the taper surface are consistent before fixing the tailstock, as shown in Fig. 2-7-19. To offset the tailstock by this method, a taper measuring rod or a standard sample of the same length as the workpiece must be selected; otherwise, the machined taper is incorrect.

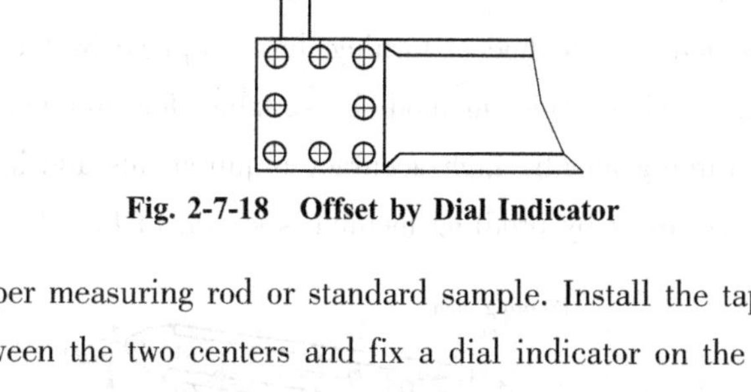

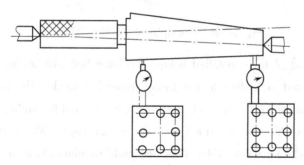

Fig. 2-7-19 Offset by Taper Measuring Rod

(5) Control of taper. In order to ensure the accuracy of taper machining, taper machining is carried out in two stages: rough turning and fine turning.

① External taper surface by rough turning: When the external taper surface by rough turning reaches 1/2 of the length, taper inspection shall be carried out first. If the taper is too large, the tailstock shall be reversely offset to reduce the offset S; if the taper is too small, the tailstock shall be offset in the same direction to increase the offset S. Rough turning can only be carried out after repeated adjustment is correct, and the allowance for semi-fine turning shall be 0.5-1.0 mm. When the external taper surface is rough turned, maneuverable feed can be used.

② The back cutting depth is determined by calculation method or moving saddle method when the external taper surface is fine turned. The external taper surface is fine turned by maneuverable feed to meet the requirements of the drawing. During mass production, the total length of the workpiece must be consistent with the size and depth of the central hole; otherwise, the taper of the workpiece machined will be inconsistent.

3) Profiling method

The profiling method is a method of turning the workpiece by the tool according to the feed of the profiling device. This method is suitable for internal and external taper workpieces with long turning length, high accuracy requirements and large production batch. The principle of turning tapers by profiling method is shown in Fig. 2-7-20.

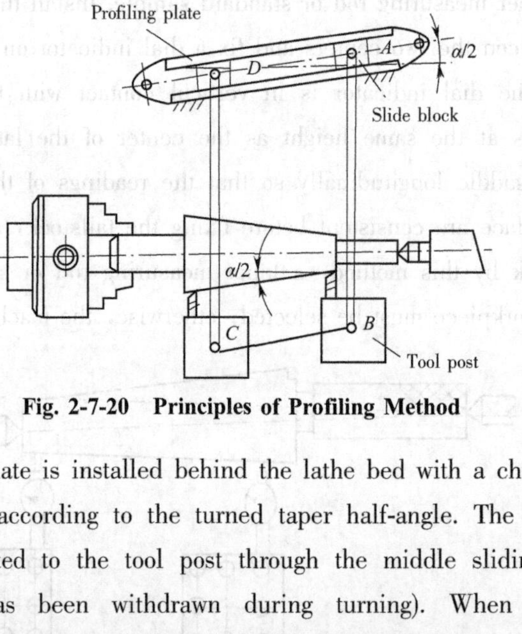

Fig. 2-7-20 Principles of Profiling Method

A fixed profiling plate is installed behind the lathe bed with a chute. The angle of the chute can be adjusted according to the turned taper half-angle. The sliding block in the chute is rigidly connected to the tool post through the middle sliding plate (the middle sliding plate screw has been withdrawn during turning). When the saddle is fed longitudinally, the sliding block slides along the profiling plate chute and drives the turning tool to move in the direction parallel to the chute, with its motion trajectory BC parallel to the chute direction AD. Therefore, the taper is turned out.

The advantages of the profiling method for turning the taper are that the taper is convenient and accurate to adjust, the central hole of the workpiece is in good contact with the center, the machining quality of the taper surface is high, and power feed turning can be carried out for the internal and external tapers with the lathe. The disadvantages are that this method can only be used on the lathe with profiling attachment, the adjustment range of profiling angle is small, and only the taper with half angle less than 12° can be turned.

4) Wide-edge turning method

Wide–edge turning method means the method of turning the external taper surface into shape once or several times with a wide-edge tool. Its working principle is essentially the molding method, so it is required that the cutting edge must be straight, and the angle between the cutting edge and the spindle axis shall be equal to the taper half-angle $\alpha/2$ of the workpiece. At the same time, the lathe is required to have good rigidity; otherwise, vibration is likely to occur.

(1) Characteristics of wide-edge turning method.

Vibration is likely to occur when turning the external taper surface with a wide-edge turning tool. Without affecting the operation, the clearance of the small sliding plate can be reduced, the extension length of the workpiece should be as short as possible, and the cutting amount should also be selected reasonably. When turning vibrates, the spindle speed should be appropriately slowed down.

(2) Selection and installation of turning tools by wide-edge turning method.

① Selection of wide-edge turning tool. For taper half-angle of 30°, 45°, 60° and 75°, the turning tool with corresponding main deflection angle can be selected, and for other taper half-angles, the turning tool with similar main deflection angle can be selected. The length of the cutting edge shall be greater than the length of the taper plain line; otherwise, grafting tool shall be used for shaping. The cutting edge shall be straight, as shown in Fig. 2-7-21. Otherwise, the taper plain line will not be straight.

② Installation of wide-edge turning tool. The clamping of the wide-edge turning tool is similar to that of the 45° end-face turning tool, and must be at the same angle as the taper surface of the taper. The clamping can be aligned with a template or a universal angle meter, as shown in Fig. 2-7-22.

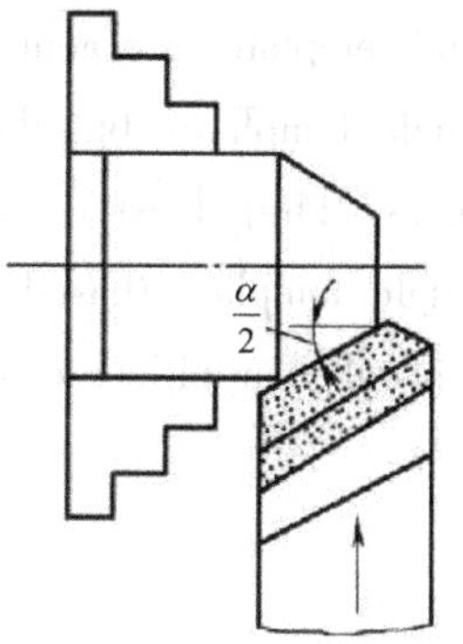

Fig. 2-7-21 Taper Turning by Wide-edge Turning Tool

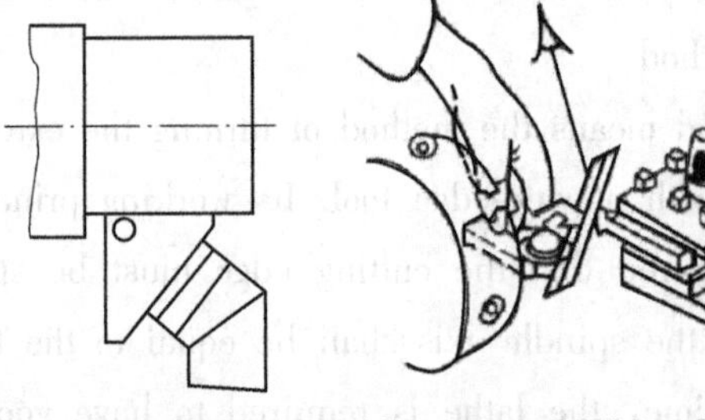

Fig. 2-7-22 Installation and Alignment of Wide-edge Turning Tool

5. Detection of external taper surface

Taper detection mainly refers to the detection of taper angle and dimensional accuracy.

1) Detection of angle or taper

(1) Measure with a vernier universal angle meter. The measurement range is 0° -320°. The measurement accuracy of this method is not high, and it is only suitable for single piece and small batch production. When the external taper angle is detected with a vernier universal angle meter, different measuring methods shall be selected according to the size of the measured angle, as shown in Fig. 2-7-23.

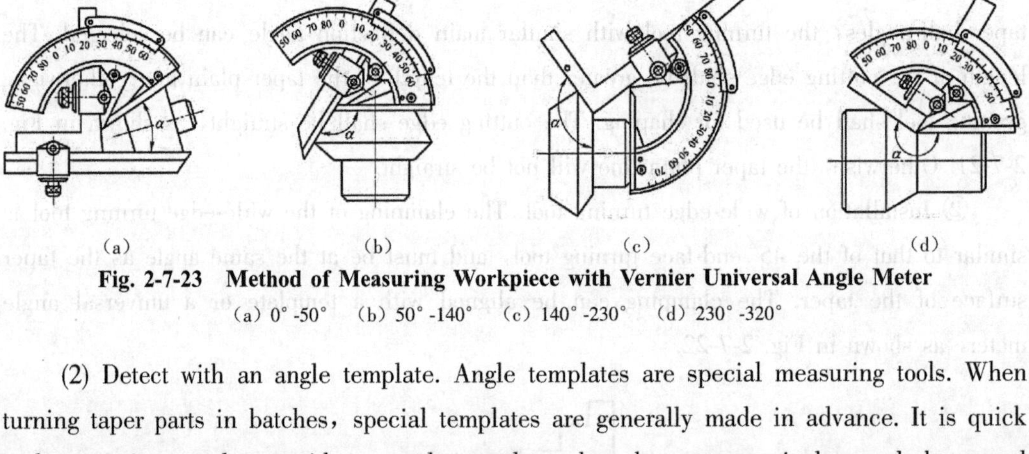

(a) (b) (c) (d)

Fig. 2-7-23 Method of Measuring Workpiece with Vernier Universal Angle Meter

(a) 0° -50° (b) 50° -140° (c) 140° -230° (d) 230° -320°

(2) Detect with an angle template. Angle templates are special measuring tools. When turning taper parts in batches, special templates are generally made in advance. It is quick and convenient to detect with an angle template, but the accuracy is low and the actual angle value cannot be measured. Fig. 2-7-24(a) shows the positive external taper angle of the bevel gear billet detected by the angle template (based on the end face); (b) shows the negative external taper angle of the bevel gear billet detected by the angle template (based on the positive external taper).

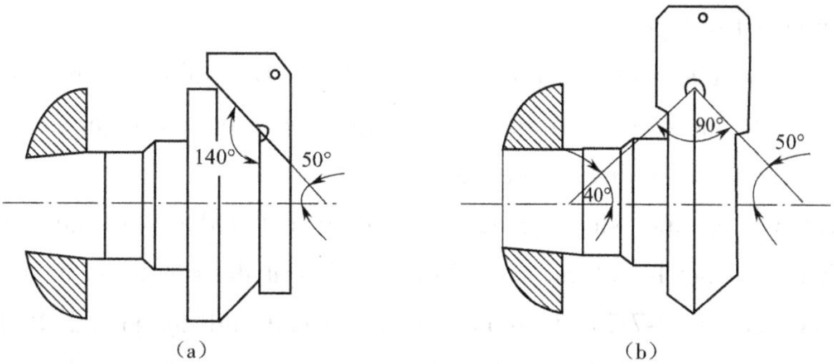

Fig. 2-7-24 Detection of Taper with Angle Template

(3) Detect by color painting. Standard taper plug gauge or taper socket gauge may be used to inspect standard tapers or workpieces with high fit accuracy requirements, as shown in Fig. 2-7-25.

The methods and steps for checking the taper with a socket gauge are as follows:

① Apply three pieces of thin display agents evenly 120° apart along the plain line on the workpiece surface, as shown in Fig. 2-7-26.

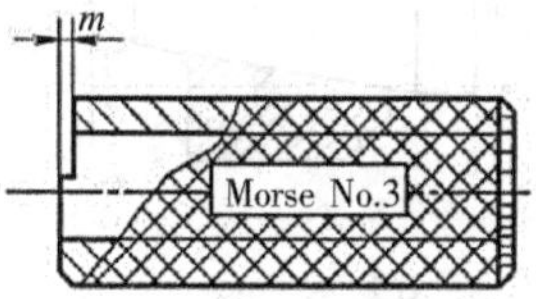

Fig. 2-7-25 Taper Socket Gauge

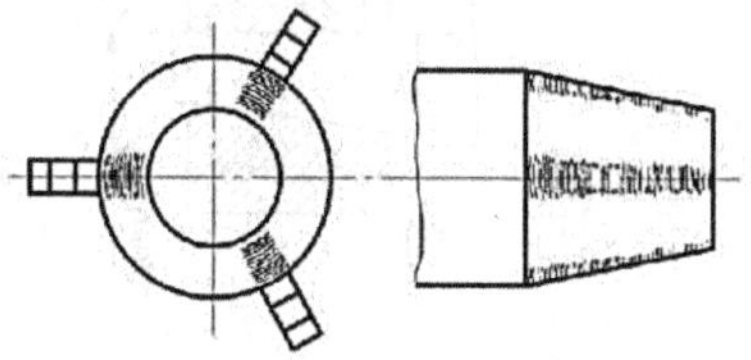

Fig. 2-7-26 Coloring Method

② Gently sleeve the socket gauge on the workpiece to rotate for not more than half a turn, as shown in Fig. 2-7-27.

③ Remove the socket gauge and observe the erasing of the display agent on the taper surface of the workpiece. If the friction trace of the display agent applied on the workpiece is uniform, it indicates that the taper hole has the correct taper, as shown in Fig. 2-7-28. If there is friction trace at the small end of the taper and there is no friction trace at the large end, it indicates that the taper has a smaller taper; otherwise, it indicates that the taper has a larger taper.

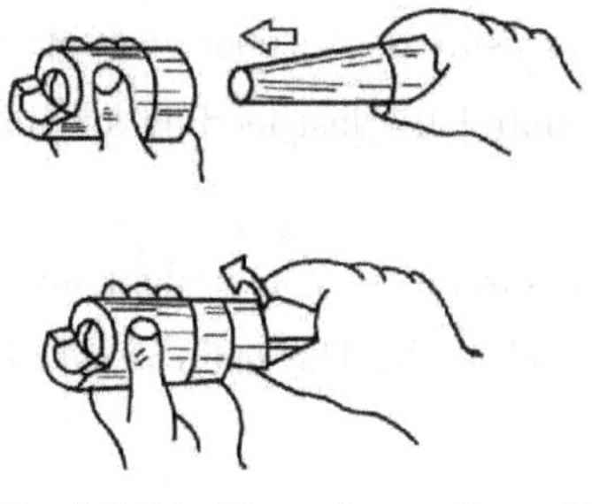

Fig. 2-7-27 Taper Inspection with Socket Gauge

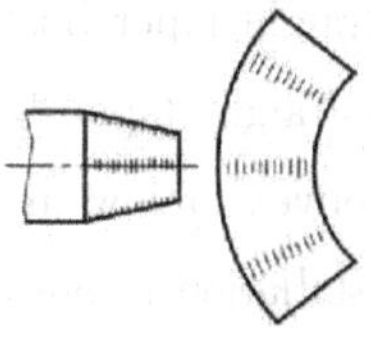

Fig. 2-7-28 Eligible Taper Expanded View

2) Measurement of taper size.

(1) Measure with a micrometer or vernier caliper. Generally, micrometers or vernier calipers are used to measure taper dimensions with low accuracy requirements or roughly measure taper dimensions during machining.

(2) Detect with a taper sleeve gauge. According to the diameter and tolerance of the workpiece, there is a notch with an axial distance of *m* at the small end of the taper sleeve gauge, as shown in Fig. 2-7-25, to represent the go end and not-go-end. If the small end face of the taper is between the notches during testing, it indicates that the small end diameter is qualified, as shown in Fig. 2-7-29(a). If the small end face of the taper fails to enter the notch, it indicates that the small end diameter is large, as shown in Fig. 2-7-29(b). If the plane of the small end of the taper exceeds the not-go-end, it indicates that the small end diameter is small, as shown in Fig. 2-7-29(c).

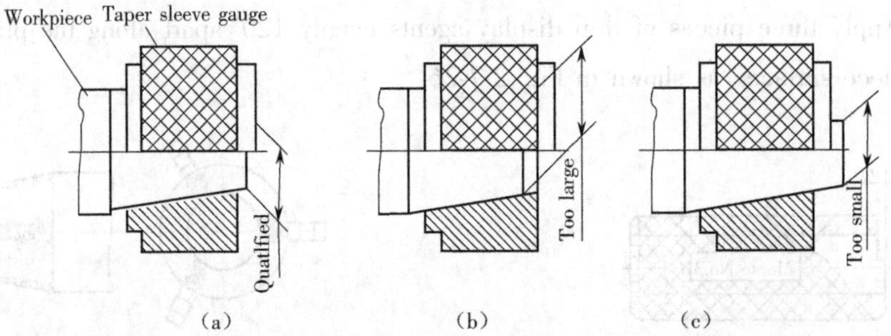

Fig. 2-7-29 Dimensions of the External Taper Detected with a Taper Socket Sleeve

6. Quality analysis of taper turning

1) Inaccurate taper

Reasons: Calculation error, or inaccurate rotation angle of small carriage and lathe end offset, which is caused by the movement of turning tool, carriage and lathe end during turning due to poor fixation; even errors in inspection and measurement may be caused by poor surface roughness of the workpiece, burrs on the gauge or workpiece or uncompleted cleaning.

2) Accurate taper but inaccurate size

Reasons: Carelessness, untimely and careless measurement, poor control of feeding amount, and especially the error caused by failing to control the last feeding amount.

3) Non-straight taper bus

The non-straight taper bus means that the taper surface is not straight and produces concave and convex or low in the middle and high at both ends. The main reason is that the turning tool installation is not aligned with the center.

4) Surface roughness not meeting the requirements

Reasons: Improper selection of cutting amount; incorrect turning tool wear or blade grinding angle; lack of surface polishing or insufficient polishing allowance; uneven manual feeding when turning the taper surface with a small trailer; in addition, large clearance of the lathe and poor rigidity of the workpiece may also affect the surface roughness of the workpiece.

[Thinking and practice]

(1) What are the methods for turning the external taper surface ?

(2) What are the methods for taper inspection of external taper surface ?

(3) What are the methods for dimensional inspection of external taper surface ?

(4) Briefly describe the method of turning the external taper surface by rotating small sliding plate method.

(5) Briefly describe the method of turning the external taper surface by tailstock offset method.

Task VIII External triangular thread machining

I. Drawings and technical requirements

As shown in Fig. 2-8-1, The blank size of the workpiece is $\phi 60$ mm×70 mm and the material is 45 steel. Scoring table see Table 2-8-1.

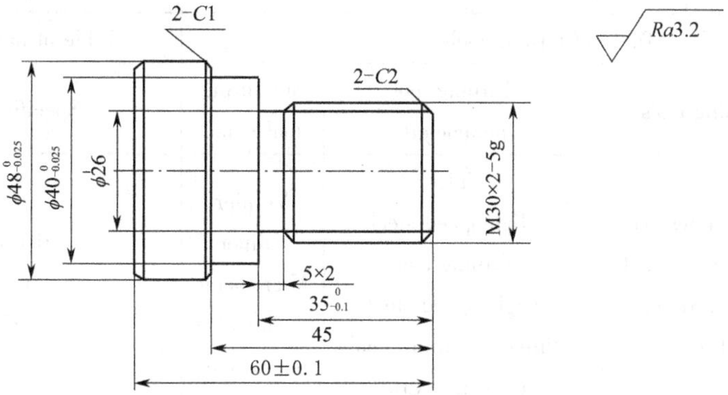

Fig. 2-8-1 External Triangular Thread Machining

Table 2-8-1 Scoring Table

Technical requirements	Assigned score	Measured results	Score
$\phi\ 48_{-0.025}^{\ 0}$ mm	10		
$\phi\ 40_{-0.025}^{\ 0}$ mm	10		
Slot 5 mm×2 mm	5		
Triangular thread large diameter ϕ30 mm	5		
Triangular thread medium diameter M30×2-5g	20		
Triangular thread angle 30°	10		
$35_{-0.1}^{\ 0}$ mm	5		
60±0.1 mm	10		
45 mm	5		
Chamfering *C*1	2		
Chamfer *C*2	2		
Surface roughness *Ra*3.2 μm	6		
Safe and civilized operation	10		

II. Analysis of processing

Refer to Table 2-8-2 for the process arrangement of each profile of the part.

Table 2-8-2 Process Card for Ordinary Lathe Machining

Part drawing No.	Fig. 2-8-1	Process card for ordinary lathe machining	Machine model	CA6140
Part name	External triangular thread		Machine tool number	
Table of turning tools			Table of measuring tools	
Name of turning tools	Turning tool parameters	Measuring tool name	Specification/mm	
90° external turning tool Slotting (cutting-off) tool Triangular thread turning tool Center drill	YT15, High-speed steel turning tool High-speed steel threaded turning tool B3 center drill	Vernier caliper Micrometer Thread micrometer	0-150/0.02 25-50/0.01 25-50	

Continue Table

Working procedure	Process content	Cutting amount			Machining Property
		S/ (r/min)	F/ (mm/r)	a_p/mm	
1	Radial turning	1 100	—	—	Manual
2	External taper rough turning to ϕ45 mm, with guaranteed length of 30 mm	450	0.33	2	Automatic
3	Turn around to clamp the turned step at ϕ45 mm×30 mm	—	—	—	Manual
4	Turning end face, drilling of center hole, supported by center	1 100	—	—	Manual
5	External taper rough turning to ϕ48 mm, with guaranteed length of 65 mm, and allowance reserved	450	0.33	2	Automatic
6	External taper rough turning to ϕ40 mm, with guaranteed length of 45 mm, and allowance reserved	450	0.33	2	Automatic
7	External taper rough turning to ϕ30 mm, with guaranteed length of 35 mm, and allowance reserved	450	0.33	2	Automatic
8	External taper fine turning of triangular thread large diameter ϕ30 mm to size	1 100	0.08	0.15	Automatic
9	External taper fine turning to $\phi 40^{0}_{-0.025}$ mm, with guaranteed length of 45 mm	1 100	0.08	0.15	Automatic
10	External taper fine turning to $\phi 48^{0}_{-0.025}$ mm, with guaranteed length of 65 mm	1 100	0.08	0.15	Automatic
11	Turning 5×2 tool retraction groove, with guaranteed relative length $35^{0}_{-0.1}$ mm	560	—	—	Manual
12	Chamfer *C*1, *C*2	450	—	—	Manual
13	M30×2-5g external triangular thread for rough and fine turning	100	—	—	Automatic
14	Cut-off, 0.5 mm allowance for total length	450	—	—	Manual
15	Turn around, pad copper sheet clamp ϕ40 mm at external taper, align with magnetic dial indicator, and tighten properly	—	—	—	Manual
16	Fine turning of end face, with guaranteed total length of 60±0.1 mm	1 120	0.1	0.1	Automatic
17	Chamfering *C*1	450	—	—	Manual

III. Knowledge and skills

1. Triangular thread

Threaded parts are widely used in mechanical products and are used for connection and transmission. For example, the connection between the lathe spindle and the chuck, the fastening of screws on the tool post to the turning tool, and the transmission of screws and nuts. There are many types of threads, which are divided into triangular threads, trapezoidal threads and rectangular threads, etc. by tooth type (as shown in Fig. 2-8-2). Threads are also divided into right-handed threads and left-handed threads (as shown in Fig. 2-8-3). Common threads (also known as metric threads) are the most widely used triangular threads. Various types and diameters of threads can be machined on the lathe, mostly for single piece and small batch production.

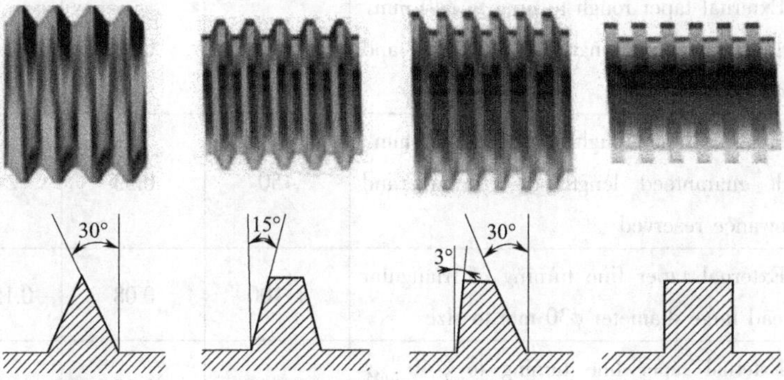

Fig. 2-8-2 Type of Threads

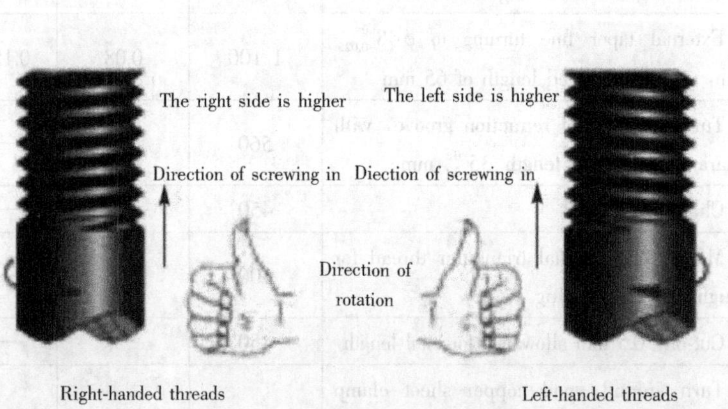

Fig. 2-8-3 Rotation Direction of Thread

2. Formation and turning principle of threaded spiral line

The spiral line can be seen as a curve formed by the beveled *AC* on the cylinder surface after the right triangle *ABC* rotates around the cylinder for one circle, as shown in Fig. 2-8-4(a). The formation of threads refers to the formation of thread profile, which is actually machined by cutting the groove of threads from the cylindrical blank, as shown in Fig. 2-8-4(b).

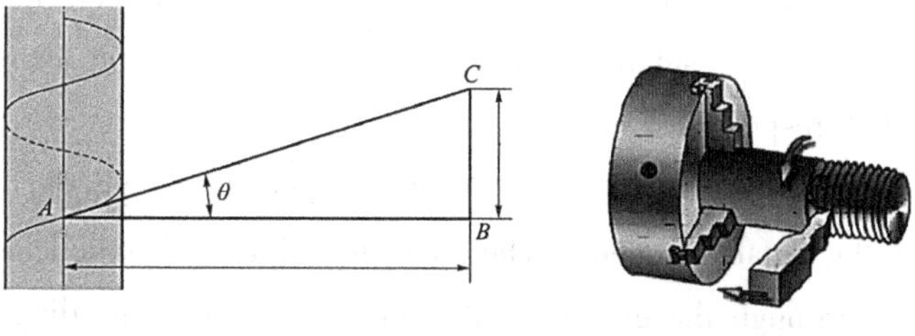

(a) (b)

Fig. 2-8-4 Formation of Thread

3. Main parameters of common triangular thread

The contour shape of the thread obtained through the thread axis profile is called the thread profile. Common profiles include triangular, trapezoidal and serrated shapes. The main parameters of thread profile are shown in Fig. 2-8-5.

1) Thread profile and thread angle

The contour shape of the thread on the cross section through the thread axis is called the thread profile. It consists of the top, bottom and both sides.

Thread angle (α): The included angle between the adjacent both sides on the profile passing through the thread axis is called the thread angle. The thread angle of most threads is symmetric to the axis vertical line, i.e. the half angle ($\alpha/2$) is equal.

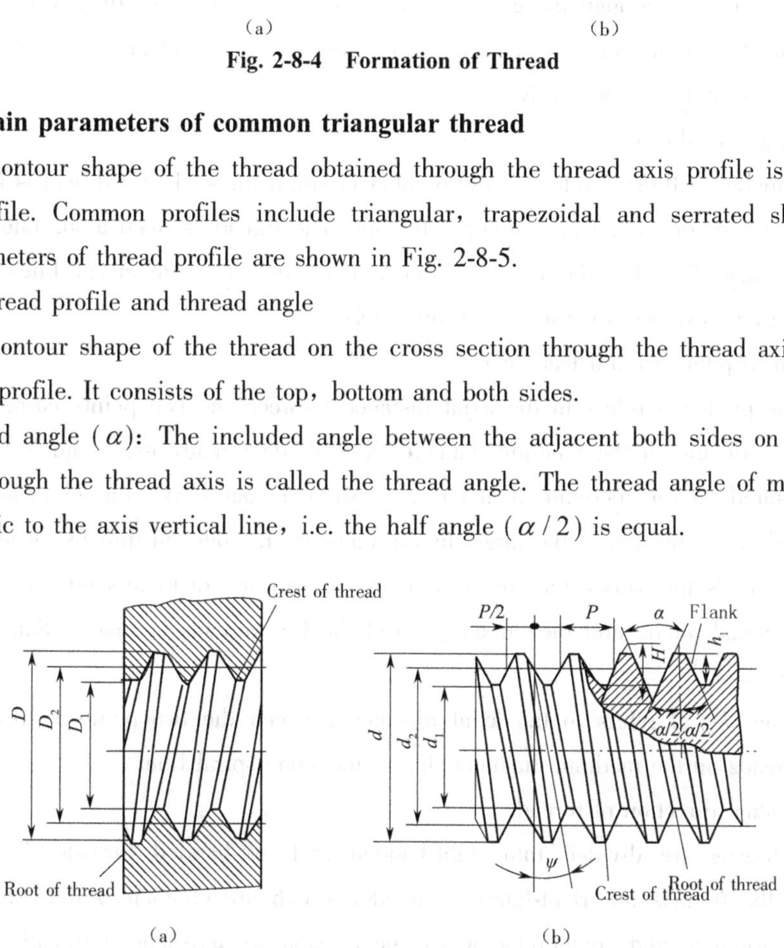

(a) (b)

Fig. 2-8-5 Main Parameters of Triangular Thread

(a) Internal thread (b) External thread

2) Thread diameters are divided

Thread diameters are divided into large, small and medium diameters.

(1) Large diameter: The diameter of the hypothetical cylindrical surface coinciding with the top of the external thread or the bottom of the internal thread is called large diameter. The large diameter is the nominal diameter. The large diameters of internal and external

threads are expressed in D and d respectively.

(2) Small diameter: The diameter of the hypothetical cylindrical surface coinciding with the bottom of the external thread or the top of the internal thread is called the small diameter of the thread. The small diameters of the internal and external threads are expressed in D_1 and d_1 respectively.

(3) Medium diameter: It is the diameter of a hypothetical cylinder, i.e. between the large diameter and the small diameter. The diameter of the hypothetical cylinder surface where the bus passes through the groove on the profile and the protruding width is equal is called the medium diameter. The medium diameters of the internal and external threads are expressed in D_2 and d_2 respectively.

3) Number of threads (n)

The number of threads refers to the number of spiral lines when a thread is formed.

There are single thread and multiple threads. The threads formed along one spiral line are called single threads; the threads formed along two or more spiral lines distributed axially at an equidistance are called multiple threads.

4) Thread pitch (P) and lead (P_h)

(1) The pitch (P) refers to the axial distance between the two points corresponding to two adjacent threads on the medium diameter line of the thread, expressed in P. P is not easily measured on the medium diameter line, so in actual work, thread measurement is often carried out at the top of the large thread diameter. In common threads, when the large thread diameter is the same, the coarse and fine threads are distinguished according to the size of the thread pitch, and the thread pitch of the fine threads is smaller than that of the coarse threads.

(2) Lead (P_h) It refers to the axial distance between the two points corresponding to adjacent threads on the medium diameter line of the same spiral line.

5) Rotation direction of thread

The threads are divided into right-handed and left-handed threads. When rotated clockwise, the threads are right-handed threads, which are characterized by low left and high right; when rotated counterclockwise, the threads are left-handed threads, which are characterized by high left and low right. Most of the threads are actually right-handed threads.

6) The height of original triangular

The height of the sharp corners intersecting the original triangle height (H) on both sides of the profile is the original triangular height.

7) The profile height (h)

The profile height (h) is the height of the profile between the top and the bottom and perpendicular to the axis of the thread.

8) The lead angle (ψ)

The lead angle (ψ) is on the medium diameter cylinder, and the included angle between the tangent of the spiral line and the plane perpendicular to the thread axis is the lead angle.

4. Dimensional calculation of common triangular thread

The common triangular thread profile is shown in Fig. 2-8-6, and the dimensional calculation of main parameters of common triangular thread is shown in Table 2-8-3.

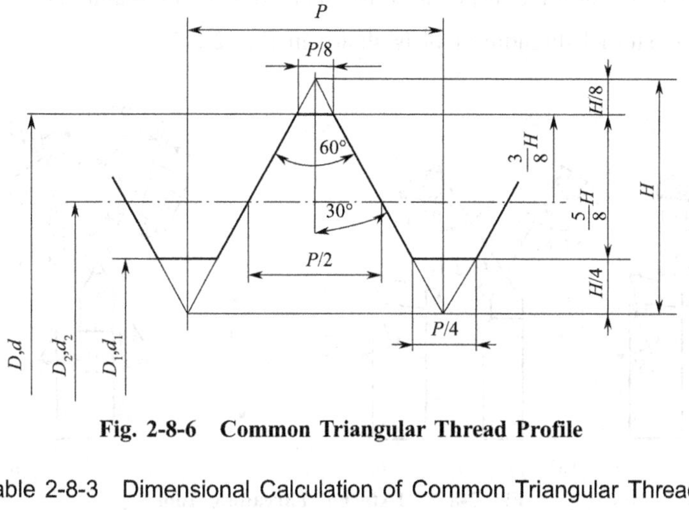

Fig. 2-8-6 Common Triangular Thread Profile

Table 2-8-3 Dimensional Calculation of Common Triangular Thread

Name		Code	Calculation formula
External thread	Thread angle	α	$\alpha = 60°$
	Original triangle height	H	$H=0.866P$
	Profile height	h	$h = \frac{5}{8}H = \frac{5}{8} \times 0.866P = 0.5413P$
	Medium diameter	d_2	$d_2 = d - 2 \times \frac{3}{8}H = d - 0.6495P$
	Small diameter	d_1	$d_1 = d - 2h = d - 1.0826P$
Internal thread	Medium diameter	D_2	$D_2 = d_2$
	Small diameter	D_1	$D_1 = d_1$
	Large diameter	D	$D=d$=nominal diameter
Lead angle		ψ	$\tan\psi = \frac{nP}{\pi d_2}$

5. Selection and installation of tools

1) Selection of threading tools

(1) External threading tool.

The thread machining must ensure the accuracy of the thread profile and pitch, and make the matching thread have the same medium diameter; otherwise, the machined thread cannot engage. In order to obtain the correct profile, the turning tool must be correctly ground. The shape of the cutting part of the thread turning tool must be completely consistent with the thread profile. The turning tool tip angle of the metric thread is 60°. When the template is used for inspection, the tool tip shall fit seamlessly with the template. The angle of the external threading tool is shown in Fig. 2-8-7.

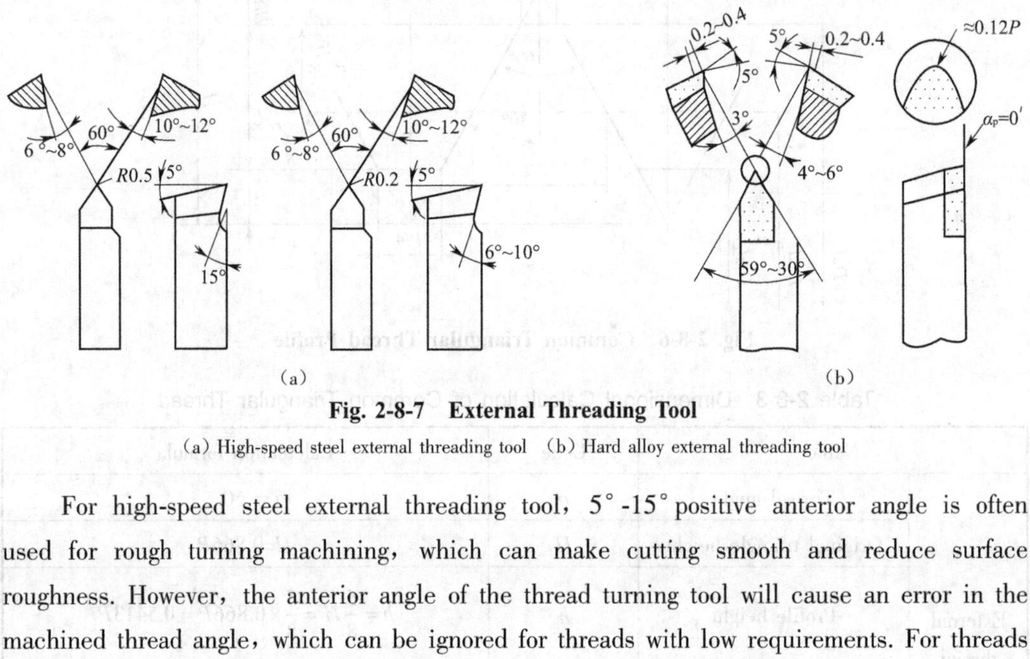

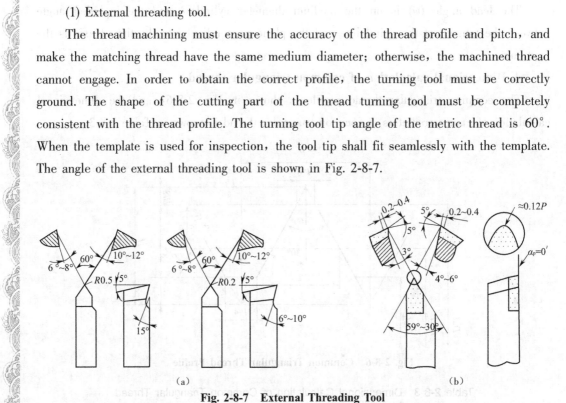

Fig. 2-8-7 External Threading Tool

(a) High-speed steel external threading tool (b) Hard alloy external threading tool

For high-speed steel external threading tool, 5°-15° positive anterior angle is often used for rough turning machining, which can make cutting smooth and reduce surface roughness. However, the anterior angle of the thread turning tool will cause an error in the machined thread angle, which can be ignored for threads with low requirements. For threads with higher accuracy requirements, the sharp angle of the threading tool needs to be ground for correction, so the thread turning tool with zero anterior angle shall be used for fine turning of threads. When the hard alloy external threading tool is cutting at high speed, the thread angle will expand. Therefore, the tool tip angle needs to be reduced by 0.5°. The working relief angle of the thread turning tool is generally 3°-5°.

Due to the influence of the thread lead angle, the relief angle of tool grinding on one side of the feed direction shall be equal to the working relief angle plus the thread lead angle. The relief angle of tool grinding on the other side shall be equal to the working relief angle minus the thread lead angle. However, the triangular thread lead angle is generally small and the impact is small, which is considered only when machining the large pitch thread.

(2) Triangular thread turning tool grinding.

Due to the limitation of the thread angle, the sharp angle of the thread turning tool is small in volume and difficult in grinding. The tool grinding requirements are as follows:

① When the anterior angle of the thread turning tool back (longitudinal) is $\gamma_0 = 0°$, the tool tip angle is equal to the thread angle; when the anterior angle of the thread turning tool back is $\gamma_0 > 0°$, the tool tip angle must be corrected.

② The cutting edges on both sides of the thread turning tool must be straight.

③ The cutting edge of the thread turning tool shall have a small surface roughness.

④ The relief angles on both sides of the thread turning tool are not equal, and a spiral lead angle shall be added or reduced on the basis of the relief angles on both sides according to the feed direction of the turning tool.

The tool grinding steps are as follows.

① Roughly grind the rake face.

② Grind the flank faces on both sides and preliminarily form the included angle between the two edges. First grind the side edge in the feed direction, and pay attention to the control of the half angle and the relief angle of the tool tip (plus a thread lead angle). Hold the tool with both hands during grinding to make the tool handle 30° to the horizontal direction of the outer circle of the grinding wheel and incline it about 8°-10° in the vertical direction, as shown in Fig. 2-8-8(a), After the turning tool contacts the grinding wheel, slightly apply pressure and move it evenly and slowly out of the flank face. Grind the side edge in the feed direction of the back, and also grasp the tool tip half angle and relief angle (minus a spiral lead angle) in the same way as the left-side edge, as shown in Fig. 2-8-8(b).

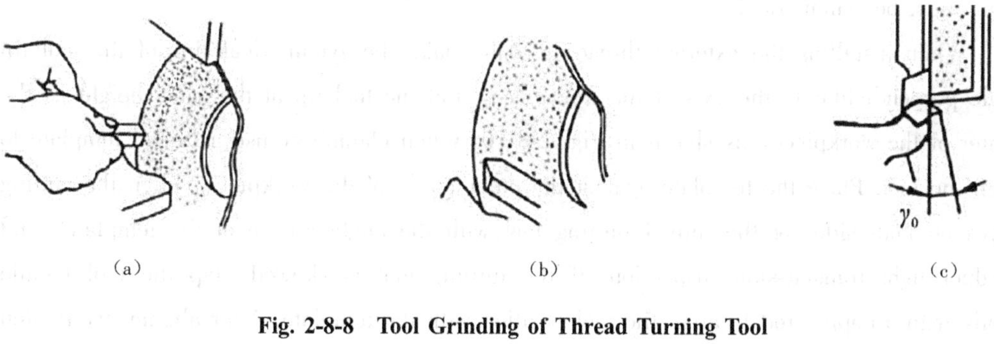

(a) (b) (c)

Fig. 2-8-8 Tool Grinding of Thread Turning Tool

(a) Left edge of tool grinding (b) Edge of tool in the feed direction of the back grinding (c) Rake face of tool fine grinding

③ Fine grind the rake face according to the anterior angle value. Tilt the rake face about 10°-15° in the horizontal direction of the grinding wheel and about 10°-15° in the vertical direction, and pay attention to making the left cutting edge slightly lower than the right cutting edge. After the rake face contacts the grinding wheel, slightly increase the pressure for tool grinding to gradually grind to the position near the tool tip, as shown in

Fig. 2-8-8(c).

④ Fine grind the flank face. The tool tip angle is measured with a thread turning tool template, as shown in Fig. 2-8-9. During measurement, the bottom plane of the tool rod shall be parallel to the plane of the template, and the correctness of the tool tip angle for tool grinding shall be judged by observing the light transmission between the tool edge and the template.

⑤ Grind the tool tip with the edge width of about 0.1*P* (pitch).

⑥ Grind the anterior and relief angles at the tool tip with petroleum and keep the cutting edge sharp.

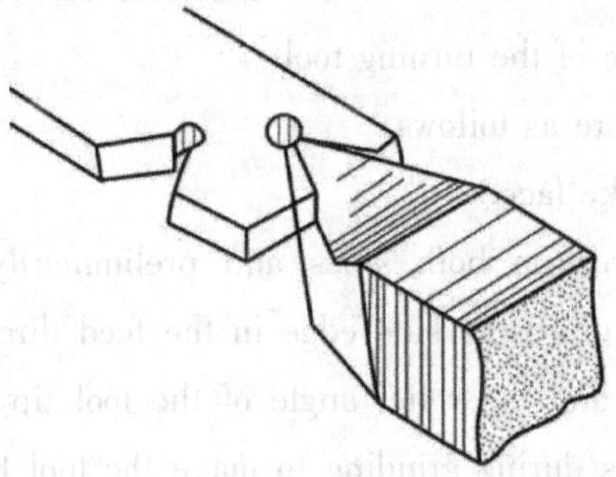

Fig. 2-8-9 Inspection of Tool Tip Angle with Templates

2) Installation of thread turning tool

The installation of thread turning tool has a great influence on the machined thread profile. If the installation is incorrect, even if the tool edge is well ground, the thread angle cannot meet the requirements of machining accuracy. For triangular threads, the profile half angle must be symmetrical.

When installing the external threading tool, make the symmetrical line of the tool tip angle perpendicular to the axis of the workpiece, and the tool tip at the same height as the center of the workpiece, as shown in Fig. 2-8-10. When clamping, use a thread template to align the tool. Place the template against the outer circle of the workpiece, align the cutting edges on both sides of the thread turning tool with the angle groove of the template, and conduct light transmission inspection. If the turning tool is skewed, tap the tool handle gently with a copper rod to align the tool position with the template. After alignment, tighten the turning tool and recheck it again to prevent the turning tool from moving when the tool post screw is tightened. After changing or grinding the thread turning tool for some reason during machining, the tool shall be re-aligned during re-clamping and the movement track of the tool tip shall be in the original spiral groove.

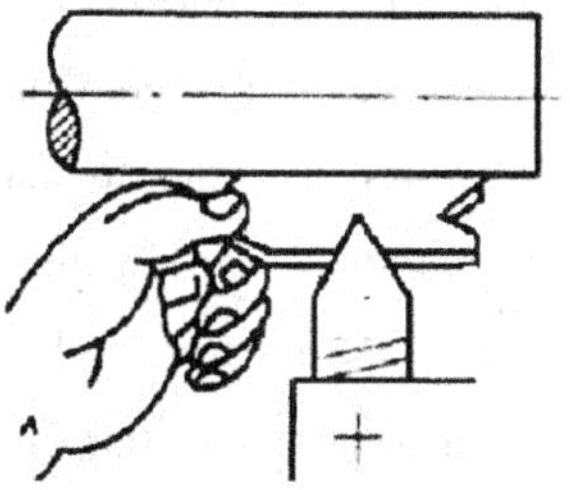
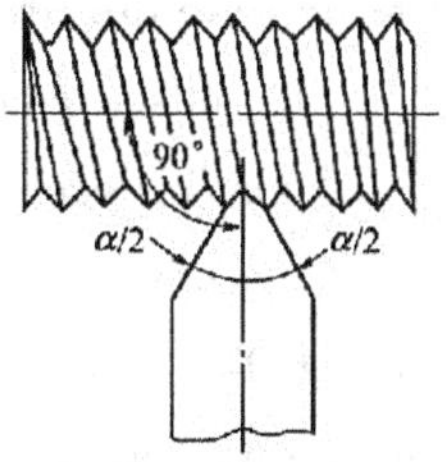

Fig. 2-8-10 Proper Installation of Thread Turning Tool

6. Method of turning triangular threads

There are two turning methods for triangular threads: low-speed turning and high-speed turning. The low-speed turning of triangular threads with high-speed steel turning tool can achieve higher thread accuracy and lower surface roughness values. However, this turning method has low production efficiency and should not be used in batch turning, which is suitable for threads of single piece or special specification. High-speed turning of threads with hard alloy turning tool is a widely used method in the machinery manufacturing industry at present, with high production efficiency and low thread surface roughness.

1) Low-speed turning of triangular thread

High–speed steel turning tool is generally used for low-speed turning of triangular threads, including rough turning and fine turning. The cutting speed can be 10-15 m/min for rough turning and 5-10 m/min for fine turning.

There are three feed methods for turning triangular threads, which shall be determined according to the material of the workpiece, the outer diameter of the threads and the size of the pitch. The three feed methods are described below.

(1) Straight feed method. Turning is performed with the straight feed method, as shown in Fig. 2-8-11. When thread is turned, the thread turning tool tip and the left and right edges are directly involved in cutting. Each feed is transversely performed by the middle sliding plate, and with the deepening of the thread depth, the back cutting depth decreases accordingly until the thread is turned. This turning method is simple to operate and the turned thread profile is correct. However, due to the simultaneous cutting of both side edges of the turning tool, it is difficult to discharge chips, the tool tip is easy to wear, the thread surface roughness value is relatively large, and the "scratches" are easy to occur when the back cutting depth is relatively deep. Therefore, this turning method is suitable for thread turning with pitch less than 2 mm or brittle material.

(2) Left and right cutting method. The left and right cutting method is shown in Fig. 2-8-12. When turning the thread, in addition to controlling the lateral feed of the thread turning tool with the middle sliding plate scale, the small sliding plate scale is used to realize the left and right trace feed of the turning tool. When the left and right cutting

method is used to turn the thread, the cutting allowance shall be reasonably distributed, and rough turning can be offset along the feed direction, generally leaving a fine turning allowance of 0.2-0.3 mm on each side. During fine turning, in order to make both sides of the thread relatively smooth, after one side is turned clean, the turning tool is moved to the other side for turning. The cutting speed is taken as 10-15 m/min for rough turning, and less than 6 m/min for fine turning and the back cutting depth is less than 0.05 mm.

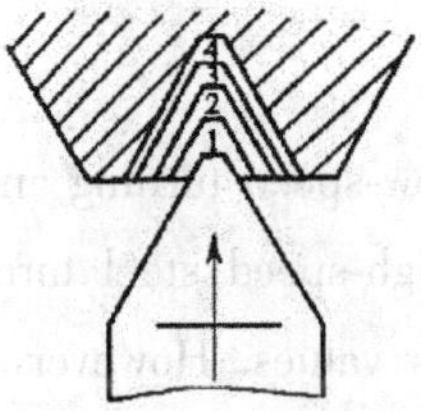

Fig. 2-8-11 Triangular Thread Turning by Straight Feed Method

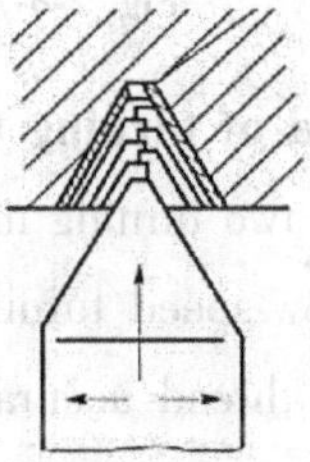

Fig. 2-8-12 Triangular Thread Turning by Left and Right Cutting Method

This turning method is more complex than the straight feed method. However, only the tool tip and one edge are involved in cutting. The chip removal is smooth, the stress and heating of the tool tip are improved, and it is not easy to produce scratches. Accordingly, the cutting amount can be increased, and a small surface roughness value can be obtained. Due to the influence of unilateral feed force, the profile error may increase. It is suitable for rough and fine turning of various threads except rectangular threads, which is conducive to increasing cutting amount and improving cutting efficiency.

(3) Inclined feed method. Compared with the left and right cutting method, the small sliding plate only feeds in one direction for turning triangular thread by inclined feed method, as shown in Fig. 2-8-13. The inclined feed method is convenient to operate, but it is only suitable for rough turning of threads because the side roughness deviating from the feed direction of the small sliding plate is large. During fine turning, the left and right cutting method must be used to obtain small surface roughness values on both sides of the thread. When high-speed steel turning tool is used for low-speed turning of threads, cutting fluid shall be filled. In order to prevent the occurrence of "scratches", it is better to use the elastic tool handle as shown in Fig. 2-8-14. When the cutting force of this handle exceeds a certain value, the turning tool can be automatically released to keep the cuttings at a proper thickness. "Scratches" can be avoided during rough turning, and the roughness value of the thread surface can be reduced during fine turning.

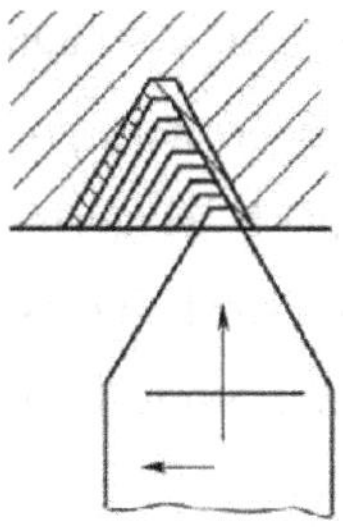

Fig. 2-8-13 Triangular Thread Turning by Inclined Feed Method

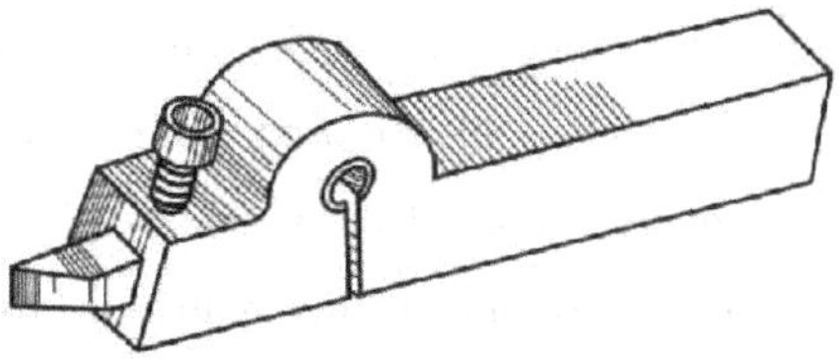

Fig. 2-8-14 Thread Turning Tool with Elastic Handle

2) High-speed turning of triangular thread

The turning tool used for high-speed turning of triangular thread is a hard alloy tool, and the cutting speed is generally taken as 50-70 m/min. During turning, only the straight feed method can be used to make the cuttings discharge perpendicular to the axis. When the thread is turned at high speed with a hard alloy tool, the back cutting depth can be larger at the beginning, and then decrease gradually. When the turning is carried out to the last time, the back cutting depth cannot be too small (generally 0.15-0.25 mm); otherwise, the surface roughness values on both sides of the thread will be large, forming a fish scale shape, and vibration will also occur in severe cases.

7. Measurement of triangular external threads

When threads are measured, different measuring methods shall be selected according to different quality requirements and production batch size. Common measurement methods include single measurement and comprehensive detection.

1) Single measurement method

(1) Measurement of thread crest diameter. The thread crest diameter refers to the large diameter of the external thread and the small diameter of the internal thread, which are generally measured with a vernier caliper or a micrometer.

(2) Measurement of pitch (or lead). Before turning the thread, first draw a shallow spiral line on the outer circle of the workpiece with a thread turning tool, and then measure the pitch (or lead) with a straight steel ruler, vernier caliper or thread template, as shown in Fig. 2-8-15.

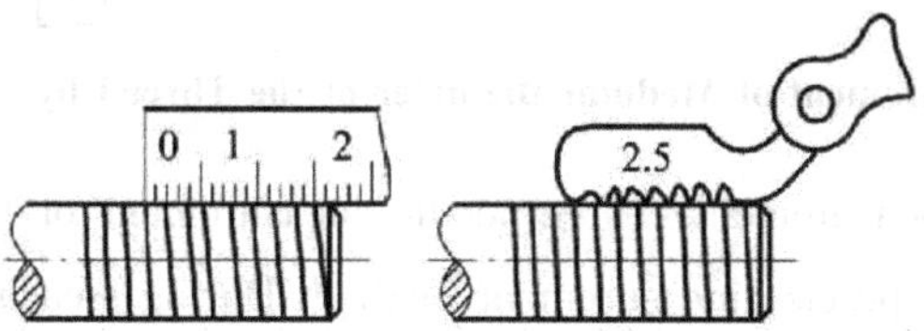

Fig. 2-8-15 Measurement of Pitch before Turning

The same method can also be used to measure the pitch (or lead) after the turning of thread is completed, as shown in Fig. 2-8-16. When measuring with a straight steel ruler or a vernier caliper, it is better to measure the pitch (or lead) of 5 or more profiles, and then take the average value. The thread template, also known as pitch gauge or profile gauge, has two types: metric and British. During measurement, the steel sheets in the thread template are embedded in the spiral groove along the direction passing through the axis of the workpiece, and if fully matched, the measured pitch is correct.

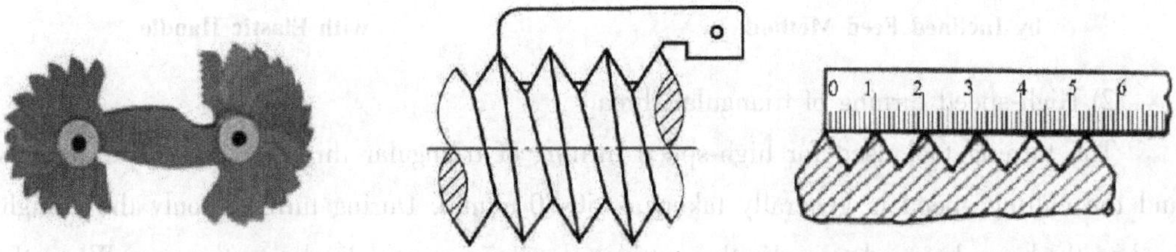

Fig. 2-8-16 Detection of Pitch and Profile after Turning

(3) Measurement of thread angle. Generally, the thread angle can be checked with a thread template or a thread angle template, as shown in Fig. 2-8-16.

(4) Measurement of medium diameter.

① Measurement with thread micrometer. The medium diameter of triangular thread can be measured with a thread micrometer. The use method is similar to that of the general micrometer. But the thread micrometer has two adjustable measuring heads (upper measuring head and lower measuring head). During measurement, two measuring heads same as the thread angle are just stuck on the thread profile side. As can be seen from Fig. 2-8-17, *ABCD* is a parallel quadrangle, so the measured size *AD* is the medium diameter of the thread. Fig. 2-8-17 Measurement of Medium Diameter of the Thread by Thread Micrometer.

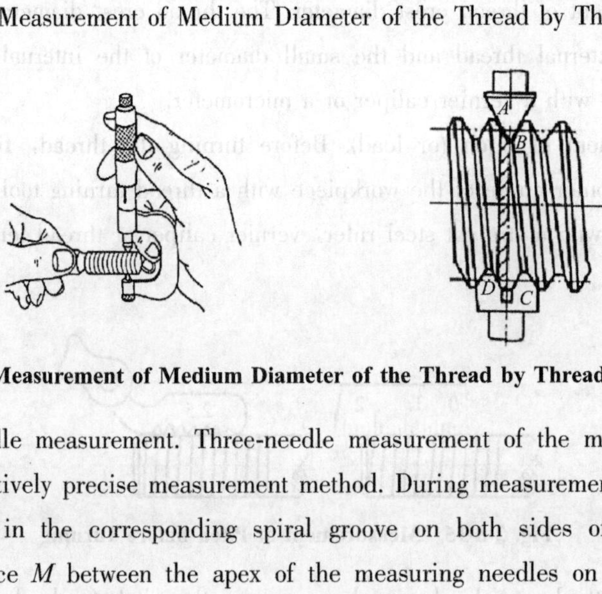

Fig. 2-8-17 Measurement of Medium Diameter of the Thread by Thread Micrometer

② Three-needle measurement. Three-needle measurement of the medium diameter of the thread is a relatively precise measurement method. During measurement, place the three measuring needles in the corresponding spiral groove on both sides of the thread, and measure the distance *M* between the apex of the measuring needles on both sides with a micrometer, as shown in Fig. 2-8-18. The actual size of the medium diameter of the thread can be calculated according to the M value. For three-needle measurement, the calculation formula for M value and measuring needle diameter is shown in Table 2-8-4.

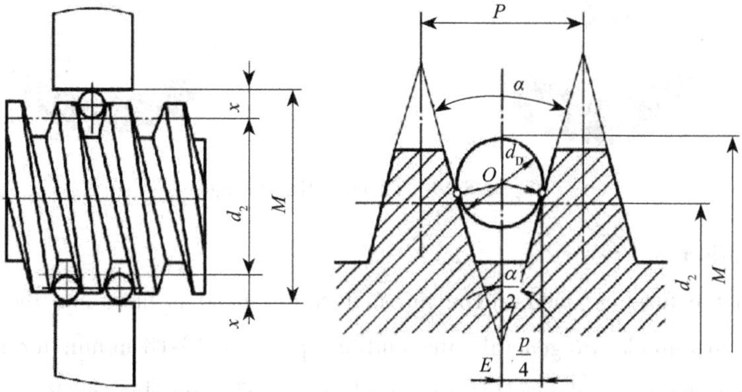

Fig. 2-8-18 Three-needle Measurement of Medium Diameter of the Thread

The three cylindrical measuring needles with the same diameter used for measurement are specially manufactured by the measuring tool manufacturer (in the absence of measuring needles, they can be replaced by the drill handle with the same diameter). The measuring needle diameter d_D cannot be too large or too small. The optimum measuring needle diameter refers to the measuring needle diameter when the cross section of the measuring needle is tangent to the profile side at the medium diameter of the thread. The maximum, optimum and minimum measuring needle diameters can be calculated from the formula in Table 2-8-4. The measuring needle shall be selected as close as possible to the optimum value to obtain higher accuracy.

Table 2-8-4 Calculation Formula for M Value and Measuring Needle Diameter d_D

Thread	Thread angle α	Calculation formula for M value	Measuring needle diameter d_D		
			Maximum value	Optimum value	Minimum value
Common thread	60°	$M = d_2 + 3d_D - 0.866P$	1.01P	0.577P	0.505P
British thread	55°	$M = d_2 + 3.1d_D - 0.961P$	0.894P–0.029	0.564P	0.481P–0.016

2) Comprehensive measurement

The triangular external thread can be comprehensively detected with a thread ring gauge. The thread ring gauge is divided into go gauge T and no-go gauge Z. Check the diameter, pitch, profile and surface roughness of the thread before checking the dimensional accuracy. When the go gauge can pass and the no-go gauge cannot pass the thread, the accuracy meets the requirements. When checking the triangular external thread with a thread ring gauge, determine whether the thread is qualified by tightening the workpiece, as shown in Fig. 2-8-19. When the thread accuracy requirement is not high, check with standard nuts.

Fig. 2-8-19 Thread Ring Gauge

8. Precautions

(1) Select the turning amount. The thread turning speed is fast, and the spindle speed should not be too quick. In general, the cutting speed is 13-18 m/min for rough turning, and the cutting depth is about 0.15 mm each time. The number of feed is calculated, leaving an allowance of 0.2 mm for fine turning. The cutting speed is 5-10 m/min for fine turning, and the cutting depth is 0.02-0.05 mm each time.

(2) The relative position of the workpiece and the spindle is fixed. When the workpiece is removed from the center for measurement, the clamp shall not be loosened; when the workpiece is reinstalled, the relative position of the clamp and the chuck must be kept unchanged.

(3) If the tool is changed during cutting, the tool must be realigned. Since there is a gap in the transmission system, the tool shall be aligned after a distance along the cutting direction and stopping. At this time, the small sliding plate is moved so that the cutting edge of the turning tool matches the thread groove.

(4) In order to ensure that the tool tip can correctly fall into the thread groove that has been turned every time turning is performed, when the pitch of the screw is not an integral multiple of the pitch of the part, the split nut cannot be opened during turning, and the forward and reverse turning method shall be adopted.

(5) When turning threads, it is not allowed to touch the workpiece by hand, or wipe the rotating threads with cotton yarn. The threads are mainly measured for pitch, pressure angle and thread medium diameter. The pitch is generally measured with a straight steel ruler, and the pressure angle is generally measured with a template, or the pitch and pressure angle can be measured at the same time with a pitch gauge. Only the medium diameter is ensured by correct operation during machining, so the medium diameter of the thread is usually measured with a thread micrometer caliper.

When machining the workpiece, the machining allowance of the workpiece needs to be cut off several times according to the drawing requirements. In order to improve the productivity and ensure the dimensional accuracy and surface roughness of the workpiece, the turning process can be divided into rough turning and fine turning. In this way, cutting parameters can be selected reasonably according to different stages of machining. Sometimes, semi-fine turning is added between rough turning and fine turning as required, and the turning parameters are between them.

[Thinking and practice]

(1) What are the types of threads ?

(2) What are the main parameters of the triangular thread ?

(3) Briefly describe the installation method of triangular thread turning tool.

(4) Briefly describe the triangular thread turning method.

(5) Briefly describe the measurement method of triangular external thread.

Item 3

Milling

[Item description]

Milling is a highly efficient machining method by cutting workpieces with rotating multi-edge tools. Machine tools for milling include horizontal or vertical milling machines and large gantry milling machines. Milling is suitable for machining planes, grooves and various forming surfaces, etc.

Through the basic training of milling, students can understand the machining principle and scope of the milling machine, master the basic skills of milling, stimulate learning interest through the implementation of tasks, cultivate engineering awareness and improve the ability to solve problems.

[Learning objective]

(1) Understand the process characteristics and scope of milling.

(2) Understand the main components and functions of common milling machines.

(3) Understand the structure, use and installation adjustment methods of common tools and accessories of milling machines.

(4) Master the operating essentials of milling and the milling method of simple parts, and understand the method of simple division with dividing head, and the dimensional accuracy and surface roughness value range that can be achieved by milling.

(5) Master the method of properly installing workpieces and tools on the milling machine, and be able to complete the operation of milling plane, inclined plane, step surface, groove and milling keyway.

Task I General knowledge of milling work

I. Function and typical job contents of milling work

Mechanical parts are generally in required shapes and dimensions by blanks through various machining methods, and milling is one of the most commonly used cutting and machining methods. The so-called milling is a cutting and machining method in which the milling tool rotates as the primary motion on the milling machine and the workpiece performs the feed motion. The main characteristics of milling are cutting with multi-edge milling tools with high efficiency; the machining range is wide, and parts with complex shapes can be machined; the machining accuracy is high, and its economic accuracy is generally IT9-IT7.

The surface roughness value is *Ra* 12.5-1.6 μm. The basic contents of milling are shown in Fig. 3-1-1.

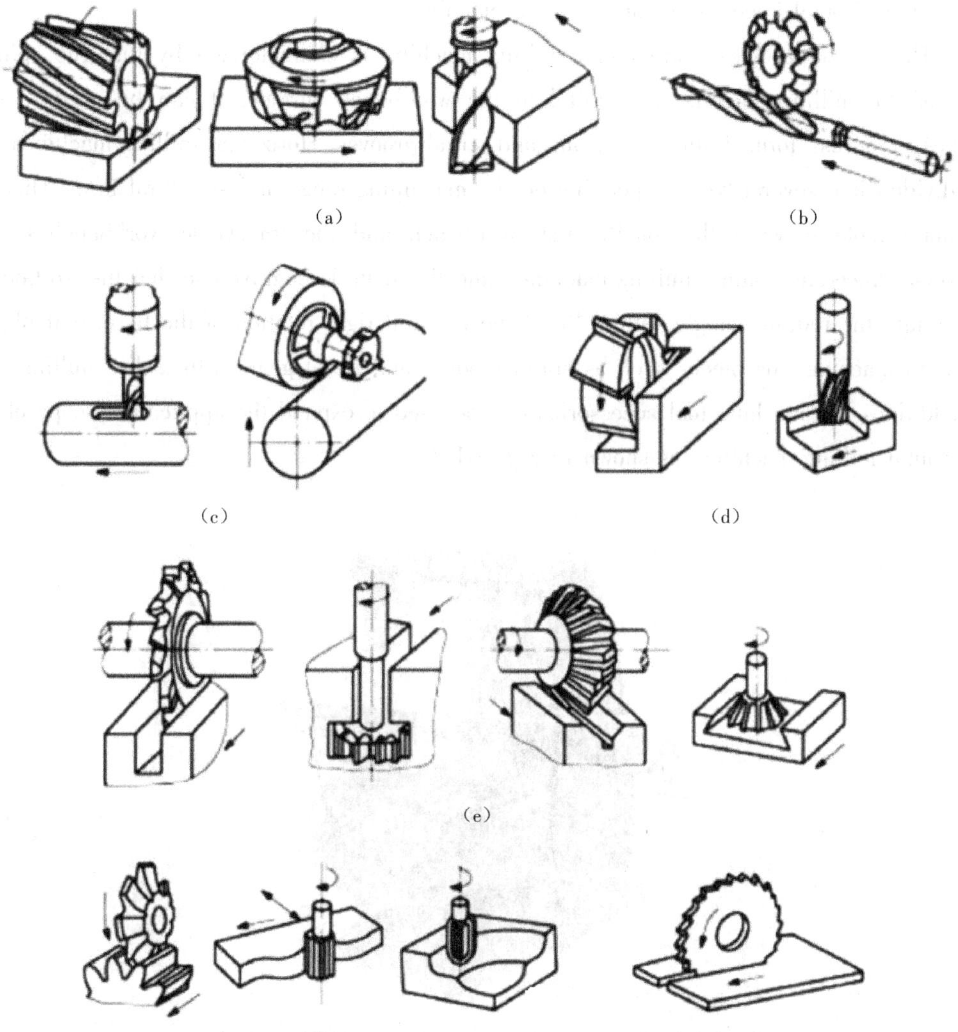

Fig. 3-1-1 Milling Contents

(a) Milling plane (b) Milling spiral groove (c) Milling step surface (d) Milling groove (e) Milling straight groove (f) Milling formed surface (g) Cutting-off

II. Basic structure of milling machine

1. Type of milling machine

Milling machine is the important equipment in the machinery manufacturing industry. It has a wide range of applications and many types. At present, horizontal knee-and-column milling machine and vertical knee-and-column milling machine are commonly used.

1) Horizontal knee-and-column milling machine

The horizontal knee-and-column milling machine is characterized by that the spindle axis of the milling machine is parallel to the workbench surface. It can be used to mill planes, grooves, formed surfaces, gears and spiral grooves. Horizontal milling machines can be divided into several types depending on the machining range and structural form. There is a rotary table between the longitudinal workbench and the transverse workbench of the universal knee-and-column milling machine, and the scale is engraved so that the workbench can rotate the required angle within 45° of the left and right rotation of the horizontal plane, so as to machine workpieces such as spiral grooves and bevel gears with a disc milling tool. In addition, the machine tool accessories can be used to expand the application scope of the horizontal milling machine, as shown in Fig. 3-1-2.

Fig. 3-1-2 Horizontal Knee-and-column Milling Machine

2) Vertical knee-and-column milling machine

The vertical knee-and-column milling machine is characterized by that the spindle axis of the milling machine is perpendicular to the workbench surface. The spindle part of the machine tool is called the vertical milling head, which can rotate the angle to make the spindle axis rotate left and right by 45°. The longitudinal workbench and the transverse workbench are connected without a rotary table, so the workbench cannot turn the angle. It can be used to mill planes, inclined planes and grooves, etc. In addition, machine tool accessories such as rotary workbench and dividing head can be used to machine complex workpieces such as arc, formed surface and gear spiral groove, as shown in Fig. 3-1-3.

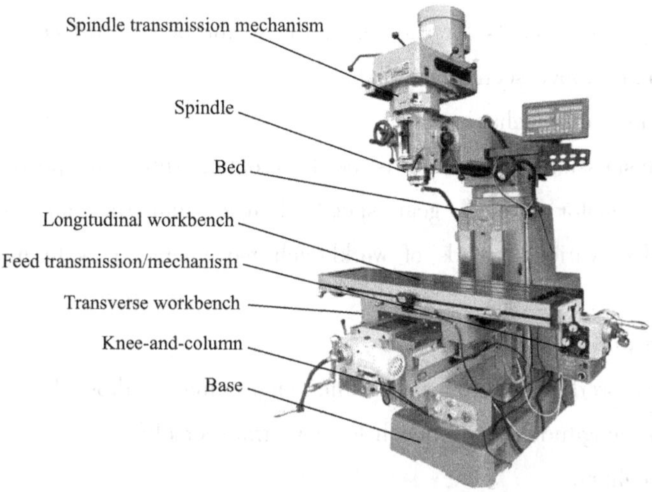

Fig. 3-1-3 Vertical Knee-and-column Milling Machine

2. Basic structure of milling machine

Although there are many types of milling machines, the basic structures of various milling machines are roughly the same. The basic components and functions of the X6330 universal vertical milling machine shown in Fig. 3-1-3 are briefly described below.

1) Spindle transmission mechanism

The spindle transmission mechanism is used to obtain different speeds from the rated speed of the main motor through gear transmission and then transmit to the spindle to meet the need of milling.

2) Spindle

The spindle is a hollow shaft with taper hole at the front end, and the milling tool shaft is installed in the taper hole. The spindle is made of high-quality structural steel, heat treated and precisely cut.

3) Bed

The bed is the main body of the machine tool, and most parts of the machine tool are installed on the bed. The bed is generally made of high-quality gray cast iron and subjected to precision cutting and aging treatment. A dovetail-shaped vertical guideway is mounted on the front wall of the bed and the knee-and-column can move upward and downward along the guideway.

4) Longitudinal workbench

The longitudinal workbench is used to install and drive the workpiece longitudinally. There are three T-shaped grooves on the workbench surface, and tools and fixtures can be installed and fixed with T-screws. There is a long groove on the front side of the workbench, which is used to install and fix the ultimate automatic stop and the automatic circulation stop. The accuracy requirements for the longitudinal workbench surface, guideway section and T-shaped straight groove section are high.

5) Feed transmission mechanism

The feed transmission mechanism is used to obtain different speeds from the rated speed of the feed motor through gear speed change, and transmit them to the feed mechanism to realize various speeds of workbench movement so as to meet the needs of milling.

6) Transverse workbench

The transverse workbench is between the longitudinal workbench and the knee-and-column to drive the longitudinal workbench to move transversely.

7) Knee-and-column

The knee-and-column is mounted on the vertical guideway on the front side of the bed, with a screw in the middle connected to the base nut. The knee-and-column is mainly used to support the workbench and drive the workbench to move up and down. The workbench and the motors and transmission devices of the feed part are installed on the knee-and-column, so the rigidity and accuracy requirements of the knee-and-column are high; otherwise, great vibration will be generated during milling, affecting the machining accuracy of the workpiece.

8) Base

The base is the supporting component of the whole machine tool and has sufficient rigidity and strength. The inner cavity of the base is filled with cutting fluid for cooling and lubrication during cutting.

III. Basic motion of milling

Milling is a cutting and machining method with the rotation motion of the milling tool as the primary motion and the straight or rotation motion of the workpiece or the straight motion of the milling tool as the feed motion. The relative motion of the workpiece and the milling tool during milling is called milling motion, including primary motion and feed motion.

1. Primary motion

The primary motion is the motion that forms the cutting speed of the machine tool or consumes the main power. The primary motion of the milling tool is rotation motion.

2. Feed motion

The feed motion is the motion required to machine the complete surface by putting the materials on the workpiece cutting layer into cutting successively. In the milling motion, both the movement or rotation of the workpiece and the movement of the milling tool are feed motions. In addition, the feed motion can be divided into longitudinal feed, transverse feed and vertical feed according to the direction of motion.

IV. Operation procedures for milling machine

(1) All parts of the milling machine must be wiped clean before operation.

(2) Check whether the mechanism and moving parts of each position of the milling machine are in good condition and whether the handles and knobs are in a reasonable position.

(3) The workbench and spindle components shall not be knocked hard with hard objects; tools, blanks and other debris shall not be placed on the workbench.

(4) When starting up, pay attention to that the working object is not in contact with the milling tool, and the workbench shall be evenly tightened back and forth; start the milling machine to make the spindle rotate and check whether the gear throws oil; check the variable speed of the milling machine spindle and feed system to make the spindle and workbench feed move from low speed to high speed and check whether the motion is normal.

(5) Do not work overload. During work, if abnormal phenomena and irregular sounds are found in the milling machine, shut down immediately and repair in time, and do not use the milling machine without repair.

(6) Check the reliability of the stroke limiter before starting the automatic feeding; when the workbench is fed automatically, disengage the manual feed clutch to prevent the handle from rotating along the shaft to hurt people; do not start automatic feeding in both feed directions at the same time; do not suddenly change the feed speed during automatic feeding; stop feeding first and then stop the rotation of the spindle (tool) after automatic feeding is completed.

(7) Do not leave the milling machine during operation, or do other things unrelated to the operation content; place the tools and measuring tools securely, neatly and reasonably, with fixed positions for easy access during operation, and return them to the original position after use.

(8) After machining, wipe the milling machine clean, remove cuttings and oil stains with soft cloth and brush, and do not blow them with compressed air to prevent fine cuttings, dust and other debris from embedding into the moving part. If there is no abnormality, lubricate each part of the lathe with oil. Carefully wipe tools, measuring tools and other accessories and place all items to original positions, clean the work site and turn off the power supply.

(9) When the milling machine is not used, each handle shall be in the neutral position, the feed fastening handle in each direction shall be released, the workbench shall be in the middle position of feed in each direction, and the guideway surface shall be properly lubricated.

(10) Blanks, semi-finished products and finished products shall be placed separately. Semi-finished products and finished products shall be stacked neatly and handled with care to prevent damage to the finished surface. Drawings and process cards shall be placed in a position convenient for reading and shall be kept clean and complete. The surroundings of the work site shall be kept clean and tidy to prevent debris stacking and tripping.

V. Safety knowledge of milling work

(1) Protective equipment shall be used as required before internship. The cuffs of the work clothes worn shall be tied up. The female students shall wear the work caps. Hair and braids shall not be left outside the caps to prevent the clothes edge, cuffs and braids from being involved in the rotating parts. It is not allowed to wear gloves when operating the milling machine.

(2) During milling, it is not allowed to touch the surface of the workpiece close to the rotating milling tool by hand; otherwise, fingers may be easily cut. Iron hooks are used to clean the cuttings. The scrap iron is thin and sharp and cannot be pulled by hand.

(3) During milling, it is forbidden to wipe the moving parts such as the workpiece or the rotating milling tool rod with cotton wire. Shut down when measuring the workpiece. Do not measure the workpiece size during cutting. It is not allowed to remove cuttings by hand grip or mouth blowing.

(4) Operators shall wear protective glasses during milling.

(5) Tools, cards and measuring tools shall be placed firmly to prevent falling from hurting people. The workpiece shall be clamped firmly and reliably to prevent accidents caused by loose workpiece during cutting.

(6) When the grinding wheel is used, operators shall stand on the side of the grinding wheel to prevent the grinding wheel from flying out and hurting people after breaking.

(7) Operators shall not stand in the direction where the cuttings flow out during operation.

(8) Protective devices such as protective covers on milling machines shall not be removed at will to prevent injury accidents due to exposure of drive belts and gears.

(9) The handwheel clutch is kept open during fast maneuvering feed.

(10) Lighting on the milling machine shall be limited to 36V low voltage and shall be powered as specified.

[Thinking and practice]

(1) What are the typical job contents of milling work ?

(2) What are the safe operation technologies of milling work ?

(3) What aspects must be done in the civilized production of milling work ?

Task II Milling tool and fixture

The milling tool used for milling is a multi-edge tool, which can carry out metal cutting and obtain workpieces of certain surface quality. There must be certain requirements for the material and geometry of the milling tool.

I. Materials for cutting part of milling tool

1. Basic requirements

The materials for manufacturing the cutting part of milling tool shall meet the following basic requirements:

1) Hardness

At normal temperature, the cutting part of the tool must have sufficient hardness to cut the workpiece. Due to the large amount of heat generated during cutting, the tool material is required to maintain its hardness at high temperature and continue cutting. This property with high temperature hardness is called red hardness.

2) Toughness and strength

The tool will bear a large impact force during cutting, so the material of the cutting part of the tool is required to have sufficient strength and toughness, and the cutting can be continued under the condition of bearing impact and vibration, which is not easy to break.

2. Commonly used materials

There are many materials with the above properties, and the commonly used milling tool materials include high-speed tool steel and hard alloy.

(1) High-speed tool steel (high-speed steel) is an alloy tool steel with tungsten (W), chromium (Cr) and vanadium (V) as the main elements, and the quenching hardness is generally 62-65 HRC. The high-speed tool steel has high strength and good toughness, which can grind sharp edges with good process performance. Forging, welding, cutting and tool grinding are easy.

Commonly used high-speed tool steel grades include W18Cr4V and W6Mo5Cr4V2, etc.

2) Hard alloy. Hard alloys are manufactured by powder metallurgical process with metallic carbide WC (tungsten carbide), TiC (titanium carbide) and Co (cobalt) as the main metallic binders. They have high normal temperature hardness, good wear resistance and high temperature resistance, and can still maintain good cutting performance at about 800℃-1 000℃. The cutting speed can be 4-8 times higher than that of high-speed steel. However, it has low bending strength and poor impact toughness, and the cutting edge is not easy to grind.

II. Type and selection of milling tool

Milling tools are multi-edge tools, which are generally manufactured by professional factories due to their complex structure. Since there are many edges involved in cutting at the same time and high cutting speed can be used, the production efficiency is high. There are many types of milling tools, which are divided by purpose, combination method and back shape. In the milling process, the milling tool shall be reasonably selected and used according to the conditions of the milling machine and the machining needs.

(1) The milling tool for milling plane is shown in Fig. 3-2-1: (a) is cylindrical milling tool for horizontal milling machine; (b) is sleeve end milling tool for horizontal milling machine or vertical milling machine; (c) is clamp milling tool for vertical milling machine.

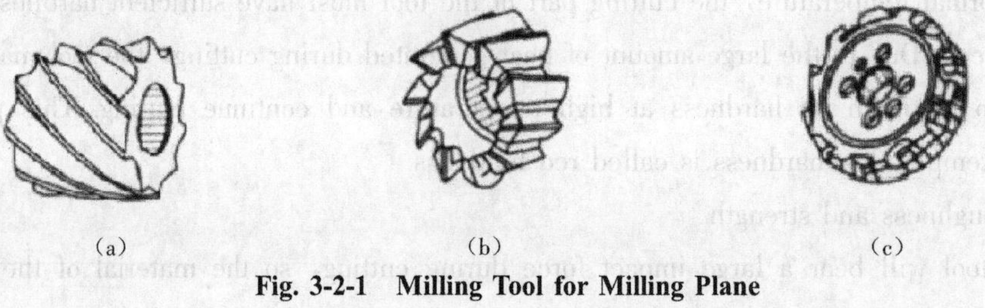

(a) (b) (c)

Fig. 3-2-1 Milling Tool for Milling Plane

(a) Cylindrical milling tool (b) Sleeve end milling tool (c) Clamp milling tool

(2) The milling tool for milling groove is shown in Fig. 3-2-2. Keyway milling tools and vertical milling tools are mostly used for vertical milling; disc slot milling tools, three-sided edge milling tools and saw blade milling tools are mostly used for horizontal milling machines.

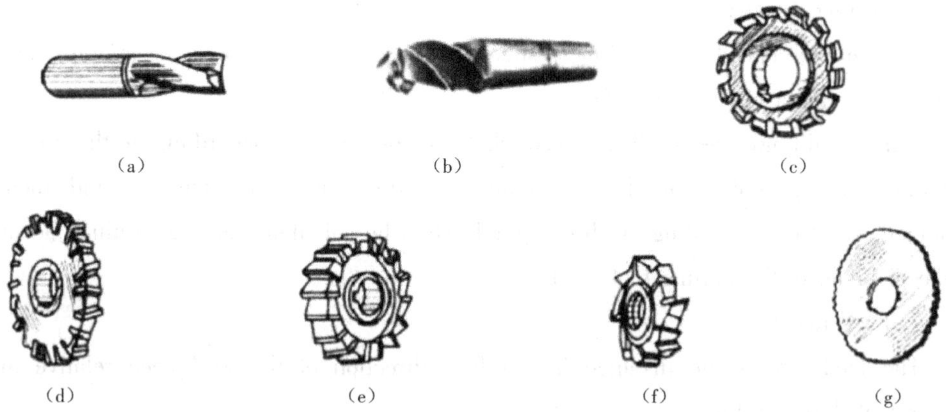

Fig. 3-2-2 Milling Tool for Milling Groove

(a) Keyway milling tool (b) Disc slot milling tool (c) Vertical milling tool (d) Inserted three-sided edge milling tool (e) Straight-toothed three-sided edge milling tool (f) Misaligned three-sided edge milling tool (g) Saw blade milling tool

(3) The milling tool for milling formed surface is shown in Fig. 3-2-3, which is commonly used to mill semi-circles, gears or other formed surfaces.

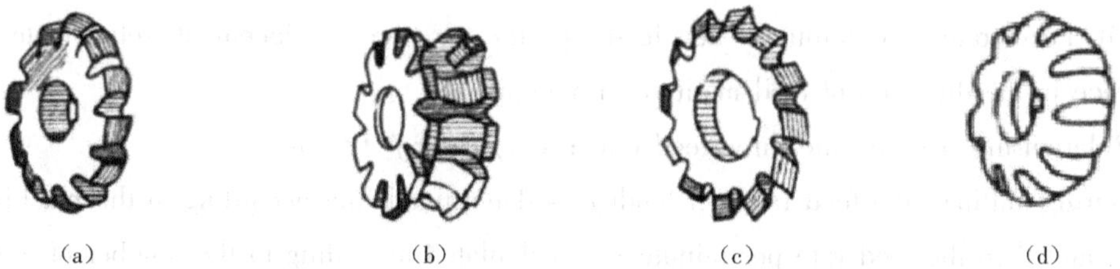

Fig. 3-2-3 Milling Tool for Milling Formed Surface

(a) Convex semicircular milling tool (b) Concave semicircular milling tool (c) Gear milling tool (d) Forming milling tool

(4) The milling tool for milling formed groove is shown in Fig. 3-2-4, which is commonly used to mill the T-shaped groove, dovetail groove and semicircular keyway, etc.

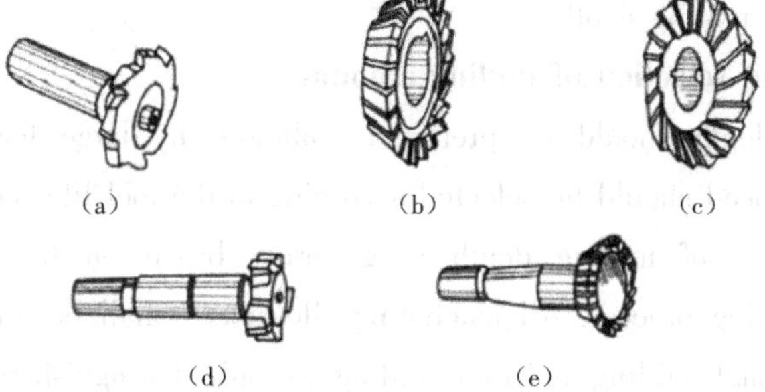

Fig. 3-2-4 Milling Tool for Milling Formed Groove

(a) T-shaped groove milling tool (b) Dovetail groove milling tool (c) Semi-circular keyway milling tool (d) Single angle milling tool (e) Double angle milling tool

III. Selection of milling amount

1. Three elements of milling

1) Milling speed v

The milling speed is the linear speed of the primary motion, in m/min, and the calculation formula is $v=\pi dn/1\ 000$.

During milling, the milling speed shall be determined according to the workpiece material, milling tool material, machining properties and other factors, and then the rotational speed of the milling machine spindle shall be calculated and determined according to the diameter of the milling tool used.

2) Feed rate f

The feed rate is the distance in the feed direction of the workpiece relative to the milling tool during milling.

Feed rate is expressed and measured in three ways according to the specific situation:

(1) Feed rate per rotation f_n: the displacement of the milling tool relative to the workpiece in the direction of feed motion per rotation, in mm/r.

(2) Feed rate per tooth f_z: the displacement of each tooth in each rotation of the milling tool relative to the workpiece in the direction of feed motion, in mm/z.

(3) Feed rate per minute (i.e. feed speed) v_f: the displacement relative to the workpiece in the direction of feed motion, in mm/min.

Relationship between the three feed rates: $v_f = f_n \cdot n = f_z \cdot z \cdot n$

During milling, the feed rate per tooth f_z is determined first according to the machining nature, and then the feed rate per minute v_f is calculated according to the number of teeth z and the rotational speed n of the milling tool, and the feed rate of the milling machine is adjusted accordingly.

3) Back cutting depth ap

Vertical distance between the surface to be machined and the machined surface during one milling feed, i.e. milling depth.

2. Sequence and selection of milling amount

Large milling depth should be preferred, followed by large feed rate, and finally appropriate milling speed should be selected according to the tool life requirements.

(1) The selection of milling depth is generally based on the cutting size of the workpiece in the milling process. All machining allowances shall be milled out at one feed as far as possible. Rough milling and Fine milling are only distinguished when the workpiece is of high machining accuracy. Refer to Table 3-2-1 for specific values.

Table 3-2-1 Rough and Fine milling Depths mm

Workpiece material	High-speed steel milling tool		Hard alloy milling tool	
	Rough milling	Fine milling	Rough milling	Fine milling
Cast iron	5-7	0.5-1	10-18	1-2
Mild steel	<5	0.5-1	<12	1-2
Medium hard steel	<4	0.5-1	<7	1-2
Hard steel	<3	0.5-1	<4	1-2

(2) Selection of feed rate per tooth. During rough machining, the feed rate shall be as large as possible. During finish machining, a smaller feed rate is generally selected. Refer to Table 3-2-2 for specific values.

Table 3-2-2 Selection of Feed Rate Per Tooth mm

Name of turning tools	High-speed steel milling tool		Hard alloy milling tool	
	Cast iron	Cast steel	Cast iron	Cast steel
Cylindrical milling tool	0.12-0.2	0.1-0.15	0.2-0.5	0.08-0.2
Vertical milling tool	0.08-0.15	0.03-0.06	0.2-0.5	0.08-0.2
Sleeve face milling tool	0.15-0.2	0.06-0.1	0.2-0.5	0.08-0.2
Three-sided edge milling tooll	0.15-0.25	0.06-0.08	0.2-0.5	0.08-0.2

(3) After the milling depth and feed rate per tooth are determined, the milling speed shall be determined on the premise of ensuring the tool life. The milling speed can be selected within the recommended range and adjusted according to the actual situation. The specific values are shown in Table 3-2-3.

Table 3-2-3 Selection of Milling Speed m/min

Workpiece material	High-speed steel milling tool	Hard alloy milling tool
20	20-45	150-190
45	20-35	120-150
40Cr	15-25	60-90
HT150	14-22	70-100
Brass	30-60	120-200
Aluminum alloy	112-300	400-600
Stainless steel	16-25	50-100

Notes: ① Take the smaller value for rough milling and the larger value for Fine milling.

② Take the smaller value when the strength and hardness of the workpiece material are high, or take the larger value.

③ Take the larger value for good thermal resistance of tool materials, or take the smaller value.

IV. Fixtures

The workpiece is installed on the milling machine with main fixtures including flat jaw pliers and three-jaw centering chucks. Generally, according to the shape and size of the workpiece, they are directly installed on the workbench with pressure plates, bolts, sizing blocks and stops. When the production volume is large, special fixtures or combined fixtures can also be used to install the workpiece.

When flat jaw pliers are installed, the bottom surface of the jaw base and the workbench surface of the milling machine shall be wiped. Generally, the position of the flat jaw pliers on the workbench surface shall be at the center of the length direction of the workbench to the left and the center of the width direction to facilitate operation.

V. Clamping of workpieces

When the vertical plane is milled by clamping the workpiece with flat jaw pliers, the workpiece is clamped with machine flat jaw pliers only by tightly fitting the workpiece reference plane with the fixing jaw. In order to make the workpiece reference plane closely fit with the fixing jaw, a round rod is often placed between the movable jaw and the workpiece. For the blank surface on the workpiece that is uneven to the reference plane or is not parallel to the reference plane, if a round rod is not placed, the reference plane of the workpiece after clamping does not necessarily fit well with the fixing jaw, as shown in Fig. 3-2-5.

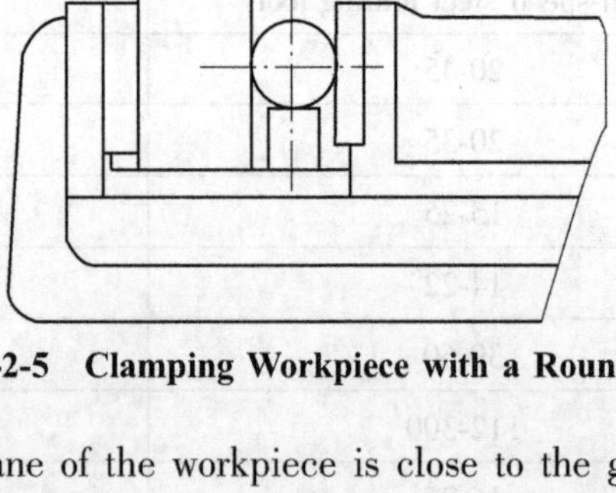

Fig. 3-2-5 Clamping Workpiece with a Round Rod

When the reference plane of the workpiece is close to the guideway surface of the plier body, a parallel sizing block shall be placed between the workpiece and the guideway. In order to make the reference plane of the workpiece parallel to the guideway surface, the aluminum or copper hammer can be used to slightly knock on the workpiece after tightening, and the sizing block can be moved manually. When it is not loose, the workpiece fits well with the sizing block and then is clamped, as shown in Fig. 3-2-6.

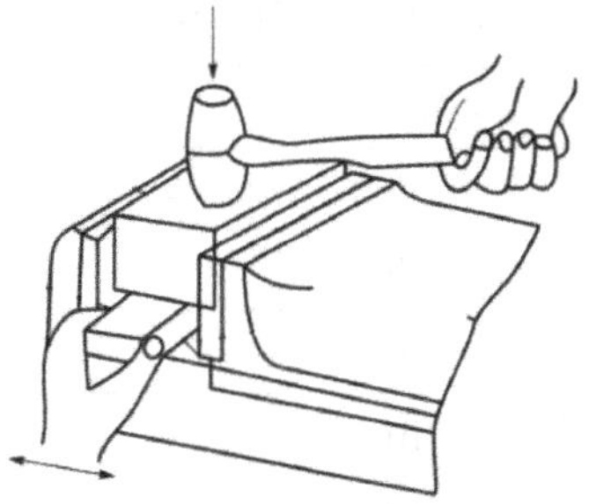

Fig. 3-2-6 Clamping Workpiece with Parallel Sizing Block

[Thinking and practice]

(1) What are the milling tool material types ?

(2) What is the type and selection method of milling tool ?

(3) Briefly describe the method for selecting milling amount.

(4) Briefly describe the use of fixtures and the clamping method of workpieces.

Task III Milling plane

I. Drawings and technical requirements

As shown in Fig. 3-3-1, The blank size of the workpiece is 120 mm×50 mm×25 mm and the material is 45 steel. Only the upper surface of the workpiece is machined. Scoring table see Table 3-3-1.

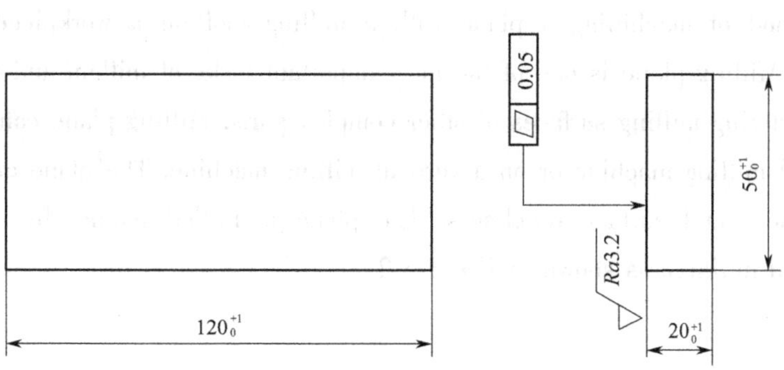

Fig. 3-3-1 Duck Beak Hammer

Table 3-3-1 Scoring Table

Technical requirements	Assigned score	Measured results	Score
Planeness 0.05 mm	30		
*Ra*3.2 μm	30		
20^{+1}_{0} mm	20		
Standardized operation	10		
Safe and civilized production	10		

II. Analysis of processing

(1) Read the part drawing (as shown in Fig. 3-3-1); check the blank size.

(2) Install flat jaw pliers.

(3) Select and install the end milling tool.

(4) Select and adjust the cutting amount (spindle speed n=118 r/min, feed speed v_f=47.5 mm/min or f=60 mm/min, and cutting depth a_p=2 mm).

(5) Install and correct the workpiece (copper gasket shall be used).

(6) Adjust the milling width (i.e. the depth of the cutting layer) and automatically feed the milling workpiece.

(7) After milling, shut down, lower the workbench and withdraw the workpiece.

(8) Measure and remove the workpiece.

(9) The teacher shall evaluate the machined workpieces and performance of the members in the team and give the scores in combination with the Scoring Table 3-3-1.

III. Knowledge and skills

1. Circumferential milling and end milling

The method of machining a plane with a milling tool on a workpiece is called as milling plane. Milling plane is one of the most important tasks of millers and the basic skill for further mastering milling surfaces of other complex parts. Milling plane can be performed on a horizontal milling machine or on a vertical milling machine. The plane quality depends on the planeness and surface roughness. The plane is milled mainly by circumferential milling and end milling, as shown in Fig. 3-3-2.

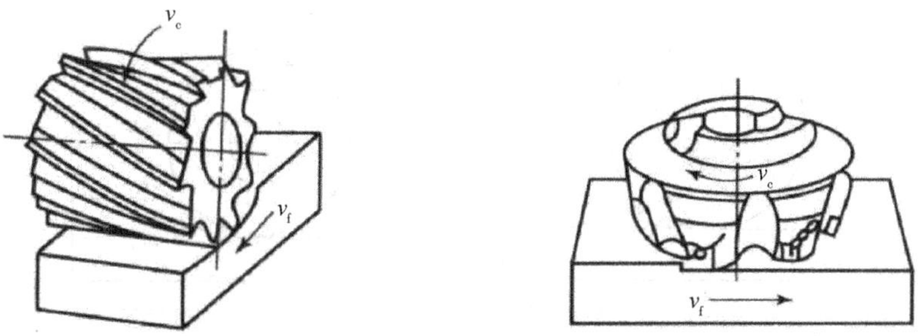

(a) (b)

Fig. 3-3-2 Circumferential Milling and End Milling

(a) Circumferential milling (b) End milling

2. Sequential milling and reverse milling

According to the relationship between the rotation direction of the milling tool and the feed direction of the workpiece, milling can be divided into sequential milling and reverse milling. When the rotation direction of the milling tool is the same as the feed direction of the workpiece, it is called sequential milling; otherwise, it is called reverse milling, as shown in Fig. 3-3-3.

Advantages of sequential milling: During sequential milling, the cutting thickness starts from the maximum, the tool wear is small and the durability is high. The surface is smooth because the vertical cutting force is downward and the clamping is reliable.

Disadvantages: During sequential milling, the component force of the milling force in the feed direction is the same as that in the feed direction of the workpiece. Due to the gap between the screw nuts of the workbench, when the feed force increases gradually, the milling force will pull the workbench to produce jumping, resulting in uneven feed, and in severe cases the milling tool will be broken.

Advantages of reverse milling: During reverse milling, the lead screw is always close to the reverse milling surface of the nut due to feed force, and the milling process is relatively smooth.

Disadvantages: During reverse milling, the cutting thickness increases gradually from zero. Due to the influence of the radius of the fillet of the cutting edge, the anterior angle is negative at the beginning of cutting, and the tool teeth are extruded and slide on the working surface, resulting in serious hardening of the workpiece surface machining and aggravating the wear of the tool teeth. Furthermore, as the vertical tangential upward direction is opposite to the clamping force of the workpiece and the gravity of the workpiece, there is a tendency to lift the workpiece from the workbench, which increases the vibration and affects the clamping and surface roughness of the workpiece.

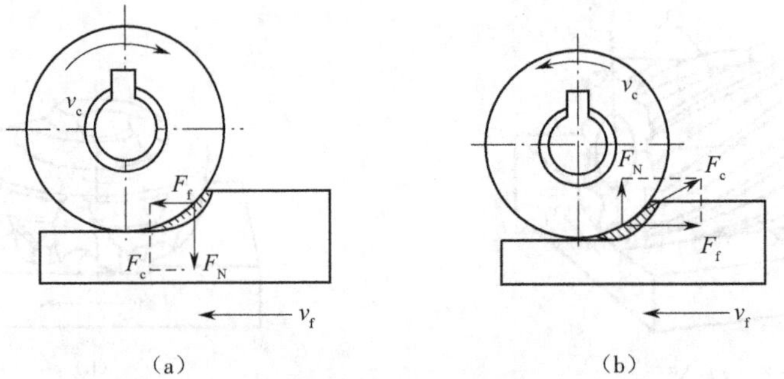

Fig. 3-3-3 Sequential Milling and Reverse Milling

(a) Sequential milling (b) Reverse milling

3. Rough milling, semi-Fine milling and Fine milling

The milling amount is selected based on the machining accuracy of the workpiece, the tool durability and the rigidity of the process system. On the premise of ensuring product quality, it is required to improve production efficiency and reduce cost as much as possible.

During rough milling, the machining accuracy of the workpiece is not high, but the milling allowance is large and the cutting force is large. The selection of milling amount shall mainly consider the milling tool durability, milling machine power, rigidity of process system and production efficiency. First, larger milling depth and width shall be selected. During milling castings and forging blanks, the tool tip shall be kept away from the surface hard layer. When machining the workpiece with small milling width, the milling depth can be appropriately increased. The milling width shall be milled out at one time as far as possible, and then a larger feed rate per tooth and a lower milling speed shall be selected.

When the surface roughness of the workpiece is required to be Ra=6.3-3.2 μm, semi-Fine milling shall be used. The semi-Fine milling depth is 0.2-2 mm. During Fine milling, in order to obtain higher dimensional accuracy and smaller surface roughness values, the milling depth should be smaller, the milling speed can be appropriately increased, and the feed rate per tooth should be smaller.

In general, the sequence of selecting the milling amount is the large milling depth first, then the feed rate per tooth, and finally the milling speed. The milling width shall be equal to the width of the workpiece machining surface as far as possible.

4. Operation method for milling plane

1) Operation method of circumferential milling plane

Move the workbench so that the workpiece is under the cylindrical milling tool and start aligning the tool. When aligning the tool, start the machine tool. After the milling tool is rotated, shake the feed handle of the knee-and-column to make the workpiece rise slowly. When the milling tool just contacts the workpiece, record the scale value of the lifting dial. Then lower the workbench, shake the longitudinal handle, withdraw the workpiece, and adjust the depth of the milling layer according to the actual size of the blank. When the allowance is not large, the reverse milling method can be used for one-time feed milling to the drawing requirements; otherwise, rough and Fine milling shall be divided.

Test workpiece: After milling, remove the workpiece, measure the dimensions of each part of the workpiece with a vernier caliper or micrometer, and compare the surface roughness of the tested workpiece with a milling roughness template.

2) Operation of end milling plane

The tool alignment method is basically the same as that of the cylindrical milling tool. The difference is that the end face cutting edge marks are used when the end milling tool is aligned, while the circumferential cutting edge marks are used when the cylindrical milling tool is aligned.

After tool alignment, asymmetric reverse milling method shall be adopted, as shown in Fig. 3-3-4.

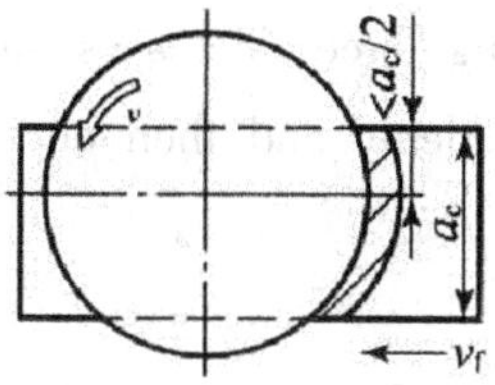

Fig. 3-3-4 Asymmetric Reverse Milling

Test workpiece: After milling, remove the workpiece, measure the dimensions of each part of the workpiece with a vernier caliper, and compare the surface roughness of the tested workpiece with a milling roughness template.

After milling the plane, the planeness is generally checked with a knife straightedge. For planes with high planeness requirements, standard flat plates can be used for inspection. During inspection, red lead powder or gentian violet solution can be applied to the planes of the standard flat plates, and then the planes of the workpieces are placed on the standard flat plates for control study. After several times of control study, the workpieces are removed to observe the coloring of the planes. If the coloring is uniform and dense, it indicates that the planeness of the planes is good, as shown in Fig. 3-3-5.

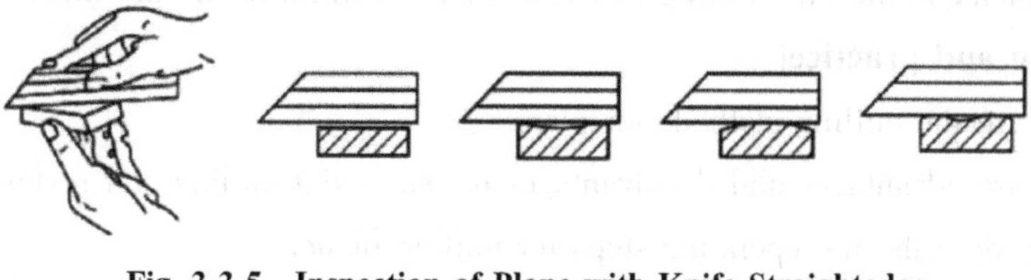

Fig. 3-3-5 Inspection of Plane with Knife Straightedge

5. Quality analysis of plane milling

The milling quality of the plane is not only related to the quality of the milling machine, fixture and milling tool used in milling, but also to many factors such as the milling amount and the selection of cutting fluid.

1) Factors affecting surface roughness

(1) The milling tool is worn and the cutting edge becomes dull.

(2) The feed rate is too large during milling.

(3) The cutting layer is too deep during milling.

(4) The selection of geometric parameters of milling tool is improper.

(5) The selection of cutting fluid during milling is improper.

(6) There is vibration during milling.

(7) The built-up edge occurs during milling, or the workbench moves due to excessive clearance of the feed drive system.

(8) Scratches occur during milling.

(9) During milling, the milling force decreases suddenly due to feed pause, which causes the milling tool to sink suddenly and then cut a pit on the workpiece machining surface (called "deep gnawing").

2) Factors affecting planeness

(1) When circumferential milling plane is used, there is an error in the cylindricality of the cylindrical milling tool.

(2) When end milling plane is used, the spindle axis of the milling machine is not perpendicular to the feed direction.

(3) The workpiece is deformed due to clamping milling force.

(4) There is internal stress on the workpiece and deformation occurs after the surface material is cut off.

(5) There is linear error of feed motion of milling machine workbench.

(6) The axial and radial clearances of the spindle bearing of the milling machine are too large.

(7) During milling, thermal deformation of the workpiece is caused by milling heat.

(8) During milling, grafting tool marks occur due to the width of the cylindrical milling tool or the diameter of the end milling tool less than the width of the machined surface.

[Thinking and practice]

(1) What are the milling methods for planes ?

(2) What are advantages and disadvantages of sequential milling and reverse milling ?

(3) Briefly describe the operating steps for milling plane.

(4) What is the method to detect planeness ?

Task IV Milling vertical and parallel surfaces

I. Drawings and technical requirements

As shown in Fig. 3-4-1, The blank size of the workpiece is 130 mm×70 mm×60 mm and the material is 45 steel. Scoring table see Table 3-4-1.

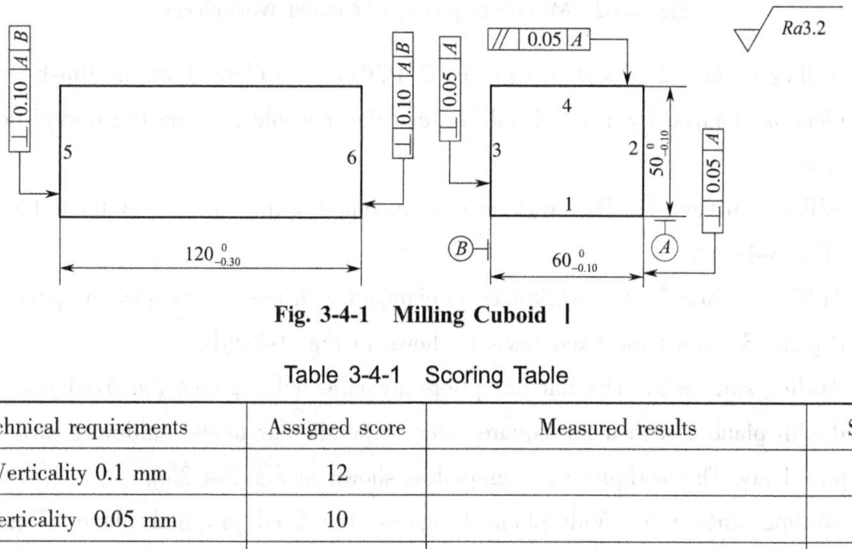

Fig. 3-4-1 Milling Cuboid Ⅰ

Table 3-4-1 Scoring Table

Technical requirements	Assigned score	Measured results	Score
Verticality 0.1 mm	12		
Verticality 0.05 mm	10		
Parallelism 0.05 mm	10		
$120_{-0.3}^{0}$ mm	10		
$60_{-0.1}^{0}$ mm	10		
$50_{-0.1}^{0}$ mm	10		
*Ra*3.2 μm	18		
Standardized operation	10		
Safe and civilized production	10		

II. Analysis of processing

(1) Milling surface 1 (reference plane A) As shown in Fig. 3-4-2(a), with plane 2 as the rough reference, the workpiece is clamped by a copper sheet between the two jaws and the workpiece close to the fixed jaw.

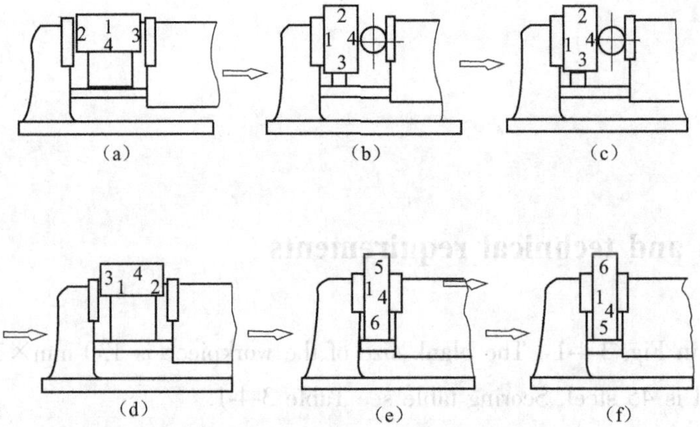

Fig. 3-4-2 Milling Sequence of Cuboid Workpieces

(2) Milling surface 2: As shown in Fig. 3-4-2(b), with plane 1 as the finish reference, the workpiece is clamped by a round rod between the movable jaw and the workpiece against the fixed jaw.

(3) Milling surface 3: The workpiece is clamped with plane 1 as the reference, as shown in Fig. 3-4-2(c).

(4) Milling surface 4: The workpiece is clamped with plane 1 against the parallel sizing block and plane 3 against the fixed jaw, as shown in Fig. 3-4-2(d).

(5) Milling surface 5: The flat jaw pliers are adjusted. Against the fixed jaw, plane 1 is aligned with plane 2 with a 90° square ruler perpendicular to the guideway surface of the flat jaw plier body. The workpiece is clamped as shown in Fig. 3-4-2(e).

(6) Milling surface 6: With plane 1 against the fixed jaw and plane 5 against the guideway surface of the flat jaw plier body, the workpiece is clamped as shown in Fig. 3-4-2(f).

III. Knowledge and skills

1. Milling of vertical plane

A plane where two adjacent surfaces intersect 90° with each other is called a vertical plane. The planes milled with cylindrical milling tool on horizontal milling machine and end milling tool on vertical milling machine are parallel to the workbench surface. Therefore, the vertical plane can be milled out only by clamping the workpiece reference plane perpendicular to the workbench surface. The machining method is basically the same as the milling plane except for the clamping requirements of the workpiece. A plane perpendicular to the reference plane is called a vertical plane. As shown in Fig. 3-4-3.

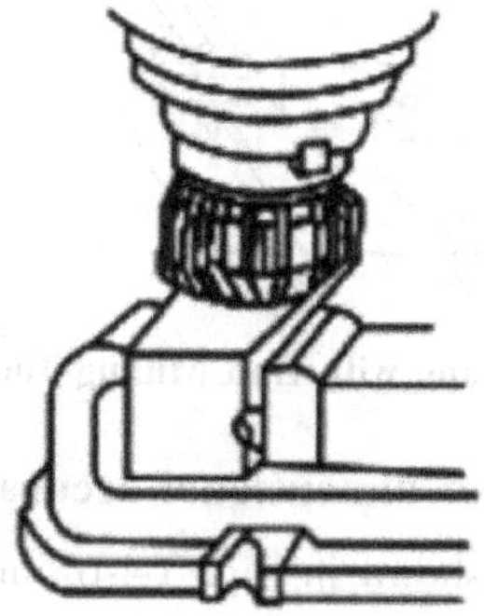

Fig. 3-4-3 Milling of Vertical Plane with End Milling Tool on Vertical Milling Machine

2. Verticality inspection method

Wide-seat right-angle rulers or 90° cylindrical angle meters or universal angle meters are often used to check the verticality of the workpiece. When checking the verticality with a right-angle ruler, make one side of the wide-seat right-angle ruler base fit with the reference plane of the workpiece, and then observe whether the contact and gap between the other side of the angle ruler and the test surface are uniform, so as to judge the correctness of the measured vertical plane, as shown in Fig. 3-4-4.

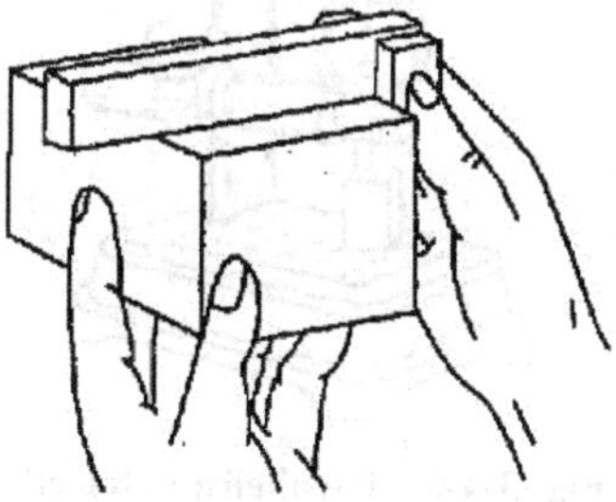

Fig. 3-4-4 Verticality Inspection by Right-angle Ruler

3. Milling of parallel plane

A parallel plane is a plane parallel to the reference plane. When milling a parallel plane, in addition to the requirements for parallelism and planeness, there are also requirements for dimensional accuracy between two parallel planes. The parallel plane is milled on the vertical milling machine with an end milling tool. When the workpiece has steps, the workpiece can be directly clamped on the workbench surface of the vertical milling machine with a pressure plate to make the reference plane fit with the workbench surface, as shown in Fig. 3-4-5.

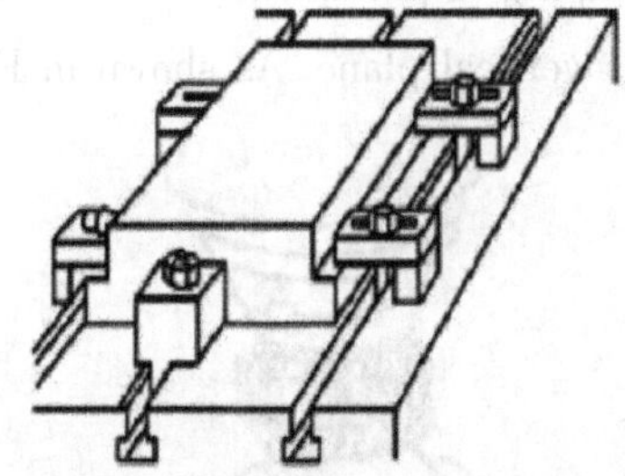

Fig. 3-4-5 Milling of Parallel Plane with End Milling Tool on Vertical Milling Machine

4. Detection of parallelism and dimensional accuracy

The parallelism is detected as shown in Fig. 3-4-6, and the reference plane is attached to the flat plate and measured with a dial indicator. When the dimensional accuracy and parallelism of the machined workpiece are detected at the same time, the four corners and the middle of the workpiece can be measured with a micrometer or a vernier caliper to check whether all dimensions are within the dimension range specified in the drawing and observe the difference between the dimensions of each part, which is the parallelism error.

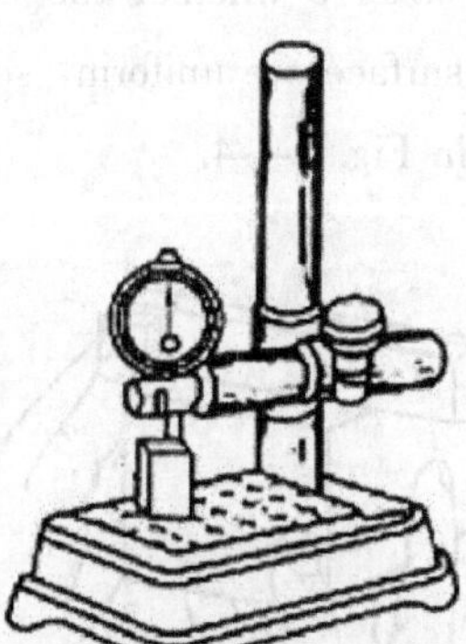

Fig. 3-4-6 Parallelism Detection

5. Quality analysis of vertical and parallel planes

1) Main factors affecting verticality

During milling, the main factors affecting verticality are as follows.

(1) The fixed jaw surface of the flat jaw pliers is not perpendicular to the workbench surface, and the milled surface is not perpendicular to the reference plane.

(2) The reference plane does not fit the fixed jaw.

(3) The planeness error of the reference plane is large, which affects the position

accuracy when the workpiece is clamped.

(4) The clamping force is too large to tilt the fixed jaw outward.

2) Factors affecting dimensional accuracy between parallel planes

During milling, the factors affecting the dimensional accuracy between parallel planes are as follows.

(1) When adjusting the depth of the milling layer, the dial is mistakenly viewed, the handle is overturned, and the clearance between the screw nut pair is not eliminated before direct withdrawal, resulting in dimensional milling error.

(2) The dimensions marked on the drawing are wrongly read and measurement is wrong.

(3) The plane of the workpiece or parallel sizing block is not cleaned and the dimension changes due to debris.

(4) The cutting marks are too deep during Fine milling and tool alignment, and the cutting marks are not removed during adjustment of the depth of the milling layer, so that the dimensions are milled small.

[Thinking and practice]

(1) Briefly describe the method of milling the vertical plane.

(2) Briefly describe the method of milling the parallel plane.

(3) Briefly describe the detection methods for parallelism and verticality.

Task V Milling inclined plane

I. Drawings and technical requirements

As shown in Fig. 3-5-1, The blank size of the workpiece is 110 mm×60 mm×30 mm and the material is 45 steel. Scoring table see Table 3-5-1.

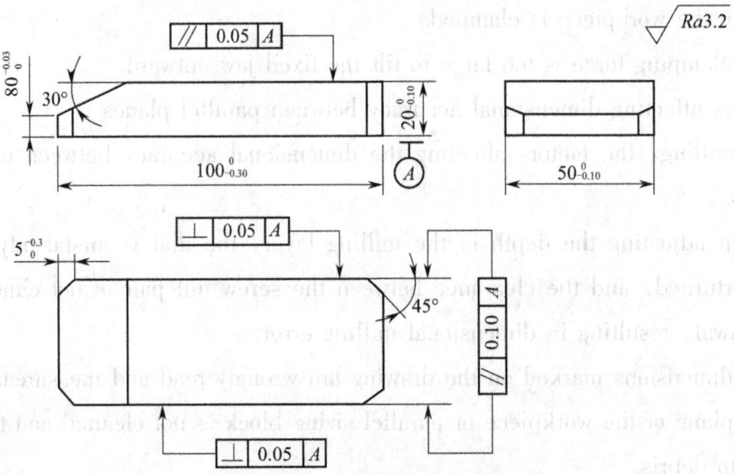

Fig. 3-5-1　Milling Cuboid Ⅱ

Table 3-5-1　Scoring Table

Technical requirements	Assigned score	Measured results	Score
Parallelism　0.05 mm	10		
Parallelism　0.1 mm	5		
Verticality　0.05 mm	5		
$20_{-0.1}^{0}$ mm	5		
$50_{-0.1}^{0}$ mm	5		
$100_{-0.3}^{0}$ mm	5		
$8_{0}^{+0.03}$ mm	5		
$5_{0}^{+0.3}$ mm	5		
30°	10		
45°	10		
*Ra*3.2 μm	20		
Standardized operation	10		
Safe and civilized production	5		

II. Analysis of processing

(1) Mill the workpiece plane to dimension.

(2) Mill 30° inclined plane.

① Align the fixed jaw of the flat jaw pliers.

② Select and install the milling tool (select the 80 mm-diameter toothed end milling tool).

③ Clamp and align the workpiece (the workpiece reference plane is parallel to the workbench surface).

④ Milling amount (take n=150 r/min, v_f=60mm/min, a_p appropriate amount in batches).

⑤ Adjust the angle of vertical milling head for α=30°.

⑥ Align the tool and mill the workpiece (align the tool, tighten the longitudinal feed after adjusting the milling depth ap, and mill the 30° inclined plane with the lateral feed in several times).

(3) Mill 45° inclined plane.

① Replace with a vertical milling tool with a diameter of 20-25 mm.

② Adjust the angle of vertical milling head for α=45°.

③ Fix the jaw with the workpiece reference plane (bottom surface) against the flat jaw pliers and clamp the workpiece.

④ Align the tool, adjust the milling width a_p (i.e. cutting depth) and mill a 45° inclined plane.

⑤ Clamp the workpiece in several times and mill the remaining three 45° inclined planes.

(4) The teacher shall evaluate the machined workpieces and performance of the members in the team and give the scores in combination with the Scoring Table 3-5-1.

III. Knowledge and skills

1. Method of milling inclined plane

The principle of milling inclined plane is consistent with that of milling plane, except that the cutting position or the installation position of the workpiece is changed accordingly so that the inclined plane can reach the accurate slope. There are mainly the following milling methods for inclined planes.

1) Convert the milling tool to the required angle to mill the inclined plane

Usually on a horizontal milling machine with a vertical milling head or on a vertical milling machine, the vertical milling head is rotated at a certain angle according to the inclination requirements of the workpiece, so that the workbench is fed laterally to machine the inclined plane, as shown in Fig. 3-5-2.

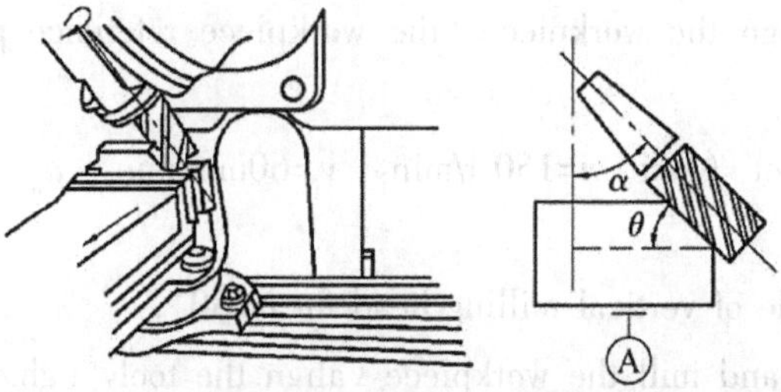

Fig. 3-5-2 Circumferential Milling of Inclined Plane for Parallelism between Reference Plane of Workpiece and Workbench Surface

2) Use an angular milling tool to mill the inclined plane

The inclined plane is directly milled with an angular milling tool, and the milling tool angle is equal to the inclined plane angle of the workpiece. However, due to the cutting edge width limitation of angular milling tool, this method is only applicable to smaller workpieces, as shown in Fig. 3-5-3.

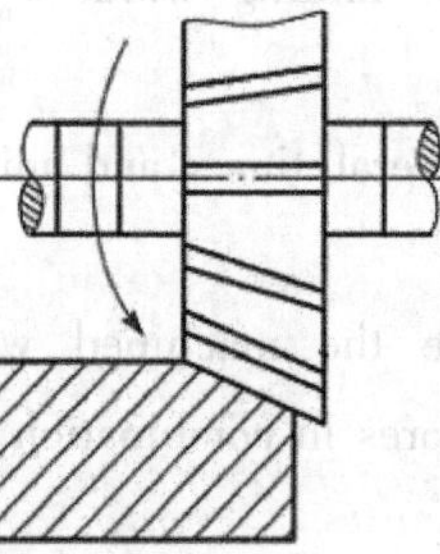

Fig. 3-5-3 Milling of Inclined Plane with Angular Milling Tool

3) Convert the workpiece to the required angle to mill the inclined plane

The installation position of the workpiece is changed, the inclined planed is turned to the horizontal position, the workpiece is clamped with auxiliary lines, dividing heads and other methods, and the inclined plane is milled in the method of milling plane, as shown in Fig. 3-5-4. The method of clamping the workpiece is as follows.

Fig. 3-5-4 Inclined Workpiece Method for Milling Inclined Plane

2. Angle detection

Check whether the included angle between the inclined plane and the reference plane meets the requirements of the drawing. There are two main detection methods.

(1) Detect with a universal angle meter: When the workpiece accuracy requirements are not high, the included angle between the inclined plane and the reference plane can be directly measured by a universal angle meter.

(2) Detect with a sine gauge: When the workpiece accuracy requirements are high, the included angle between the inclined plane and the reference plane can be detected with a sine gauge and a dial indicator and a measuring block.

3. Quality analysis of inclined plane

1) Factors affecting dimensional accuracy of inclined plane

(1) The scale is mistakenly viewed or the number of rotations of the handle is mistakenly made, and the clearance of the screw nut pair is not eliminated.

(2) Measurement is inaccurate, and dimensions are wrongly milled.

(3) The workpiece is loose during milling.

2) Factors affecting angle of inclined plane

(1) The rotation angle of vertical milling head is inaccurate.

(2) The scribing is inaccurate when clamping the workpiece according to the scribing, or displacement of the workpiece occurs during milling.

(3) When circumferential milling is used, the cylindricality error of the milling tool is large (if tapered) .

(4) When milling with an angular milling tool, the angle of the milling tool is not accurate.

(5) When the workpiece is clamped, the jaws of the flat jaw pliers, the guideway surface of the jaw body and the workpiece surface are not cleaned.

3) Factors affecting surface roughness of inclined plane

(1) The feed rate is too large.

(2) The milling tool is not sharp.

(3) The machine tool and fixture have poor rigidity and vibration during milling.

(4) During milling, the surface of the workpiece is "gnawed" when the workbench feeding or the spindle rotating suddenly stops.

(5) Cutting fluid is not used when milling steel parts, or the selection of cutting fluid is improper.

[Thinking and practice]

(1) What are the milling methods for inclined planes ?

(2) What measuring tools are used for the angle detection ?

(3) What are the factors affecting the angle of the inclined plane during milling inclined plane ?

Task VI　Milling steps

I. Drawings and technical requirements

As shown in Fig. 3-6-1, The blank size of the workpiece is 90 mm×50 mm×40 mm and the material is 45 steel. Scoring table see Table 3-6-1.

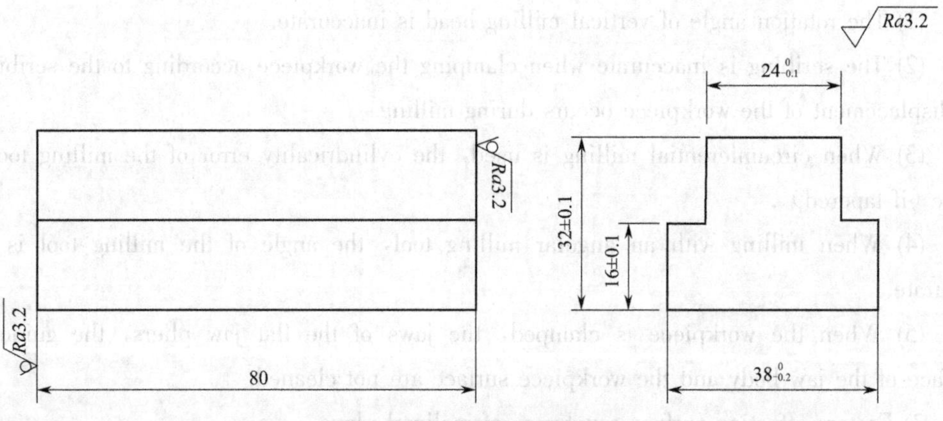

Fig. 3-6-1　Milling Steps

Table 3-6-1　Scoring Table

Technical requirements	Assigned score	Measured results	Score
32±0.1mm	20		
16±0.1 mm	20		
$24^{\ 0}_{-0.1}$ mm	15		
$38^{\ 0}_{-0.2}$ mm	15		
*Ra*3.2 μm	20		
Standardized operation	5		
Safe and civilized production	5		

II. Analysis of processing

(1) Install flat jaw pliers and correct the fixed jaw parallel to the longitudinal feed direction of the workbench.

(2) Select and install the end milling tool (sleeve end milling tool 63 mm×40 mm).

(3) Adjust the cutting amount (n=118 r/min, v_f=60 mm/min).

(4) Mill four sides to size.

(5) Mill two steps to size.

(6) Measure and remove the workpiece.

(7) The teacher shall evaluate the machined workpieces and performance of the members in the team and give the scores in combination with the Scoring Table 3-6-1.

III. Knowledge and skills

1. Form and process requirements of steps

The form of common step parts includes single step and double steps (T-key) and other forms of steps. The process requirements for steps include.

(1) Dimensional accuracy: the dimensions matching other parts on the steps generally require higher dimensional accuracy.

(2) Accuracy of shape and position: such as planeness of each surface, parallelism between the side of the step and the reference plane, and symmetry of the centering lines of double steps.

(3) Surface roughness: The mating surfaces on both sides of the step generally require low surface roughness.

2. Method of milling steps

1) Scribe and align the tool

Before clamping, scribe the dimension line of the step with a height gauge, install it in flat jaw pliers, start the machine tool, visually check whether the side edge of the milling tool falls on the edge line, slowly rise the vertical workbench, make the milling tool cut the cutting mark on the workpiece surface, and observe whether the cutting mark is between the edge lines. If there is any deviation, adjust the workbench again.

2) Use end milling tool to mill steps

Wide and shallow steps are often milled on vertical milling machines with end milling tools. The end milling tool rod has large stiffness, small change in cutting thickness during milling, smooth cutting, good machining surface quality and high production efficiency.

During milling, the diameter of the selected end milling tool shall be greater than the width of the step, which can be generally selected as $D=(1.4\text{-}1.6)\times B$, as shown in Fig. 3-6-2.

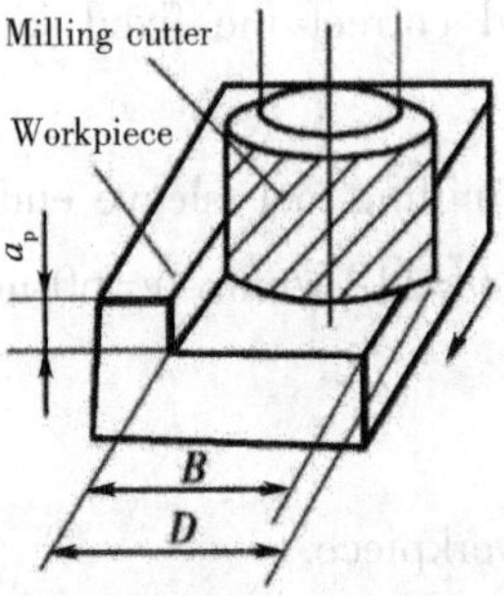

Fig. 3-6-2 Milling Steps with End Milling Tools

3) Mill the deeper steps with vertical milling tools

When milling the steps with the vertical milling tool by circumferential milling, it is necessary to first adjust to the required depth of the steps, and the width of the steps can be milled in several times. Due to the weak strength of the vertical milling tool, the allowable cutting amount shall be smaller, as shown in Fig. 3-6-3.

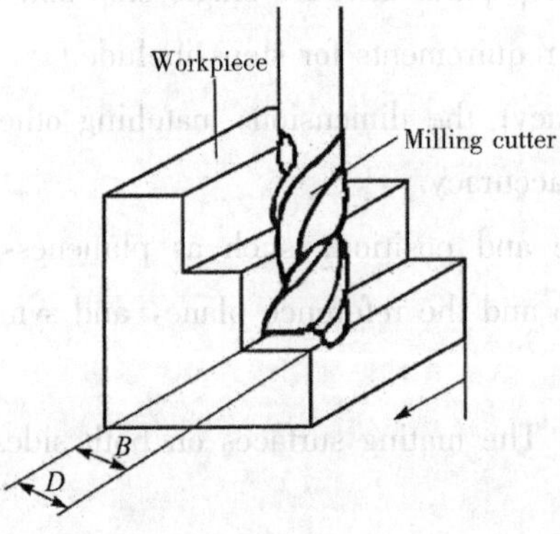

Fig. 3-6-3 Milling Steps with Vertical Milling Tools

3. Measurement of steps

The width and depth of the steps can generally be measured with a vernier caliper and a depth vernier caliper. Symmetrical steps on both sides can be measured with a micrometer when the steps are deep, as shown in Fig. 3-6-4(a); when the steps are shallow, they can be measured with a limit gauge, as shown in Fig. 3-6-4(b).

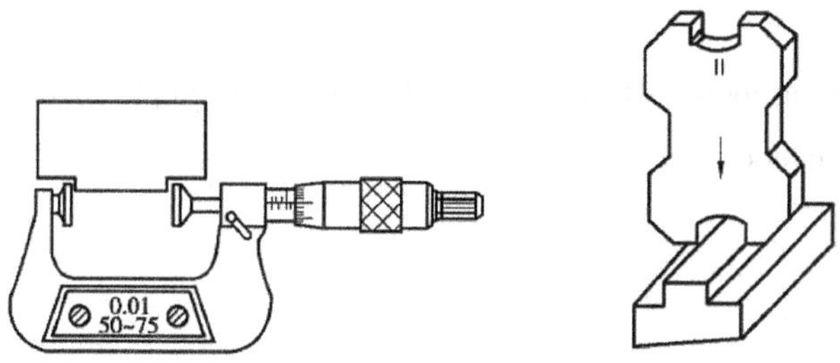

(a) (b)

Fig. 3-6-4 Measurement of Steps

(a) Measurement of steps with a common normal micrometer (b) Measurement of steps with a limit gauge

4. Quality analysis of steps

1) Factors affecting step dimensions

(1) The workbench is not shaken accurately when moving to adjust the dimension.

(2) Measurement is inaccurate.

(3) During milling, the force on the milling tool is not uniform to cause "back-off".

(4) The milling tool swing is too large.

(5) The zero position of the workbench is inaccurate, which will reduce the upper dimension of the step when milling the step with a three-sided edge milling tool.

2) Factors affecting the shape and position accuracy of steps

(1) Fixed jaws of flat jaw pliers are not corrected, or the workpiece is not corrected when clamping with pressure plates, and the milled steps are skewed.

(2) The zero position of the workbench is not accurate. When milling with a three-sided edge milling tool, the steps will be milled into the upper narrow and lower wide, and the side will be milled into the concave surface.

(3) The zero position of the vertical milling head is not accurate. When the longitudinal feed is milled with a vertical milling tool, the bottom surface of the step will have a concave surface.

3) Factors affecting surface roughness of steps

(1) The milling tool becomes dull.

(2) The milling tool swing is too large.

(3) The selection of milling amount is improper, especially excessive feed rate.

(4) Cutting fluid is not used when milling steel parts, or the use of cutting fluid is improper.

(5) The vibration during milling is too large, the unused feed mechanism is not fastened, and the workbench moves.

[Thinking and practice]

(1) What is the tool diameter determination method when milling steps with end milling

tool ?

(2) What measuring tools can be used for the width and depth of the steps ?

(3) How to mill deeper steps ?

Task VII Milling groove

I. Drawings and technical requirements

As shown in Fig. 3-7-1, The blank size of the workpiece is 110 mm×60 mm×30 mm and the material is 45 steel. Scoring table see Table 3-7-1.

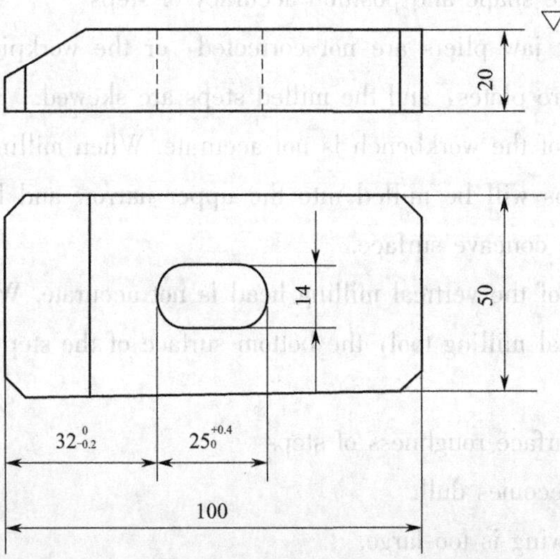

Fig. 3-7-1 Milling Enclosed Groove

Table 3-7-1 Scoring Table

Technical requirements	Assigned score	Measured results	Score
$32_{-0.2}^{0}$ mm	20		
$25_{0}^{+0.4}$ mm	20		
14 mm	20		

Continue Table

Technical requirements	Assigned score	Measured results	Score
*Ra*3.2 μm	20		
Standardized operation	10		
Safe and civilized production	10		

II. Analysis of processing

(1) Fix the flat jaw pliers and correct the jaw perpendicular to the spindle axis of the milling machine.

(2) Install the universal vertical milling head so that the spindle axis of the vertical milling head is perpendicular to the workbench surface.

(3) Mark the dimension, position line and drilling position line of the groove on the workpiece.

(4) Install and correct workpieces.

(5) Install the ϕ13 mm drill bit and drill the tool hole.

(6) Select and install milling tool (ϕ14 mm vertical milling tool).

(7) Align the tool and lock the lateral feed.

(8) Mill the enclosed groove in several times.

(9) Measure and remove the workpiece.

(10) The teacher shall evaluate the machined workpieces and performance of the members in the team and give the scores in combination with the Scoring Table 3-7-1.

III. Knowledge and skills

There are many types of grooves that can be machined on the milling machine, commonly including right-angle grooves, V-grooves, dovetail grooves, T-grooves, arc grooves and various keyways. Right-angle groove machining is common and can be divided into open, semi-enclosed and enclosed types, as shown in Fig. 3-7-2.

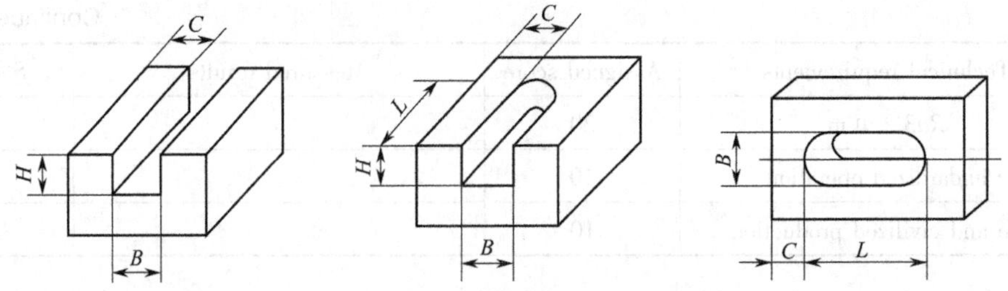

(a)　　(b)　　(c)

Fig. 3-7-2　Types of Right-angle Grooves

(a) Open　(b) Semi-enclosed　(c) Enclosed

1. Milling of right-angle grooves

1) Milling semi-enclosed groove

Generally, when the semi-enclosed groove is milled with a vertical milling tool, the diameter of the vertical milling tool selected shall not be greater than the width of the groove. Due to poor rigidity of the vertical milling tool, the milling tool is easily broken due to deflection and excessive stress during milling. Therefore, when machining deep grooves, the milling shall be carried out in several times to achieve the required depth. After the depth of the groove is milled, both sides of the groove shall be expanded. During the expanded milling, sequential milling shall be avoided to avoid damaging the milling tool and the workpiece, as shown in Fig. 3-7-3.

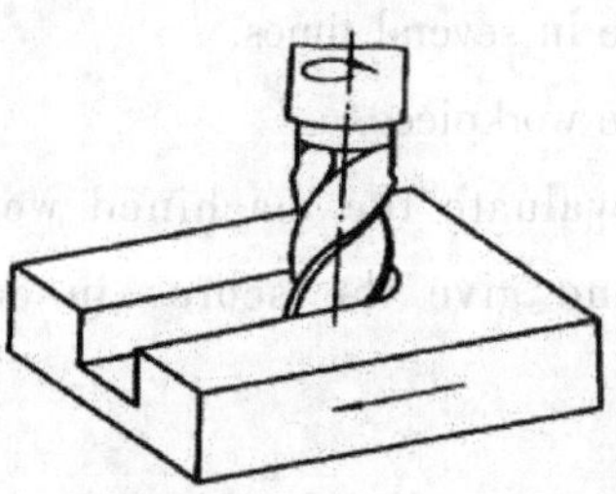

Fig. 3-7-3　Milling Semi-enclosed Groove

2) Milling enclosed groove

Keyway milling tool or vertical milling tool may be used for milling enclosed keyway. When milling with a keyway milling tool, the vertical feed of the workpiece shall be moved to the milling tool, the longitudinal feed of the workpiece shall be cut to the full length of the keyway with a certain cutting depth, and then the cutting shall be vertically fed, and finally the reverse longitudinal feed shall be carried out repeatedly until the machining of the keyway is completed, as shown in Fig. 3-7-4. When milling the enclosed groove with a vertical milling tool, since there is no cutting surface near the end face center of the vertical milling tool and the cutting workpiece cannot be fed vertically, the cutting hole shall be pre-drilled, as shown in Fig. 3-7-5. The depth of the cutting hole shall be slightly greater than the depth of the groove, and its diameter shall be less than 0.5-1mm of the width of the milled groove. During milling, feeding shall be carried out in several times. Each feeding shall be milled from one end of the cutting hole to the other end. After the groove depth reaches the requirement, both sides shall be expanded milled. During milling, the unused feed mechanism shall be tightened (if longitudinal milling is used, the transverse feed mechanism shall be locked; otherwise, the longitudinal feed mechanism shall be locked), and the sequential milling shall be avoided at both sides of the expanded milling.

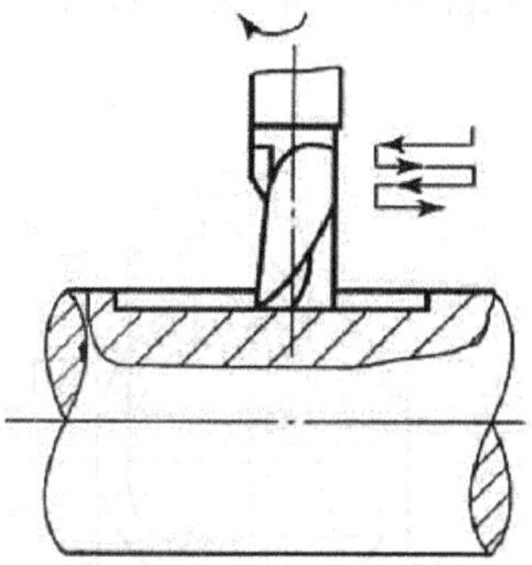

Fig. 3-7-4 Milling Enclosed Groove

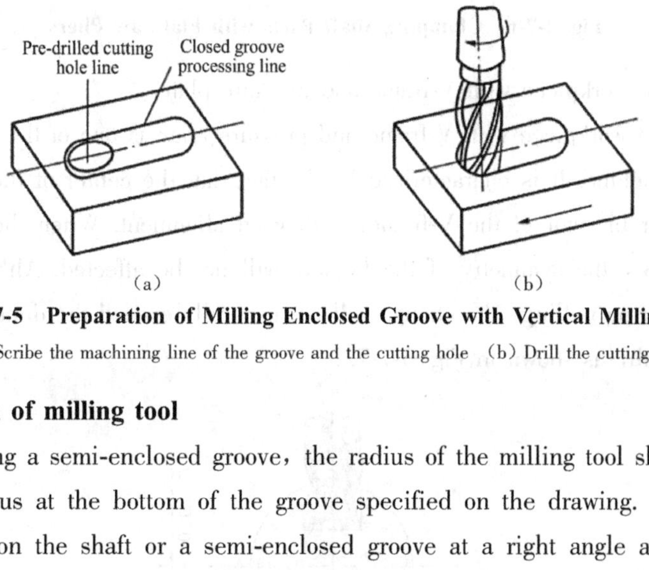

Fig. 3-7-5 Preparation of Milling Enclosed Groove with Vertical Milling Tool

(a) Scribe the machining line of the groove and the cutting hole (b) Drill the cutting hole

2. Selection of milling tool

When milling a semi-enclosed groove, the radius of the milling tool shall be consistent with the arc radius at the bottom of the groove specified on the drawing. When milling an enclosed groove on the shaft or a semi-enclosed groove at a right angle at one end of the groove bottom, a keyway milling tool is generally used, and the width of the keyway is guaranteed by the diameter of the milling tool.

3. Clamping and correction of workpiece

When milling the keyway, there are many clamping methods for the workpiece, which are usually clamped with flat pliers or V-shaped frames, so that the axis of the workpiece is consistent with the feed direction of the workbench and parallel to the workbench surface.

1) Clamp the workpiece with flat jaw pliers

The workpiece is clamped with flat jaw pliers, which is easy to clamp and suitable for single piece production. In order to ensure that both sides and bottom surface of the milled shaft groove are parallel to the axis of the workpiece, the axis of the workpiece must be parallel to both the longitudinal feed direction of the workbench and the workbench surface. When the workpiece is clamped with flat jaw pliers, a dial indicator shall be used to align the fixed jaw parallel to the feed direction of the workbench, and when the workpiece is clamped, the busbar on the workpiece shall also be aligned parallel to the workbench surface, as shown in Fig. 3-7-6.

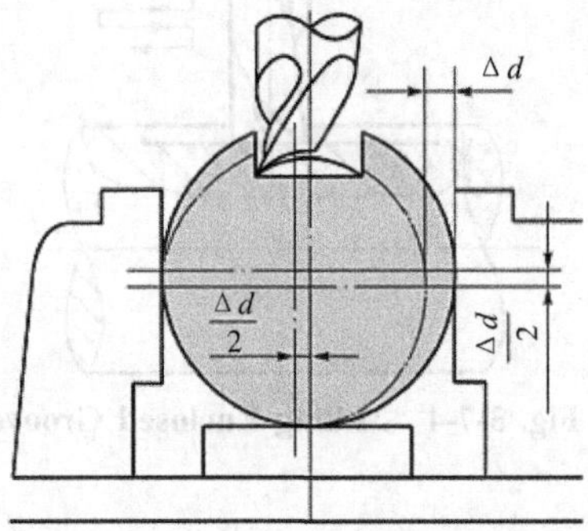

Fig. 3-7-6 Clamping Shaft Parts with Flat Jaw Pliers

2) Clamp the workpiece with V-frame and pressure plate

Clamping the workpiece with V-frame and pressure plate is one of the common methods on the milling machine. It is characterized by the fact that the center of the workpiece must be on the angular bisector of the V-frame, with good alignment. When the diameter of the workpiece changes, the symmetry of the keyway will not be affected. Although the milling depth changes during milling, the change will not generally exceed the dimensional tolerance of the groove depth, as shown in Fig. 3-7-7.

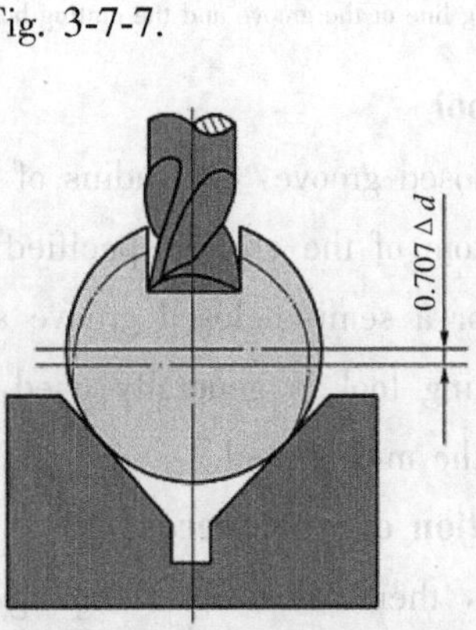

Fig. 3-7-7 Clamping of Shaft Workpiece with V-frame

4. Adjustment of tool

In order to ensure the symmetry of the keyway on the shaft, the position of the milling tool must be adjusted so that the axis of the keyway milling tool passes through the axis of the workpiece. Common adjustment methods include:

(1) Adjust the symmetrical center of the workpiece according to the cutting marks. The principle and method of adjusting the cutting marks of the keyway milling tool are the same as that of the disc groove milling tool, except that the cutting marks milled by the keyway milling tool are a small quadrangle plane with a side length equal to the diameter of the milling tool.

During centering, it is enough to place the milling tool in the middle of the small plane when rotating. This method has a low centering accuracy, but is simple to use and is the most commonly used method, as shown in Fig.3-7-8.

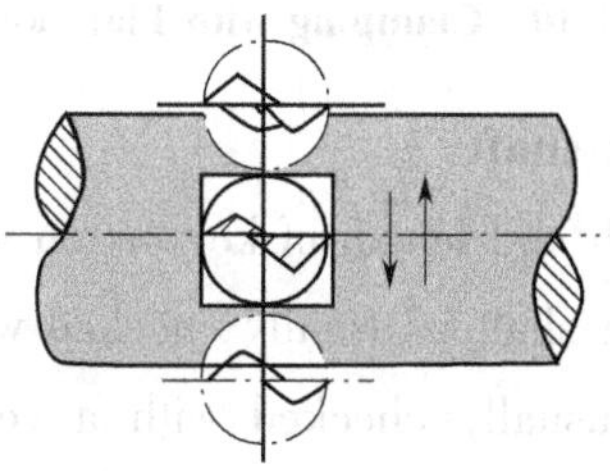

Fig. 3-7-8 Centering of Cutting Marks of Keyway Milling Tool

(2) Adjust the symmetrical center of the workpiece according to the side. During adjustment, thin paper shall be pasted on the side of the workpiece first, and the machine tool shall be started to make the rotating milling tool gradually towards the workpiece. After the cutting edge of the milling tool contacts the thin paper, the workbench shall be lowered to withdraw the workpiece, and then transversely moved for a distance A. This method has a high centering accuracy and is suitable for a long keyway milling tool with a larger diameter. As shown in Fig. 3-7-9.

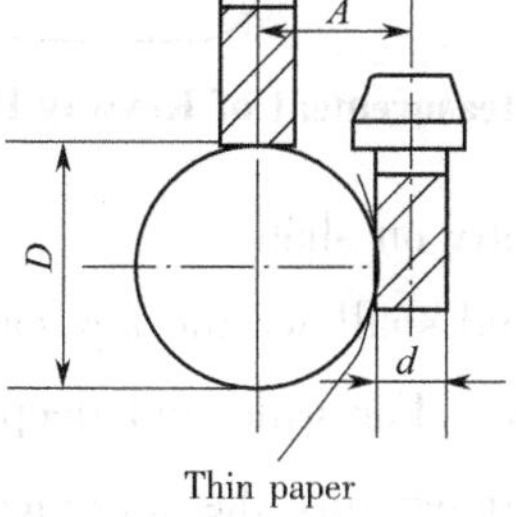

Fig. 3-7-9 Centering of Cutting Marks of Keyway Milling Tool

(3) Adjust the milling tool position center with a lever dial indicator. During

adjustment, fix the lever dial indicator on the spindle of the vertical milling head, rotate the spindle by hand, observe the readings of the dial indicator on both sides of the jaw, on both sides of the V-frame and on both sides of the angle meter, and move the workbench horizontally to make the readings on both sides the same. This method has high centering accuracy, as shown in Fig. 3-7-10.

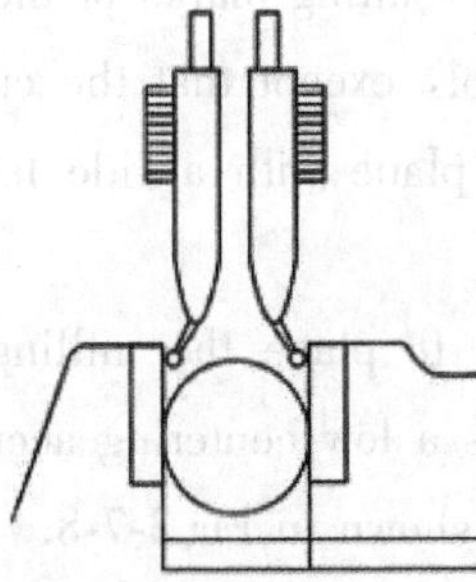

Fig. 3-7-10 Clamping with Flat Jaw Pliers

5. Detection of keyway on shaft

1) Detection of width, depth and length of keyway on shaft

The width of the keyway on shaft is usually checked with a plug gauge, and the depth and length of the keyway are usually checked with a vernier caliper. The depth of the enclosed keyway can be measured with a vernier caliper, and a rectangular measuring block slightly larger than the keyway depth is placed in the keyway. The measured size minus the size of the rectangular measuring block is the size from the keyway bottom to the cylindrical surface. The keyway with a width greater than the diameter of the micrometer measuring rod can be directly measured with a micrometer, as shown in Fig. 3-7-11.

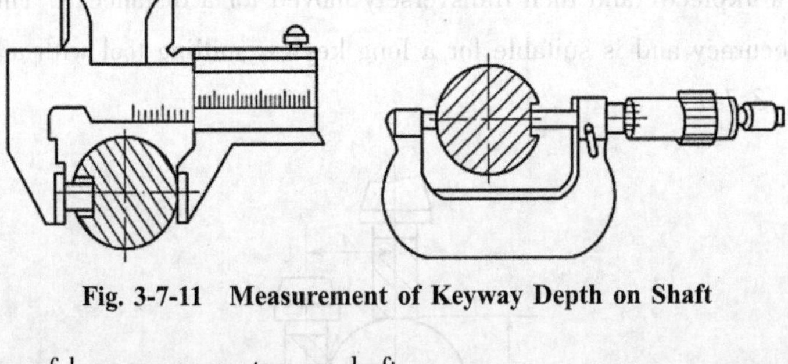

Fig. 3-7-11 Measurement of Keyway Depth on Shaft

2) Detection of keyway symmetry on shaft

When the keyway is narrow and shallow, the key can be inserted tightly in the keyway to increase the measurement surface. For wide and deep keyway, the sides of the keyway can be measured directly. During detection, the workpiece is placed between the V-frames or the two centers and together on the flat plate to make it possible to rotate around the fixed axis. Place the keyway on one side, calibrate the upper plane of the key parallel to the plate with a dial indicator, record the reading of the dial indicator, then turn the workpiece by 180°, and detect it in the same way to obtain another reading. The difference between the two readings is the symmetry error, as shown in Fig. 3-7-12.

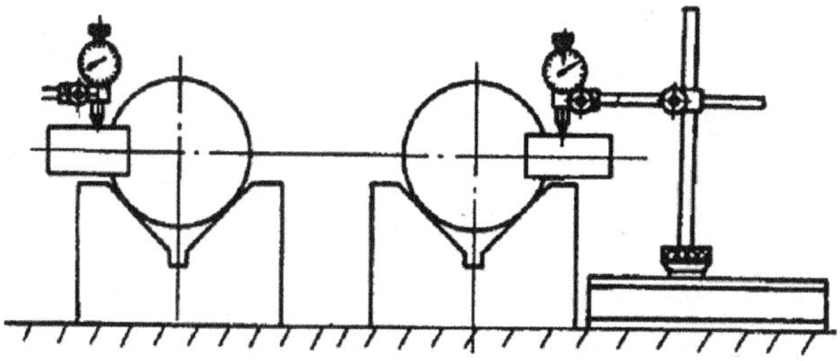

Fig. 3-7-12 Detection of Keyway Symmetry

6. Quality analysis of keyway on shaft

1) Factors affecting dimensional accuracy

(1) The milling tool size is not checked by trial milling, and the workpiece is directly milled, causing dimensional error.

(2) The keyway is milled with a keyway milling tool, and the radial circular runout of the milling tool is too large.

(3) During milling, the cutting depth is too large and the feed rate is too large, resulting in "back-off", and the keyway is milled wide.

2) Factors affecting symmetry

(1) The symmetrical center of the milling tool is not adjusted correctly. During milling, the milling tool back-off is too large.

(2) In batch production, the dimensional error of the workpiece outer circle is too large.

(3) The allowance for expanded milling on both sides of the shaft groove is inconsistent.

3) Factors affecting the parallelism of both sides of the keyway to the axis

(1) The cylindricality of the outer circle diameter of the workpiece is out of tolerance.

(2) When the workpiece is clamped with flat jaw pliers or V-frames, the flat jaw pliers or V–frames are not aligned.

[Thinking and practice]

(1) What are the types of grooves that can be machined on the milling machine ?

(2) Briefly describe the method of milling the enclosed keyway with a keyway milling tool.

(3) What are the clamping methods for the workpiece when milling the keyway ?

(4) What are detected for keyway on shaft ?